AF806250

Cancer Neuroscience

Moran Amit • Nicole N. Scheff
Editors

Cancer Neuroscience

Springer

Editors
Moran Amit
The University of Texas MD Anderson Cancer Center
Houston, TX, USA

Nicole N. Scheff
Hillman Cancer Center
University of Pittsburgh
Pittsburgh, PA, USA

ISBN 978-3-031-33840-3 ISBN 978-3-031-32429-1 (eBook)
https://doi.org/10.1007/978-3-031-32429-1

Cover credit: David Aten, MA, CMI

This Springer imprint is published by the registered company Springer Nature Switzerland AG
The registered company address is: Gewerbestrasse 11, 6330 Cham, Switzerland

Dedicated to my wife Heli Kimhi, friend,
Dov G Cohen and Mentor, Jeffrey N Myers.

- Moran Amit

Preface

The nervous system intimately intersects with cancer, resulting in profound consequences for patients. Cancer neurobiology has emerged as a new research area from the now prominent field of tumor microenvironment research; it offers a new angle for developing innovative clinical strategies for the diagnosis, prognosis, and treatment of cancer. Preclinical and clinical evidence is mounting quickly that the peripheral nerves are a pivotal component of the tumor microenvironment; inhibition of peripheral nerve signaling is thought to slow tumor progression. Therapeutic strategies to target peripheral neurons for cancer treatment are currently being evaluated in combination with standard-of-care approaches as well as immunotherapy drugs. Recent accumulating evidence has demonstrated the diagnostic and prognostic value of nerves and neurotrophic factors in the cancer microenvironment and the functional implications of the presence and outgrowth of nerves in human tumors. Here we will comprehensively discuss the role of nerves in tumorigenesis and cancer progression and how the innervation of tumors can serve as both prognostic and predictive biomarkers as well as actionable therapeutic targets. In this book, we collected all foundational resources for the budding Hallmark of Cancer, i.e., cancer neuroscience. Pioneers in the field, both cancer biologists and neuroscientists, provide an expansive review of data on the mechanisms that govern neuron-cancer trophic relationships, as novel insights continue to be published, offering new understanding of clinical phenomena as well as avenues for treatment, targeting the nervous system as the fifth pillar of cancer therapy. We compiled this comprehensive resource for the scientists and health-care workers entering this field to not only summarize our current understanding of these mechanisms but to also point to further reading for additional learning.

Houston, TX, USA — Moran Amit
Pittsburgh, PA, USA — Nicole N. Scheff

Preface

Contents

Part I Cancer and the Central Nervous System

1 **Neuronal Activity in Brain Tumor Pathogenesis: Adding to the Complexities of Central Nervous System Neoplasia** 3
Khushboo Irshad, Nicole Brossier, and Yuan Pan

2 **Neuron-Cancer Synaptic and Other Electrical Signaling** 27
Humsa S. Venkatesh

3 **Mechanisms of Cancer-Induced Remodeling of the Central Nervous System** 35
Saritha Krishna, Vardhaan Ambati, and Shawn L. Hervey-Jumper

Part II Cancer and the Peripheral Nervous System

4 **Neural Influence on Cancer Invasion and Metastasis** 51
Ligia B. Schmitd, Cindy Perez-Pacheco, and Nisha J. D'Silva

5 **Cancer Induced Remodeling of the Peripheral Nervous System** 69
Anthony C. Restaino and Paola D. Vermeer

6 **Neuroimmune Interactions and Their Role in Carcinogenesis** 83
Shahrukh Ali, Dan Yaniv, and Moran Amit

7 **Neuroimmunoregulation of Cancer: The Case for Multiple Myeloma** 101
Sheeba Ba Aqeel, Caitlin James, Jens Hillengass, and Elizabeth Repasky

8 **Nervous System Interactions with Nonimmune Elements in Cancer Microenvironment: A Missing Piece?** 117
Kaan Çifcibaşı, Carmen Mota Reyes, Rouzanna Istvanffy, and Ihsan Ekin Demir

9 Reprogrammed Schwann Cells in Cancer 131
Sylvie Deborde and Richard J. Wong

Part III Perspectives of the Field

10 Systemic Interactions Between Cancer and the Nervous System 147
Yue Wu and Jeremy C. Borniger

11 Tools and Model Systems to Study Nerve-Cancer Interactions 169
Peter L. Wang, Nicole A. Lester, Jimmy A. Guo, Jennifer Su, Carina Shiau, and William L. Hwang

12 Future Direction of Cancer Neuroscience 185
Jami L. Saloman, Nicole N. Scheff, and Brian M. Davis

Correction to: Tools and Model Systems to Study Nerve-Cancer Interactions C1
Peter L. Wang, Nicole A. Lester, Jimmy A. Guo, Jennifer Su, Carina Shiau, and William L. Hwang

Index 201

Introduction

Traditionally, the nervous and immune systems have been viewed as functionally and anatomically distinct even though both systems detect and respond to internal and external threats. Whereas the role of the immune system in cancer development and progression is well established and whose functions in cancer are considered hallmarks, the nervous system typically has been viewed as being *affected* by cancer and as a conduit for the transmission of cancer-related pain and perineural invasion of tumors. This by-stander viewpoint was most likely because, despite penetrating every tissue of the body, nerves were largely ignored by pathologists, oncologists, and basic scientists alike. If nerves were studied in the context of cancer, it was usually due to cancer-induced growth of neural processes that made them macroscopic, symptomatic, and hard to ignore. But even then, this proliferation was mostly explored as a mechanism explaining cancer pain or cancer dissemination along nerves (i.e., perineural invasion) and to some extent angiogenesis. However, in 1998, Gustavo Ayala, PhD, published the first instance of peripheral sensory neurons interacting with prostate cancer cells in vitro. This seminal work set the stage for the field of neuroscience and for Paul Frenette, PhD, who, in 2013, published the results of a 5-year study in *Science* which demonstrated that ablation or chemical blockade of different portions of the peripheral nervous system slowed or even prevented tumorigenesis in an in vivo mouse model of prostate cancer (Science 341: 1236361). This study attracted considerable attention as it was conducted in a genetic model that mimicked many of the features of human disease and because the signaling processes identified in the model appeared to be conserved in humans. Shortly after its publication, a number of studies were published that showed that ablation or silencing of different aspects of the nervous system could slow or block cancers as diverse as gastric cancer, pancreatic ductal adenocarcinoma, and basal cell carcinoma. It was even shown that neural activity could modulate the growth of primary and metastatic brain cancers.

The primary impediment to advancing our understanding of the complex crosstalk between nerves and cancer cells in the past has been the wide chasm in everything from animal models to lexicon for biological description that existed between oncology and neuroscience. Over the past decade, however, emerging technologies have enabled investigators to begin to bridge the gap between cancer and neuroscience research to unveil the interactive roles of nerves in cancer. We are now beginning to define how the nervous system contributes to cancer initiation, growth, spread, recurrence, and even resistance to oncologic therapeutic strategies. Indeed, neuromodulation with both genetic and pharmacological approaches has been shown to affect not only tumor growth but also antitumor immune response. Collectively, studies have demonstrated a significant and pressing need to recognize cancer neuroscience interplay as a hallmark of cancer to (1) establish the neoneurogenic process as a highly relevant therapeutic target for both the prevention and treatment of cancer, (2) foster interdisciplinary cross-talk among experts in cancer biology and neuroscience, which have traditionally progressed along parallel paths, and (3) integrate human variables such as biological sex, age, race, and gender as underlying factors that can impact cancer neuroscience.

This book is the first and current viewpoint of the burgeoning cancer neuroscience field provided by scientists responsible for the most seminal papers over the last decade, as well as promising up-and-coming scientists in the field.

Moran Amit
Nicole N. Scheff

Part I
Cancer and the Central Nervous System

Chapter 1
Neuronal Activity in Brain Tumor Pathogenesis: Adding to the Complexities of Central Nervous System Neoplasia

Khushboo Irshad, Nicole Brossier, and Yuan Pan

Cellular Origins of Glioma

Gliomas are one of the most common primary CNS tumors in both adults and children [93]. Tumors within this class have histologic features reminiscent of glial origin cells (e.g., astrocytes and oligodendrocytes) and range from the World Health Organization (WHO) Grade 1–4 [80, 81]. Neural stem cells (NSCs), multipotent (neuroglial) progenitors, glial restricted progenitors (GRPs), and oligodendrocyte precursor cells (OPCs) have all been suggested as potential cells of origin for different types of glial-histology tumors [1, 2, 5, 36, 44, 70, 84, 87, 88, 90, 105, 106]. The most well-studied of these is adult glioblastoma, one of the most lethal subtypes of brain tumor. While gliomas can be found in all regions of the brain, glioblastoma most often arises in close proximity to the subventricular zone (SVZ) of the lateral ventricles [10, 60]. Interestingly, this area is home to a prominent pool of NSCs, which are capable of giving rise to glial cells throughout mammalian life [66, 96, 100]. Self-renewing stem-like cells displaying NSC markers (e.g., nestin and CD133; Fig. 1.1) have been identified in glioblastoma (denoted as "glioma stem-like cells" or GSCs) [40, 103], and orthotopic transplantation of these GSCs can give rise to new tumors whose histology recapitulates that of the primary brain

K. Irshad
Department of Symptom Research, University of Texas MD Anderson Cancer Center, Houston, TX, USA

Department of Biochemistry, All India Institute of Medical Sciences, New Delhi, India

N. Brossier
Department of Pediatrics, Washington University in St. Louis, St. Louis, MO, USA

Y. Pan (✉)
Department of Symptom Research, University of Texas MD Anderson Cancer Center, Houston, TX, USA
e-mail: Ypan4@mdanderson.org

M. Amit, N. N. Scheff (eds.), *Cancer Neuroscience*,
https://doi.org/10.1007/978-3-031-32429-1_1

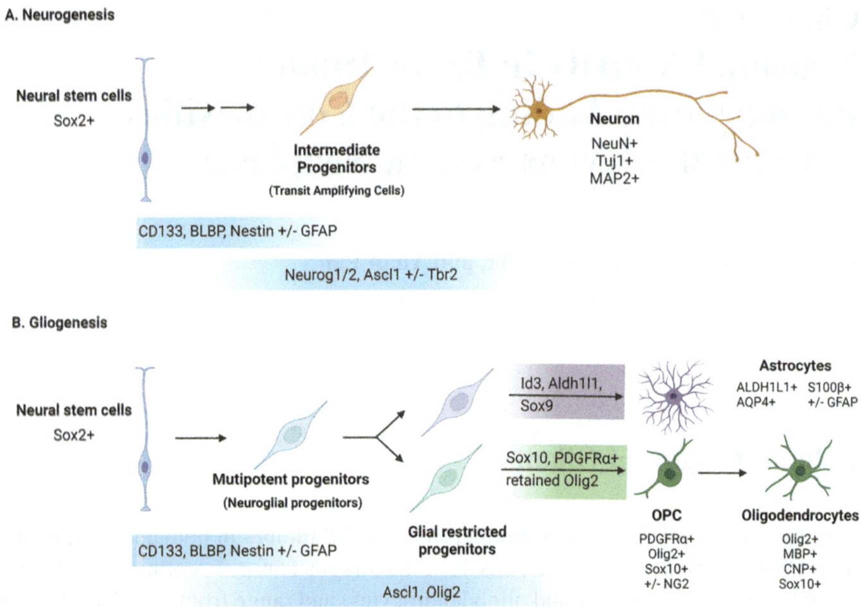

Fig. 1.1 Neurogenesis and gliogenesis from neural stem cells. *OPC* oligodendrocyte precursor cell. Molecular markers for each lineage are indicated

tumor [40]. In support of NSCs as a potential cell of origin for glioblastoma, analysis of matched tumor-free and tumor-bearing tissue specimens from human glioblastoma patients revealed NSCs in the tumor-free SVZ region that contained low-level GBM driver mutations. In addition, NSCs containing glioblastoma driver mutations migrate from the SVZ to distal sites to form brain tumors in mice [63]. In murine models, conditional loss of tumor suppressor genes (e.g., *Trp53*, *Nf1*, and *Pten*) in NSCs induces high-grade glioma formation in genetically engineered mouse models [1]. Furthermore, orthotopic transplantation of mutant NSCs, but not mutant astrocytes, results in malignant brain tumor formation [49].

Outside of NSCs, murine modeling data have implicated other cells capable of giving rise to glioma. Indeed, certain mutations of *Trp53*, *Nf1*, and *Pten* can be used to drive murine high-grade glioma formation from either adult Nestin+ NSCs, Ascl1+ neuroglial progenitors/OPCs, or neuron-glial antigen 2-positive (NG2+) OPCs (Fig. 1.1), with the cell of origin dictating where the tumor forms in the brain as well as its transcriptional profile [2]. Consistent with that observation, pediatric high-grade gliomas can be induced by oncogenic mutations in either NSCs or OPCs, with mutations in NSCs resulting in a higher average tumor grade and decreased survival [109]. Moreover, mutation in either NSCs or Olig2+ pre-OPCs can also produce diffuse midline gliomas (DMG) [61, 115], aggressive gliomas seen in pediatric and young adult patients which are typically histone-mutated (H3K27M). However, histone mutation appears to provide a growth advantage and decrease survival rate compared to *Trp53* mutation alone only in NSC-derived, not OPC-derived, tumors. This is in seeming contradiction to the presence of abundant

OPC-like cells in DMG that display enhanced proliferation and tumor-inducing capacity [36, 84, 87, 88]. This discrepancy may be partially explained by a prior study using mosaic analysis with double markers (MADM), which demonstrated that oncogenic mutations in NSCs result in clonal growth prior to malignancy only in derivative OPCs, suggesting that OPCs may be the cell to undergo neoplastic transformation even if in some cases mutations are introduced in earlier progenitors [70]. Overall, these studies are consistent with multiple cellular origins for glioma, with increasing lineage restriction accompanied by decreasing susceptibility to malignant transformation [73].

Fewer studies have been devoted to elucidating the origin of lower grade gliomas, in part due to the challenges of modeling these tumors. Much of the work that has been done has focused on transgenic modeling of optic pathway gliomas (OPG) in the context of neurofibromatosis type 1 (NF1) [69], a cancer predisposition syndrome. Similar to higher-grade tumors, $CD133^{+}$ GSCs can be identified in NF1-OPG [25], suggesting a potential NSC origin. In support of this hypothesis, conditional inactivation of *Nf1* in NSCs results in NF1-OPG formation [7, 45, 106]. Although these tumors characteristically form in the optic pathway, several studies support their origination from NSCs in the embryonic third ventricle (TV) [16, 26, 50, 62, 114], which then migrate into the optic nerve according to normal developmental programming [91, 104]. As with higher-grade gliomas, the cell undergoing neoplastic transformation may be a migrating glial progenitor rather than the NSC itself [50]. Transgenic studies also suggest a similarly decreased susceptibility to transformation with increasing lineage specification, as loss of *Nf1* in Olig2+ pre-OPCs (Fig. 1.1) results in tumor formation with an increased latency compared to *Nf1* loss in NSCs [106], and NG2+ glia do not form OPG upon *Nf1* loss [105]. Recent low-grade glioma modeling with induced pluripotent stem cells (iPSCs) derived from human cell types further supports this hypothesis, as tumor growth occurred after transplantation of *Nf1*-null iPSC-derived NSCs, GRPs, and GRPs with either astrocytic or oligodendroglial differentiation, but not astrocytes [5].

Studies on the cellular origins of non-NF1 low-grade gliomas are even more limited. *KIAA1549-BRAF*, the most common driver mutation in pediatric low-grade glioma, increases the proliferation and glial differentiation of cerebellar NSCs with glioma-like lesions formed upon orthotopic injection of mutant NSCs but not astrocytes [55]. Similar to other gliomas, however, NSCs are not the only cell type capable of neoplastic transformation upon *KIAA1549-BRAF* mutation. iPSC modeling demonstrates that more differentiated progenitors (GRPs, OPCs) can also generate tumors upon orthotopic transplantation, with decreased susceptibility to transformation observed in fully differentiated cells [5]. In human tumor specimens, single-cell sequencing has identified developmental gene expression overlapping with progenitors around the fourth ventricle that display evidence of MAPK activation [126]. As observed in NF1-OPG, this may be consistent with mutations in the NSC compartment followed by expansion of a MAPK-activated migrating glial progenitor [50, 126]. In total, the data suggest that gliomas can arise from any one of series of less differentiated cell types, ranging from NSCs to GRPs to OPCs, under the appropriate conditions. Furthermore, these different cellular origins may contribute to phenotypic variability within a given tumor type.

Effect of Neuronal Activity on NSC Dynamics

The electrical activity originating from neurons (neuronal activity) and activity-dependent factors (e.g., neurotransmitters) are responsible for many aspects of CNS development, including modulating NSC dynamics [13, 28, 120, 124]. Neurotransmitters, released from neurons in an activity-dependent manner, carry out neuron stimulatory, inhibitory, and modulatory functions. Depending on the chemical structures, neurotransmitters are classified into three categories: (1) amino acids such as glutamate, gamma-aminobutyric acid (GABA), glycine, and acetylcholine (Ach); (2) biogenic amines such as dopamine, norepinephrine, epinephrine, and serotonin; and (3) neuropeptides such as substance P (SP), neuropeptide Y (NPY), opioids, calcitonin gene related peptide (CGRP), vasoactive intestinal polypeptide (VIP), bombesin, and neurotensin [51]. Neurotransmitters exhibit a variety of effects on neural stem and progenitor cells, depending on the developmental stage, stem/progenitor cell type, and the types of neurotransmitters (reviewed in [12, 112, 116]). For example, glutamate decreases NSC proliferation in the embryonic neocortex [74] but increases the neural progenitor proliferation in the developing telencephalon [76, 77].

Advanced neuromodulatory techniques such as optogenetics and chemogenetics have started to elucidate the relationship between neuronal activity and NSC dynamics in specific brain regions in vivo. Optogenetic techniques employ engineered light-sensing proteins (most often ion channels) to control the electrical property of living cells, whereas chemogenetic techniques use engineered ligand-receptor pairs to manipulate intracellular signaling [31, 107]. Recent studies leveraged optogenetics and chemogenetics to induce neuronal activity in specific neuronal populations and investigated their regulation of NSC behavior. For example, the activity of dentate gyrus parvalbumin (PV)-positive interneurons inhibits NSC activation through $GABA_A$ receptors [108]. Leveraging chemogenetics, a study identified that the activity of hypothalamic proopiomelanocortin (POMC) neurons regulates NSC dynamics in a subtype-dependent manner (e.g., by specifically innervating part of the ventricular-subventricular zone and inducing proliferation and neurogenesis of $Nkx2.1^+$ NSCs) [97]. These findings demonstrate that NSC dynamics are tightly regulated by local neuronal activity.

Effect of Neuronal Activity on OPC Dynamics

OPCs are progenitors that give rise to oligodendrocytes, the glial cells that myelinate neuronal axons. OPC proliferation and differentiation are regulated by neuronal activity, an adaptive process that contributes to the plasticity of the oligodendroglial lineage. Recent in vivo studies have uncovered that experience modulates OPC proliferation and differentiation in mice. Sensory enrichment induces OPC differentiation into oligodendrocytes that support de novo myelination and fine-tuning of neuronal circuitry [47]. Similarly, motor learning using a complex wheel test can increase OPC proliferation and oligodendrogenesis (OPC differentiation) in corresponding brain regions [79, 125]. Importantly, genetic inhibition of OPC differentiation during the motor learning experience decreases oligodendrogenesis and impairs motor learning [79]. Like motor learning, memory consolidation during spatial learning also depends on OPC differentiation in response to experience [110]. On the other hand, certain experiences like social isolation decrease oligodendrogenesis and myelin sheath thickness [71, 72, 78]. These findings demonstrate that OPC proliferation and differentiation are dynamic processes influenced by experiences.

Experience is a complex interplay of various types of neurons and their activity. To demonstrate a direct relationship, optogenetic approaches were used to induce neuronal activity specifically in the motor planning region (M2) to observe OPC dynamics in the neuronal projections [41]. Optogenetically induced physiological firing of layer V neurons enhances OPC proliferation and differentiation within the stimulated circuit [41]. This change results in increased myelin thickness in the M2 subcortical white matter and changes in motor function. Further, pharmacological inhibition of OPC differentiation reduces the activity-induced oligodendroglial responses and changes in swing speed (motor function). Similarly, chemogenetically induced neuronal activity in somatosensory cortical neurons increases OPC proliferation, oligodendrogenesis, and myelination within the active brain region [83]. These studies revealed that neuronal activity directly modulates OPC proliferation and differentiation in certain brain regions and consequential de novo myelination.

Neuron-OPC Synapses

Synapses are junctions that form between the end of a neuron and another cell. Although synapses were originally thought to form exclusively between neurons in the CNS, OPCs were later found to form bona fide synapses with neurons [14, 68] (Fig. 1.2). OPCs express a variety of neurotransmitter receptors [19, 111] and were first found to form excitatory synapses with neurons [14]. In the hippocampus, the activity of excitatory neurons induces inward currents in OPCs in an AMPA receptor-dependent manner via the glutamatergic neuron-OPC synapses [14, 27, 58]. Later, OPCs were found to form synapses with GABAergic neurons [68, 117]. GABA depolarizes OPC membrane potential through the GABAergic synapses due to a high intracellular Cl^- concentration in OPCs [68]. Moreover, GABA induces transient inhibition of AMPA receptor-mediated currents in OPCs, suggesting a feedback mechanism between GABAergic and glutamatergic synapses in OPCs.

Investigation of brain-wide neuron-OPC connections using monosynaptically restricted rabies virus tracing of OPC afferents [85] confirmed that OPCs in the same brain region can form synapses with both excitatory and inhibitory neurons.

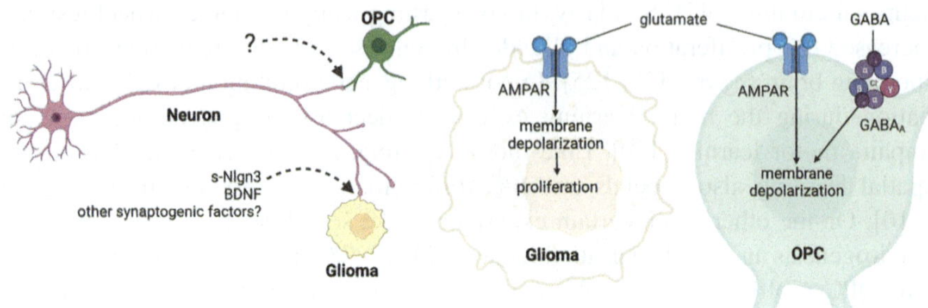

Fig. 1.2 Neuron-glial synapses in glioma cells and OPCs. Both excitatory and inhibitory synapses have been identified in OPCs. To date, only excitatory synapses are reported in glioma cells

Neurons that synapse on OPCs in the same brain region were found spread out in multiple brain regions. These findings reveal the expansive synaptic input on OPCs and suggest that OPCs integrate information from both excitatory and inhibitory inputs which may be important for OPC dynamics. The biological function of neuron-OPC synapses is still under intensive investigation and could be highly context-dependent (reviewed in [86]).

Activity-Dependent Pathogenesis of Gliomas

Given that primary brain tumor cells resemble features in NSCs and OPCs, the tumor cells could subvert the activity-dependent mechanisms to promote tumor progression and therapy resistance. To date, activity-dependent regulation of glioma has been demonstrated by several groups (reviewed in [53, 95, 120]).

As mentioned in the previous section, experience plays substantial roles in modulating OPC proliferation and differentiation. The regulation of glioma proliferation by experiences has been recently reported [23, 94] (Box 1.1). For example, visual experience plays a role in the pathogenesis of optic pathway glioma (OPG), a type of low-grade glioma that develops in the optic pathway. OPG is commonly observed in patients with NF1. To model NF1-associated OPG, genetically engineered mice (*Nf1*$^{fl/-}$*;Gfap-Cre,* or *Nf1*^{OPG} mice) were generated [8]. These mice develop gliomas specifically in the prechiasmatic region, where the only neuronal component of the tumor microenvironment is the axons of retinal ganglion cells (RGC). Dark-rearing, which suppresses visual experience induced RGC activity, inhibits NF1-OPG initiation and growth, demonstrating that visual experience is required for gliomagenesis in the optic nerve in this model [94].

Box 1.1: Animal Models Used to Study Experience-Dependent Gliomagenesis

- *Nf1*$^{fl/-}$*;Gfap-Cre* mice develop low-grade glioma in the optic nerve around 9 weeks of age. Light, which evokes visual and nonimaging forming experience in the eye, drives initiation and maintenance of glioma in this model via neuronal activity-dependent ADAM10-mediated neuroligin-3 shedding.
- *Nf1*$^{fl/fl}$*;Trp53*$^{fl/fl}$*;NG2-Cre* mice develop high-grade glioma throughout the brain with the olfactory bulb being the hotspot of gliomagenesis. Olfactory experience drives the proliferation of both preneoplastic *Nf1−/−Trp53−/−* OPCs and neoplastic glioma cells via activity-dependent release of IGF1.

Similar to visual experience, olfactory experience drives glioma pathogenesis in the olfactory bulb. In a genetically engineered mouse model (*Trp53* and *Nf1* knockout in adult OPCs) where glioma constantly transforms from mutant OPCs and develops in the olfactory bulb, naris occlusion (suppresses olfactory experience) reduces tumor volume [23]. These data indicate that olfactory experience is required for gliomagenesis in this model.

Experience induces complex neuronal interactions as well as changes in nonneuronal cells. For this reason, optogenetics was used to specifically induce neuronal activity and study the effect on glioma pathogenesis. In human high-grade glioma xenograft models, optogenetically induced neuronal activity increases glioma proliferation specifically within the active circuitry [121]. Similarly, optogenetic stimulation of RGC firing increases the volume of optic glioma in the *Nf1*^{OPG} mice [94]. In the olfactory glioma model, chemogenetically silencing and activating the activity of olfactory receptor neurons (ORNs) decreases and increases glioma volume, respectively [23].

Activity-Dependent Mitogens for Glioma

Optogenetics have also been used to identify secretory molecules responsible for activity-dependent glioma pathogenesis. The ectodomain of neuroligin-3 (shed-NLGN3) and brain-derived neurotrophic factor (BDNF) were identified as key activity-dependent secretory factors in optogenetically stimulated brain slices and retinal explants [94, 121]. In the culture, NLGN3 increases the growth of both high- and low-grade gliomas cells; and genetically ablating *Nlgn3* inhibits high-grade glioma growth in xenograft models and low-grade glioma initiation in *Nf1*^{OPG} mice [94, 121].

NLGN3 is normally a postsynaptic membrane protein and is expressed by both neurons and OPCs. Interestingly, OPCs are responsible for the majority of the activity-dependent release of shed-NLGN3, as evidenced by comparing NLGN3 shedding in neuron- vs. OPC-specific *Nlgn3* knockout brain slices [123]. ADAM10 is a sheddase that cleaves the extracellular domain of multiple membrane proteins, including NLGN3 (reviewed in [33]). While shed-NLGN3 primarily originates from OPCs, ADAM10 secreted from neurons is primarily responsible for cleaving NLGN3 [123]. ADAM10 secretion is increased upon neuronal activation. Pharmacologically blocking ADAM10 is sufficient to reduce NLGN3 shedding in forebrain slices and optic nerve, and inhibit glioma progression [94, 123].

In the olfactory glioma mouse model, insulin-like growth factor 1 (IGF1) is the key activity-dependent gliomagenesis factor [23]. Olfactory experience excites mitral and tufted (M/T) cells, a type of output neuron of the olfactory bulb, to secrete IGF1 in an activity-dependent manner. *Igf1* knockout in M/T cells suppresses OPC transformation; and genetically ablating *Igf1r* (encoding IGF1 receptor) in OPCs inhibits activity-induced proliferation of preneoplastic OPCs. IGF1 seems to drive gliomagenesis in this system independent of the neuron-glioma synapses [23].

Activity-Dependent Glioma Initiation in Cancer Predisposition Syndromes

Patients with certain cancer predisposition syndromes (e.g., Li-Fraumeni syndrome, neurofibromatosis, tuberous sclerosis, and Turcot's syndrome) have increased susceptibility to developing brain tumors [34]. In an attempt to recapitulate both the *NF1* germline mutation and the "second-hit" somatic *NF1* mutations in patients with NF1-OPG, the *NF1*^{OPG} (*Nf1*$^{fl/-}$;*Gfap-Cre*) mice were generated in which the *NF1* homologous deletion in NSCs is incorporated into the *NF1*-heterozygous background [37, 43]. Another mouse model that harbors the same *Nf1* homologous deletion in NSCs, but in a wild-type background (*Nf1*$^{fl/fl}$;*Gfap-Cre*), does not develop optic glioma, indicating that the *Nf1*-heterozygous background (germline *Nf1* mutation) is required for optic glioma initiation [8]. It was unclear why the germline *Nf1* mutation is required for gliomagenesis until it was recently shown that only *Nf1*$^{+/-}$, but not wild-type, optic nerves exhibit increased NLGN3 shedding in response to neuronal activity [94], demonstrating that the germline *Nf1* mutation synergizes with neuronal activity to increase NLGN3 shedding that drives optic gliomagenesis (Fig. 1.3).

A recent study uncovered that the germline *Nf1* mutation increases NLGN3 shedding by inducing neuronal hyperexcitability via inhibiting the HCN (hyperpolarization-induced cyclic nucleotide-gated) channel [6]. Neuronal hyperexcitability also increases neuronal production of midkine, which is required for activating the T cell/microglial axis to promote optic glioma growth (Fig. 1.3). Importantly, not all germline *Nf1* mutations lead to neuronal hyperexcitability, and the *Nf1* germline mutation (e.g., Arg1809Cys) that is not optic glioma-prone does not induce neuronal hyperexcitability [6]. Such observations suggest one potential contributor to the incomplete penetrance of OPG in individuals with NF1.

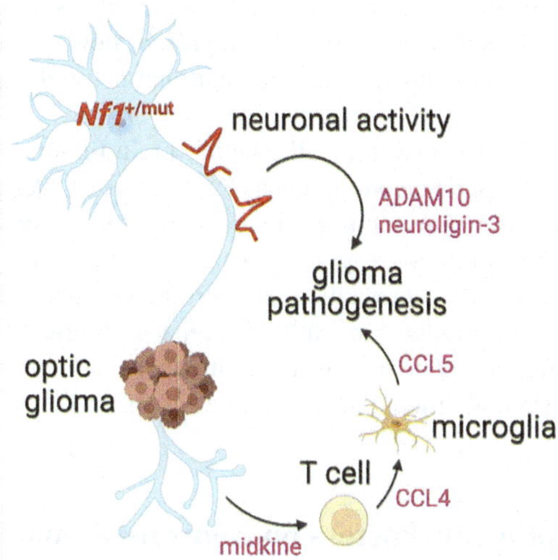

Fig. 1.3 Molecular and cellular mechanisms governing neuronal regulation of NF1 optic glioma pathogenesis. *Nf1* mutations induce hyperexcitability in neurons and increase midkine secretion that stimulates T cells to produce Ccl4 production. Ccl4 reprograms microglia to secrete Ccl5 that increases glioma growth. In addition, the activity of *Nf1*-mutant neurons increases ADAM10-dependent cleavage of neuroligin-3, which is required for optic glioma initiation and maintenance

Whereas certain *NF1* germline mutations could synergize with neuronal activity to drive glioma development, whether germline mutations in other tumor suppressor genes (e.g., *TP53*, *PTEN*, *TSC*) influence neuronal activity and brain tumor progression remained to be understood.

Neuron-Glioma Synapses

Similar to OPCs, ultrastructural studies found *bona fide* synapses between glioma cells (postsynaptic) and neurons (presynaptic) [118, 122] (Fig. 1.2). Neuronal activity induces fast inward currents in glioma cells which results in membrane depolarization; and depolarizing glioma cell membrane is sufficient to increase proliferation [118, 122]. The activity-induced inward currents in glioma cells are reminiscent of the excitatory postsynaptic currents (EPSCs) in OPCs and can be inhibited by tetrodotoxin, a sodium channel blocker. The neuron-tumor synapses can be found on the tumor microtubes (actin-rich, mm wide, hundreds of mm long, long-lived [>60 min] membrane extensions) [92, 118]. Pharmacologically (perampanel) or genetically targeting AMPA receptors in glioma-bearing mice inhibits tumor growth [118, 122].

In breast-to-brain metastases, tumor cells also communicate to neurons but by approaching the existing brain neuron-neuron synapses in a perisynaptic fashion [128], reminiscent of the tripartite configuration between neurons and astrocytes

[98]. These metastatic tumor cells leverage NMDA receptors to promote metastatic colonization [128]. Together, these findings highlight how brain tumor (both primary and secondary) cells integrate into the neuronal networks electrically to promote malignancy.

It remains to be investigated in detail whether brain tumors other than gliomas also form synaptic or perisynaptic contacts with neurons. Perhaps more refined knowledge regarding how these tumor cells interact with neurons could be devised from previous findings in tumor responses to activity-dependent factors. For example, medulloblastoma cells express the metabotropic glutamate receptor mGluR4; and mGluR4 stimulation reduces medulloblastoma proliferation [48]. These findings suggest that glutamatergic neuronal activity could influence medulloblastoma pathogenesis; however, additional in vivo studies are needed.

Effect of Synaptogenic Factors on Neuron-Glioma Synapses

It has been well-established that BDNF levels are mediated by neuronal activity, and BDNF serves as a mitogen of glioma cells [46, 121]. In the healthy CNS, BDNF modulates synaptic connectivity and strength to facilitate neurodevelopment and neuroplasticity [20, 22, 54, 65, 75, 82, 89]. A recent study revealed that BDNF also functions as a synaptogenic factor in glioma by increasing the number of neuron-glioma synapses [113]. In addition, BDNF increases the strength of neuron-glioma synapses by elevating AMPA receptor trafficking to the glioma cell surface. Consistently, genetically inhibiting activity-dependent BDNF expression improves the survival of glioma-bearing mice [113]. The effect of BDNF on glioma cells is mediated through TrkB, as *Ntrk2* (TrkB encoding gene; BDNF receptor) knockdown in glioma cells blocks the BDNF-induced changes in glioma cells and inhibits glioma growth in vivo.

Shed-NLGN3 also carries out a synaptogenic effect on neuron-glioma synapses as NLGN3 upregulates the transcription of synaptic genes within glioma cells, and *Nlgn3* knockout reduces the number of neuron-glioma synapses [121, 122]. Together, these findings underline the synaptogenic and mitogenic significance of BDNF and NLGN3 in driving glioma pathophysiology.

Synaptic/synaptogenic signatures are present in glioma cells and are associated with glioma-associated seizures [52]. For example, in the healthy brain, the formation of excitatory synapses is in part facilitated by astrocyte-derived glypicans (glypican 4 and 6) [3]. Glypican 3 is instead utilized by glioma cells harboring the *PIK3CA*^{C420R} mutation to drive synaptogenesis and network hyperexcitability in the tumor microenvironment [127]. Another study showed that glioblastoma cells that are functionally connected to existing neuronal network exhibit high levels of thrombospondin-1, a synaptogenic factor normally secreted by astrocytes [57]. These findings suggest that certain glioma-derived synaptogenic factors may enhance synaptic connections (neuron-neuron and/or neuron-glioma synapses) in the tumor microenvironment to further induce cancer progression and seizures.

Gap Junctions in Brain Tumors

Before neurons attain maturity or functional synapses have been produced, patterned waves of electrical activity persist in the developing CNS. Earlier studies have shown that during neurodevelopment, NSCs/NPCs arrange into adjacent clusters through gap junctions consisting of connexin-26 (Cx26) and connexin-43 (Cx43) [15]. These coupled clusters sense microenvironmental stimuli, which induce synchronized membrane depolarization and subsequent calcium transients in the coupled cells. Likewise, glioma cells form interconnected networks via Cx43 gap junctions [92, 118, 122].

Neuronal activity induces prolonged potassium currents in some glioma cells [118, 122]. These currents are non-synaptic and have low-input resistance, and their amplitude is positively regulated by neuronal activity. Gap junction coupling is responsible for the low-input resistance of these currents and also for inducing synchronized calcium transients among glioma cells, which can be reduced by gap junction blockers. In addition, gap junction blockers inhibit glioma growth in vivo and sensitize gliomas for therapies in xenograft models [99, 118, 122]. Gap junction blockers that disrupt tumor-astrocyte connectivity in brain metastases also inhibit brain metastatic outgrowth [24, 56, 67]. These findings indicate that brain tumor cells leverage gap junction-mediated self- and non-self-connectivity to sustain tumor growth and therapy resistance.

While the gap junction-mediated glioma network plays vital roles in glioma growth and therapy resistance, recent findings elucidated the role of non-gap junction connected glioblastoma cells in brain invasion [119]. These cells are neuron- and neural progenitor-like in transcriptomic signatures, and they invade the brain parenchyma in a way that resembles neuronal migration during development. This invasion behavior is regulated by neuronal activity, which increases the migration speed of the non-gap junction connected glioma cells. Together, gap junction connected and unconnected tumor cells play important and yet different roles in the pathogenesis of glioblastoma, which contribute to the malignant nature of this brain cancer.

Extracellular Vesicles as Potential Activity-Dependent Factors in Brain Tumor Pathogenesis

Extracellular vesicles (e.g., exosomes) are emerging as important cell-cell communication mediators in cancer and neurodegenerative disorders [42], and their secretion can be modulated by neuronal activity [59]. Exosomes released by brain cells can specifically target other cells distally or locally [11]. Recent findings demonstrated a crucial role of exosomes in neural circuit development by facilitating NPC proliferation, neurogenesis, and circuit connectivity [101]; the authors identified proteins that regulate neurodevelopment in exosomes isolated from induced pluripotent stem cell (iPSC)-derived neurons. These exosomes can rescue deficits in the *MECP2*-mutant (known to cause aberrant neural circuit development in the Rett syndrome) neural cultures and promote hippocampal neurogenesis in vivo, highlighting the importance of neuronal exosomes in facilitating neurodevelopment and potentially brain tumor development [35].

Exosomes released from glial cells (e.g., oligodendrocytes, astrocytes, and microglia) also contribute to neuronal function and brain tumor pathogenesis [18]. Glial exosomal release (e.g., from oligodendrocytes) could be modulated by neuronal activity [39]. Oligodendrocyte-derived exosomes can enter neurons and alter neuronal behaviors, including increasing firing and improving survival under stress conditions (e.g., oxygen/glucose deprivation) [38]. The oligodendrocyte-derived exosome-like vesicles can also inhibit myelination in oligodendrocyte-neuron cocultures [9]. Astrocyte-derived extracellular vesicles carry miRNAs that downregulate *PTEN* in brain metastatic tumor cells and increase tumor proliferation [129]. Microglia, when exhibiting the M2-polarization, increase glioma proliferation and migration by inhibiting *Bmal1* expression via exosomal miR-7239-3p [64].

Extracellular vesicles are also capable of intercellular delivery of oncogenic proteins. For example, EGFRvIII, an oncogenic receptor tyrosine kinase residing in the plasma membrane, is incorporated into microvesicles derived from glioma cells. Internalization of these exosomes into naive cells inserts the oncogenic EGFRvIII into the plasma membrane and activates tumorigenic signaling pathways such as MAPK and AKT [4]. These findings demonstrate important contributions of extracellular vesicles to brain tumor progression, yet whether neuronal activity regulates brain tumor pathogenesis via neuronal and glial extracellular vesicles remains to be determined.

Targeting Neuron-Tumor Interactions in the CNS

Recent discoveries in the interaction between neurons and tumor cells in the CNS provide new avenues for treating brain tumors (reviewed in [95, 119]). For instance, preclinical studies suggest that synaptic communications can be targeted by AMPA receptor inhibitors [118, 122]. Perampanel is a Food and Drug Administration (FDA)-approved noncompetitive AMPA receptor antagonist for treating seizures. The effect of perampanel on brain tumor associated seizures/hyperexcitability is currently being tested in clinical trials (NCT02363933, NCT03636958, NCT04497142), but the effect on glioma progression needs to be determined. NMDA receptors could be the target for inhibiting the tripartite synaptic signaling in breast cancer brain metastases [128]. The effect of NMDA receptor antagonists on treating brain metastases remains to be determined clinically.

Targeting the glioma-induced neuronal hyperexcitability could potentially break the malignant neuron-glioma interaction cycles. Glioma cells can produce glutamate via the x_c^- cystine glutamate exchanger and induce neuronal hyperexcitability [17, 21, 29]. In preclinical mouse models, pharmacologically targeting x_c^- with sulfasalazine, a drug used to treat ulcerative colitis, inhibits glioma-associated epilepsy [17, 21]. Likewise, lamotrigine, an HCN channel agonist used to treat seizures, suppresses neuronal hyperexcitability and tumor progression in mouse models of NF1-OPG [6]. These data suggest that x_c^- and HCN could serve as therapeutic targets for glioma and glioma-associated neuronal hyperexcitability.

Gap junctions can be targeted to reduce activity-induced synchronized intracellular events among glioma cells. Gap junction inhibitor meclofenamate reduces the frequency and amplitude of the slow inward currents in glioma cells and inhibits glioma growth in vivo [118, 122]. Meclofenamate is currently under clinical trial for brain metastases (NCT02429570). Tonabersat, a clinically well-tolerated gap junction inhibitor [102], exhibits an additional glioma-inhibitory effect when used as an adjuvant in combination with radio-chemotherapy in xenograft models [30]; the effect on treating glioma clinically remains to be determined.

Shed-NLGN3 is a glioma mitogen and a synaptogenic factor for the neuron-glioma synapses; NLGN3 shedding is thus an important therapeutic target for preventing neuron-glioma interactions. Inhibitors of ADAM10 have also demonstrated tumor-inhibitory effect in human glioma xenograft models and genetically engineered mouse models of NF1-OPG [94, 123]. INCB7839, an ADAM10/17 inhibitor, is currently being tested in a clinical trial for pediatric high-grade gliomas (NCT04295759).

Like the ADAM10/NLGN3 axis, the BDNF/TrkB axis is also being considered to prevent neuron-glioma communications. Entrectinib, a pan-Trk inhibitor, reduces glioma growth in multiple glioma xenograft models [113]. Entrectinib has good blood-brain barrier penetrance [32] and has been approved by the FDA for *NTRK*-fusion cancers. Entrectinib is currently being tested in patients with CNS tumors (NCT02650401, NCT02568267).

Conclusion

The newly identified activity-dependent regulation of brain tumors has opened the gate for novel therapeutic strategies, yet there is much remained in the field to be understood. For example, how are glioma cells regulated by other types of neurons in addition to the glutamatergic neurons? How do other experiences influence brain tumor progression? How does neuronal activity regulate the pathophysiology of non-glioma brain tumors? Do various oncogenic mutations differentially influence neuron-brain tumor interaction? Is there a region-specific effect of neuron-brain tumor interactions? How are other glial cells affected by neuron-tumor interactions? Future investigations are urgently needed for the field to better understand the mechanisms and deduce therapeutic strategies.

Acknowledgments The authors gratefully acknowledge the Alex's Lemonade Stand Foundation (19-16681 to Y.P.; 18-12558 to N.B.), Hyundai Hope on Wheels (DR-2019-672 to N.B.), the Neurofibromatosis Therapeutic Acceleration Program (210112 to N.B.), the National Institute of

Child Health and Human Development (K12HD076244 to N.B.), the Cancer Prevention and Research Institute of Texas (RR210085 to Y.P. as the CPRIT Scholar), the Gilbert Family Foundation (622030 to Y.P.), and the DHR Young Scientist Grant from the Department of Health Research (DHR-ICMR, India, No.R.12014/11/2019-HR to K.I.).

References

1. Alcantara Llaguno, S., Chen, J., Kwon, C.-H., Jackson, E.L., Li, Y., Burns, D.K., Alvarez-Buylla, A., Parada, L.F., 2009. Malignant Astrocytomas Originate from Neural Stem/Progenitor Cells in a Somatic Tumor Suppressor Mouse Model. Cancer Cell 15 1 , 45–56. doi:https://doi.org/10.1016/j.ccr.2008.12.006
2. Alcantara Llaguno, S.R., Wang, Z., Sun, D., Chen, J., Xu, J., Kim, E., Hatanpaa, K.J., Raisanen, J.M., Burns, D.K., Johnson, J.E., Parada, L.F., 2015. Adult Lineage Restricted CNS Progenitors Specify Distinct Glioblastoma Subtypes. Cancer Cell 28 4 , 429–440. doi:https://doi.org/10.1016/j.ccell.2015.09.007
3. Allen, N.J., Bennett, M.L., Foo, L.C., Wang, G.X., Chakraborty, C., Smith, S.J., Barres, B.A., 2012. Astrocyte glypicans 4 and 6 promote formation of excitatory synapses via GluA1 AMPA receptors. Nature 486 7403 , 410–414. doi:https://doi.org/10.1038/nature11059
4. Al-Nedawi, K., Meehan, B., Micallef, J., Lhotak, V., May, L., Guha, A., Rak, J., 2008. Intercellular transfer of the oncogenic receptor EGFRvIII by microvesicles derived from tumour cells. Nat Cell Biol 10 5 , 619–624. doi:https://doi.org/10.1038/ncb1725
5. Anastasaki, C., Chatterjee, J., Cobb, O., Sanapala, S., Scheaffer, S.M., De Andrade Costa, A., Wilson, A.F., Kernan, C.M., Zafar, A.H., Ge, X., Garbow, J.R., Rodriguez, F.J., Gutmann, D.H., 2022a. Human induced pluripotent stem cell engineering establishes a humanized mouse platform for pediatric low-grade glioma modeling. Acta Neuropathol Commun 10 1 , 120. doi:https://doi.org/10.1186/s40478-022-01428-2
6. Anastasaki, C., Mo, J., Chen, J.-K., Chatterjee, J., Pan, Y., Scheaffer, S.M., Cobb, O., Monje, M., Le, L.Q., Gutmann, D.H., 2022b. Neuronal hyperexcitability drives central and peripheral nervous system tumor progression in models of neurofibromatosis-1. Nat Commun 13, 2785. doi:https://doi.org/10.1038/s41467-022-30466-6
7. Bajenaru, M.L., Garbow, J.R., Perry, A., Hernandez, M.R., Gutmann, D.H., 2005. Natural history of neurofibromatosis 1–associated optic nerve glioma in mice. Annals of Neurology 57 1 , 119–127. doi:https://doi.org/10.1002/ana.20337
8. Bajenaru, M.L., Hernandez, M.R., Perry, A., Zhu, Y., Parada, L.F., Garbow, J.R., Gutmann, D.H., 2003. Optic nerve glioma in mice requires astrocyte Nf1 gene inactivation and Nf1 brain heterozygosity. Cancer Res 63 24 , 8573–8577.
9. Bakhti, M., Winter, C., Simons, M., 2011. Inhibition of myelin membrane sheath formation by oligodendrocyte-derived exosome-like vesicles. J Biol Chem 286 1 , 787–796. doi:https://doi.org/10.1074/jbc.M110.190009
10. Barami, K., Sloan, A.E., Rojiani, A., Schell, M.J., Staller, A., Brem, S., 2009. Relationship of gliomas to the ventricular walls. J Clin Neurosci 16 2 , 195–201. doi:https://doi.org/10.1016/j.jocn.2008.03.006
11. Basso, M., Bonetto, V., 2016. Extracellular Vesicles and a Novel Form of Communication in the Brain. Front Neurosci 10, 127. doi:https://doi.org/10.3389/fnins.2016.00127
12. Berg, D.A., Belnoue, L., Song, H., Simon, A., 2013. Neurotransmitter-mediated control of neurogenesis in the adult vertebrate brain. Development 140 12 , 2548–2561. doi:https://doi.org/10.1242/dev.088005
13. Bergles, D.E., Richardson, W.D., 2016. Oligodendrocyte Development and Plasticity. Cold Spring Harb Perspect Biol 8 2 , a020453. doi:https://doi.org/10.1101/cshperspect.a020453

14. Bergles, D.E., Roberts, J.D., Somogyi, P., Jahr, C.E., 2000. Glutamatergic synapses on oligodendrocyte precursor cells in the hippocampus. Nature 405 6783 , 187–191. doi:https://doi.org/10.1038/35012083
15. Bittman, K.S., LoTurco, J.J., 1999. Differential regulation of connexin 26 and 43 in murine neocortical precursors. Cereb Cortex 9 2 , 188–195. doi:https://doi.org/10.1093/cercor/9.2.188
16. Brossier, N.M., Thondapu, S., Cobb, O.M., Dahiya, S., Gutmann, D.H., 2021. Temporal, spatial, and genetic constraints contribute to the patterning and penetrance of murine neurofibromatosis-1 optic glioma. Neuro Oncol 23 4 , 625–637. doi:https://doi.org/10.1093/neuonc/noaa237
17. Buckingham, S.C., Campbell, S.L., Haas, B.R., Montana, V., Robel, S., Ogunrinu, T., Sontheimer, H., 2011. Glutamate release by primary brain tumors induces epileptic activity. Nat Med 17 10 , 1269–1274. doi:https://doi.org/10.1038/nm.2453
18. Budnik, V., Ruiz-Cañada, C., Wendler, F., 2016. Extracellular vesicles round off communication in the nervous system. Nat Rev Neurosci 17 3 , 160–172. doi:https://doi.org/10.1038/nrn.2015.29
19. Cahoy, J.D., Emery, B., Kaushal, A., Foo, L.C., Zamanian, J.L., Christopherson, K.S., Xing, Y., Lubischer, J.L., Krieg, P.A., Krupenko, S.A., Thompson, W.J., Barres, B.A., 2008. A Transcriptome Database for Astrocytes, Neurons, and Oligodendrocytes: A New Resource for Understanding Brain Development and Function. J Neurosci 28 1 , 264–278. doi:https://doi.org/10.1523/JNEUROSCI.4178-07.2008
20. Caldeira, M.V., Melo, C.V., Pereira, D.B., Carvalho, R., Correia, S.S., Backos, D.S., Carvalho, A.L., Esteban, J.A., Duarte, C.B., 2007. Brain-derived neurotrophic factor regulates the expression and synaptic delivery of alpha-amino-3-hydroxy-5-methyl-4-isoxazole propionic acid receptor subunits in hippocampal neurons. J Biol Chem 282 17 , 12619–12628. doi:https://doi.org/10.1074/jbc.M700607200
21. Campbell, S.L., Buckingham, S.C., Sontheimer, H., 2012. Human glioma cells induce hyperexcitability in cortical networks. Epilepsia 53 8 , 1360–1370. doi:https://doi.org/10.1111/j.1528-1167.2012.03557.x
22. Chapleau, C.A., Pozzo-Miller, L., 2012. Divergent roles of p75NTR and Trk receptors in BDNF's effects on dendritic spine density and morphology. Neural Plast 2012, 578057. doi:https://doi.org/10.1155/2012/578057
23. Chen, P., Wang, W., Liu, R., Lyu, J., Zhang, L., Li, B., Qiu, B., Tian, A., Jiang, W., Ying, H., Jing, R., Wang, Q., Zhu, K., Bai, R., Zeng, L., Duan, S., Liu, C., 2022. Olfactory sensory experience regulates gliomagenesis via neuronal IGF1. Nature 606 7914 , 550–556. doi:https://doi.org/10.1038/s41586-022-04719-9
24. Chen, Q., Boire, A., Jin, X., Valiente, M., Er, E.E., Lopez-Soto, A., Jacob, L., Patwa, R., Shah, H., Xu, K., Cross, J.R., Massagué, J., 2016. Carcinoma-astrocyte gap junctions promote brain metastasis by cGAMP transfer. Nature 533 7604 , 493–498. doi:https://doi.org/10.1038/nature18268
25. Chen, Y.-H., McGowan, L.D., Cimino, P.J., Dahiya, S., Leonard, J.R., Lee, D.Y., Gutmann, D.H., 2015. Mouse low-grade gliomas contain cancer stem cells with unique molecular and functional properties. Cell Rep 10 11 , 1899–1912. doi:https://doi.org/10.1016/j.celrep.2015.02.041
26. Dahiya, S., Lee, D.Y., Gutmann, D.H., 2011. Comparative Characterization of the Human and Mouse Third Ventricle Germinal Zones. J Neuropathol Exp Neurol 70 7 , 622–633. doi:https://doi.org/10.1097/NEN.0b013e31822200aa
27. De Biase, L.M., Nishiyama, A., Bergles, D.E., 2010. Excitability and synaptic communication within the oligodendrocyte lineage. J Neurosci 30 10 , 3600–3611. doi:https://doi.org/10.1523/JNEUROSCI.6000-09.2010
28. de Faria, O., Gonsalvez, D.G., Nicholson, M., Xiao, J., 2019. Activity-dependent central nervous system myelination throughout life. J Neurochem 148 4 , 447–461. doi:https://doi.org/10.1111/jnc.14592

29. de Groot, J., Sontheimer, H., 2011. Glutamate and the Biology of Gliomas. Glia 59 8 , 1181–1189. doi:https://doi.org/10.1002/glia.21113
30. De Meulenaere, V., Bonte, E., Verhoeven, J., Kalala Okito, J.-P., Pieters, L., Vral, A., De Wever, O., Leybaert, L., Goethals, I., Vanhove, C., Descamps, B., Deblaere, K., 2019. Adjuvant therapeutic potential of tonabersat in the standard treatment of glioblastoma: A preclinical F98 glioblastoma rat model study. PLoS One 14 10 , e0224130. doi:https://doi.org/10.1371/journal.pone.0224130
31. Deisseroth, K., 2015. Optogenetics: 10 years of microbial opsins in neuroscience. Nat Neurosci 18 9 , 1213–1225. doi:https://doi.org/10.1038/nn.4091
32. Doebele, R.C., Drilon, A., Paz-Ares, L., Siena, S., Shaw, A.T., Farago, A.F., Blakely, C.M., Seto, T., Cho, B.C., Tosi, D., Besse, B., Chawla, S.P., Bazhenova, L., Krauss, J.C., Chae, Y.K., Barve, M., Garrido-Laguna, I., Liu, S.V., Conkling, P., John, T., Fakih, M., Sigal, D., Loong, H.H., Buchschacher, G.L., Garrido, P., Nieva, J., Steuer, C., Overbeck, T.R., Bowles, D.W., Fox, E., Riehl, T., Chow-Maneval, E., Simmons, B., Cui, N., Johnson, A., Eng, S., Wilson, T.R., Demetri, G.D., 2020. Entrectinib in patients with advanced or metastatic NTRK fusion-positive solid tumours: integrated analysis of three phase 1–2 trials. Lancet Oncol 21 2 , 271–282. doi:https://doi.org/10.1016/S1470-2045(19)30691-6
33. Edwards, D.R., Handsley, M.M., Pennington, C.J., 2008. The ADAM metalloproteinases. Mol Aspects Med 29 5 , 258–289. doi:https://doi.org/10.1016/j.mam.2008.08.001
34. Farrell, C.J., Plotkin, S.R., 2007. Genetic causes of brain tumors: neurofibromatosis, tuberous sclerosis, von Hippel-Lindau, and other syndromes. Neurol Clin 25 4 , 925–946, viii. doi:https://doi.org/10.1016/j.ncl.2007.07.008
35. Feldman, D., Banerjee, A., Sur, M., 2016. Developmental Dynamics of Rett Syndrome. Neural Plast 2016, 6154080. doi:https://doi.org/10.1155/2016/6154080
36. Filbin, M.G., Tirosh, I., Hovestadt, V., Shaw, M.L., Escalante, L.E., Mathewson, N.D., Neftel, C., Frank, N., Pelton, K., Hebert, C.M., Haberler, C., Yizhak, K., Gojo, J., Egervari, K., Mount, C., van Galen, P., Bonal, D.M., Nguyen, Q.-D., Beck, A., Sinai, C., Czech, T., Dorfer, C., Goumnerova, L., Lavarino, C., Carcaboso, A.M., Mora, J., Mylvaganam, R., Luo, C.C., Peyrl, A., Popović, M., Azizi, A., Batchelor, T.T., Frosch, M.P., Martinez-Lage, M., Kieran, M.W., Bandopadhayay, P., Beroukhim, R., Fritsch, G., Getz, G., Rozenblatt-Rosen, O., Wucherpfennig, K.W., Louis, D.N., Monje, M., Slavc, I., Ligon, K.L., Golub, T.R., Regev, A., Bernstein, B.E., Suvà, M.L., 2018. Developmental and oncogenic programs in H3K27M gliomas dissected by single-cell RNA-seq. Science 360 6386 , 331–335. doi:https://doi.org/10.1126/science.aao4750
37. Fisher, M.J., Jones, D.T.W., Li, Y., Guo, X., Sonawane, P.S., Waanders, A.J., Phillips, J.J., Weiss, W.A., Resnick, A.C., Gosline, S., Banerjee, J., Guinney, J., Gnekow, A., Kandels, D., Foreman, N.K., Korshunov, A., Ryzhova, M., Massimi, L., Gururangan, S., Kieran, M.W., Wang, Z., Fouladi, M., Sato, M., Øra, I., Holm, S., Markham, S.J., Beck, P., Jäger, N., Wittmann, A., Sommerkamp, A.C., Sahm, F., Pfister, S.M., Gutmann, D.H., 2021. Integrated molecular and clinical analysis of low-grade gliomas in children with neurofibromatosis type 1 (NF1). Acta Neuropathol 141 4 , 605–617. doi:https://doi.org/10.1007/s00401-021-02276-5
38. Fröhlich, D., Kuo, W.P., Frühbeis, C., Sun, J.-J., Zehendner, C.M., Luhmann, H.J., Pinto, S., Toedling, J., Trotter, J., Krämer-Albers, E.-M., 2014. Multifaceted effects of oligodendroglial exosomes on neurons: impact on neuronal firing rate, signal transduction and gene regulation. Philos Trans R Soc Lond B Biol Sci 369 1652 , 20130510. doi:https://doi.org/10.1098/rstb.2013.0510
39. Frühbeis, C., Fröhlich, D., Kuo, W.P., Amphornrat, J., Thilemann, S., Saab, A.S., Kirchhoff, F., Möbius, W., Goebbels, S., Nave, K.-A., Schneider, A., Simons, M., Klugmann, M., Trotter, J., Krämer-Albers, E.-M., 2013. Neurotransmitter-Triggered Transfer of Exosomes Mediates Oligodendrocyte–Neuron Communication. PLoS Biol 11 7 , e1001604. doi:https://doi.org/10.1371/journal.pbio.1001604
40. Galli, R., Binda, E., Orfanelli, U., Cipelletti, B., Gritti, A., De Vitis, S., Fiocco, R., Foroni, C., Dimeco, F., Vescovi, A., 2004. Isolation and Characterization of Tumorigenic, Stem-

like Neural Precursors from Human Glioblastoma. Cancer Research 64 19 , 7011–7021. doi:https://doi.org/10.1158/0008-5472.CAN-04-1364

41. Gibson, E.M., Purger, D., Mount, C.W., Goldstein, A.K., Lin, G.L., Wood, L.S., Inema, I., Miller, S.E., Bieri, G., Zuchero, J.B., Barres, B.A., Woo, P.J., Vogel, H., Monje, M., 2014. Neuronal Activity Promotes Oligodendrogenesis and Adaptive Myelination in the Mammalian Brain. Science 344 6183 , 1252304. doi:https://doi.org/10.1126/science.1252304
42. Guo, M., Hao, Y., Feng, Y., Li, H., Mao, Y., Dong, Q., Cui, M., 2021. Microglial Exosomes in Neurodegenerative Disease. Front Mol Neurosci 14, 630808. doi:https://doi.org/10.3389/fnmol.2021.630808
43. Gutmann, D.H., McLellan, M.D., Hussain, I., Wallis, J.W., Fulton, L.L., Fulton, R.S., Magrini, V., Demeter, R., Wylie, T., Kandoth, C., Leonard, J.R., Guha, A., Miller, C.A., Ding, L., Mardis, E.R., 2013. Somatic neurofibromatosis type 1 (NF1) inactivation characterizes NF1-associated pilocytic astrocytoma. Genome Res 23 3 , 431–439. doi:https://doi.org/10.1101/gr.142604.112
44. Haag, D., Mack, N., Benites Goncalves da Silva, P., Statz, B., Clark, J., Tanabe, K., Sharma, T., Jäger, N., Jones, D.T.W., Kawauchi, D., Wernig, M., Pfister, S.M., 2021. H3.3-K27M drives neural stem cell-specific gliomagenesis in a human iPSC-derived model. Cancer Cell 39 3 , 407–422.e13. doi:https://doi.org/10.1016/j.ccell.2021.01.005
45. Hegedus, B., Yeh, T.-H., Lee, D.Y., Emnett, R.J., Li, J., Gutmann, D.H., 2008. Neurofibromin regulates somatic growth through the hypothalamic–pituitary axis. Hum Mol Genet 17 19 , 2956–2966. doi:https://doi.org/10.1093/hmg/ddn194
46. Hong, E.J., McCord, A.E., Greenberg, M.E., 2008. A biological function for the neuronal activity-dependent component of Bdnf transcription in the development of cortical inhibition. Neuron 60 4 , 610–624. doi:https://doi.org/10.1016/j.neuron.2008.09.024
47. Hughes, E.G., Orthmann-Murphy, J.L., Langseth, A.J., Bergles, D.E., 2018. Myelin remodeling through experience-dependent oligodendrogenesis in the adult somatosensory cortex. Nat Neurosci 21 5 , 696–706. doi:https://doi.org/10.1038/s41593-018-0121-5
48. Iacovelli, L., Arcella, A., Battaglia, G., Pazzaglia, S., Aronica, E., Spinsanti, P., Caruso, A., De Smaele, E., Saran, A., Gulino, A., D'Onofrio, M., Giangaspero, F., Nicoletti, F., 2006. Pharmacological Activation of mGlu4 Metabotropic Glutamate Receptors Inhibits the Growth of Medulloblastomas. J Neurosci 26 32 , 8388–8397. doi:https://doi.org/10.1523/JNEUROSCI.2285-06.2006
49. Jacques, T.S., Swales, A., Brzozowski, M.J., Henriquez, N.V., Linehan, J.M., Mirzadeh, Z., O'Malley, C., Naumann, H., Alvarez-Buylla, A., Brandner, S., 2010. Combinations of genetic mutations in the adult neural stem cell compartment determine brain tumour phenotypes. EMBO J 29 1 , 222–235. doi:https://doi.org/10.1038/emboj.2009.327
50. Jecrois, E.S., Zheng, W., Bornhorst, M., Li, Y., Treisman, D.M., Muguyo, D., Huynh, S., Andrew, S.F., Wang, Y., Jiang, J., Pierce, B.R., Mao, H., Krause, M.K., Friend, A., Nadal-Nicolas, F., Stasheff, S.F., Li, W., Zong, H., Packer, R.J., Zhu, Y., 2021. Treatment during a developmental window prevents NF1-associated optic pathway gliomas by targeting Erk-dependent migrating glial progenitors. Developmental Cell 56 20 , 2871-2885.e6. doi:https://doi.org/10.1016/j.devcel.2021.08.004
51. Jiang, S.-H., Hu, L.-P., Wang, X., Li, J., Zhang, Z.-G., 2020. Neurotransmitters: emerging targets in cancer. Oncogene 39 3 , 503–515. doi:https://doi.org/10.1038/s41388-019-1006-0
52. John Lin, C.-C., Yu, K., Hatcher, A., Huang, T.-W., Lee, H.K., Carlson, J., Weston, M.C., Chen, F., Zhang, Y., Zhu, W., Mohila, C.A., Ahmed, N., Patel, A.J., Arenkiel, B.R., Noebels, J.L., Creighton, C.J., Deneen, B., 2017. Identification of diverse astrocyte populations and their malignant analogs. Nat Neurosci 20 3 , 396–405. doi:https://doi.org/10.1038/nn.4493
53. Jung, E., Alfonso, J., Osswald, M., Monyer, H., Wick, W., Winkler, F., 2019. Emerging intersections between neuroscience and glioma biology. Nat Neurosci 22 12 , 1951–1960. doi:https://doi.org/10.1038/s41593-019-0540-y

54. Kang, H., Schuman, E.M., 1995. Long-lasting neurotrophin-induced enhancement of synaptic transmission in the adult hippocampus. Science 267 5204 , 1658–1662. doi:https://doi.org/10.1126/science.7886457
55. Kaul, A., Chen, Y.-H., Emnett, R.J., Dahiya, S., Gutmann, D.H., 2012. Pediatric glioma-associated KIAA1549:BRAF expression regulates neuroglial cell growth in a cell type-specific and mTOR-dependent manner. Genes Dev 26 23 , 2561–2566. doi:https://doi.org/10.1101/gad.200907.112
56. Kim, S.-J., Kim, J.-S., Park, E.S., Lee, J.-S., Lin, Q., Langley, R.R., Maya, M., He, J., Kim, S.-W., Weihua, Z., Balasubramanian, K., Fan, D., Mills, G.B., Hung, M.-C., Fidler, I.J., 2011. Astrocytes upregulate survival genes in tumor cells and induce protection from chemotherapy. Neoplasia 13 3 , 286–298. doi:https://doi.org/10.1593/neo.11112
57. Krishna S, Choudhury A, Keough MB, Seo K, Ni L, Kakaizada S, Lee A, Aabedi A, Popova G, Lipkin B, Cao C, Nava Gonzales C, Sudharshan R, Egladyous A, Almeida N, Zhang Y, Molinaro AM, Venkatesh HS, Daniel AGS, Shamardani K, Hyer J, Chang EF, Findlay A, Phillips JJ, Nagarajan S, Raleigh DR, Brang D, Monje M, Hervey-Jumper SL. Glioblastoma remodelling of human neural circuits decreases survival. Nature. 2023 May;617(7961):599-607. https://doi.org/10.1038/s41586-023-06036-1. Epub 2023 May 3. PMID: 37138086; PMCID: PMC10191851.
58. Kukley, M., Nishiyama, A., Dietrich, D., 2010. The fate of synaptic input to NG2 glial cells: neurons specifically downregulate transmitter release onto differentiating oligodendroglial cells. J Neurosci 30 24 , 8320–8331. doi:https://doi.org/10.1523/JNEUROSCI.0854-10.2010
59. Lachenal, G., Pernet-Gallay, K., Chivet, M., Hemming, F.J., Belly, A., Bodon, G., Blot, B., Haase, G., Goldberg, Y., Sadoul, R., 2011. Release of exosomes from differentiated neurons and its regulation by synaptic glutamatergic activity. Mol Cell Neurosci 46 2 , 409–418. doi:https://doi.org/10.1016/j.mcn.2010.11.004
60. Larjavaara, S., Mäntylä, R., Salminen, T., Haapasalo, H., Raitanen, J., Jääskeläinen, J., Auvinen, A., 2007. Incidence of gliomas by anatomic location. Neuro Oncol 9 3 , 319–325. doi:https://doi.org/10.1215/15228517-2007-016
61. Larson, J.D., Kasper, L.H., Paugh, B.S., Jin, H., Wu, G., Kwon, C.-H., Fan, Y., Shaw, T.I., Silveira, A.B., Qu, C., Xu, R., Zhu, X., Zhang, Junyuan, Russell, H.R., Peters, J.L., Finkelstein, D., Xu, B., Lin, T., Tinkle, C.L., Patay, Z., Onar-Thomas, A., Pounds, S.B., McKinnon, P.J., Ellison, D.W., Zhang, Jinghui, Baker, S.J., 2019. Histone H3.3 K27M Accelerates Spontaneous Brainstem Glioma and Drives Restricted Changes in Bivalent Gene Expression. Cancer Cell 35 1 , 140–155.e7. doi:https://doi.org/10.1016/j.ccell.2018.11.015
62. Lee, C., Hu, J., Ralls, S., Kitamura, T., Loh, Y.P., Yang, Y., Mukouyama, Y., Ahn, S., 2012. The Molecular Profiles of Neural Stem Cell Niche in the Adult Subventricular Zone. PLOS ONE 7 11 , e50501. doi:https://doi.org/10.1371/journal.pone.0050501
63. Lee, Joo Ho, Lee, Jeong Ho, 2018. The origin-of-cell harboring cancer-driving mutations in human glioblastoma. BMB Rep 51 10 , 481–483. doi:https://doi.org/10.5483/BMBRep.2018.51.10.233
64. Li, X., Guan, J., Jiang, Z., Cheng, S., Hou, W., Yao, J., Wang, Z., 2021. Microglial Exosome miR-7239-3p Promotes Glioma Progression by Regulating Circadian Genes. Neurosci Bull 37 4 , 497–510. doi:https://doi.org/10.1007/s12264-020-00626-z
65. Li, X., Wolf, M.E., 2011. Brain-derived neurotrophic factor rapidly increases AMPA receptor surface expression in rat nucleus accumbens. Eur J Neurosci 34 2 , 190–198. doi:https://doi.org/10.1111/j.1460-9568.2011.07754.x
66. Lim, D.A., Alvarez-Buylla, A., 2016. The Adult Ventricular–Subventricular Zone (V-SVZ) and Olfactory Bulb (OB) Neurogenesis. Cold Spring Harb Perspect Biol 8 5 , a018820. doi:https://doi.org/10.1101/cshperspect.a018820
67. Lin, Q., Balasubramanian, K., Fan, D., Kim, S.-J., Guo, L., Wang, H., Bar-Eli, M., Aldape, K.D., Fidler, I.J., 2010. Reactive astrocytes protect melanoma cells from chemotherapy by sequestering intracellular calcium through gap junction communication channels. Neoplasia 12 9 , 748–754. doi:https://doi.org/10.1593/neo.10602

68. Lin, S.-C., Bergles, D.E., 2004. Synaptic signaling between GABAergic interneurons and oligodendrocyte precursor cells in the hippocampus. Nat. Neurosci 7:24–32.
69. Listernick, R., Louis, D.N., Packer, R.J., Gutmann, D.H., 1997. Optic pathway gliomas in children with neurofibromatosis 1: consensus statement from the NF1 Optic Pathway Glioma Task Force. Ann Neurol 41 2 , 143–149. doi:https://doi.org/10.1002/ana.410410204
70. Liu, C., Sage, J.C., Miller, M.R., Verhaak, R.G.W., Hippenmeyer, S., Vogel, H., Foreman, O., Bronson, R.T., Nishiyama, A., Luo, L., Zong, H., 2011. Mosaic Analysis with Double Markers (MADM) Reveals Tumor Cell-of-Origin in Glioma. Cell 146 2 , 209–221. doi:https://doi.org/10.1016/j.cell.2011.06.014
71. Liu, J., Dietz, K., DeLoyht, J.M., Pedre, X., Kelkar, D., Kaur, J., Vialou, V., Lobo, M.K., Dietz, D.M., Nestler, E.J., Dupree, J., Casaccia, P., 2012. Impaired adult myelination in the prefrontal cortex of socially isolated mice. Nat Neurosci 15 12 , 1621–1623. doi:https://doi.org/10.1038/nn.3263
72. Liu, J., Dupree, J.L., Gacias, M., Frawley, R., Sikder, T., Naik, P., Casaccia, P., 2016. Clemastine Enhances Myelination in the Prefrontal Cortex and Rescues Behavioral Changes in Socially Isolated Mice. J Neurosci 36 3 , 957–962. doi:https://doi.org/10.1523/JNEUROSCI.3608-15.2016
73. Llaguno, S.A., Sun, D., Pedraza, A., Vera, E., Wang, Z., Burns, D.K., Parada, L.F., 2019. Cell of Origin Susceptibility to Glioblastoma Formation Declines with Neural Lineage Restriction. Nat Neurosci 22 4 , 545–555. doi:https://doi.org/10.1038/s41593-018-0333-8
74. LoTurco, J.J., Owens, D.F., Heath, M.J., Davis, M.B., Kriegstein, A.R., 1995. GABA and glutamate depolarize cortical progenitor cells and inhibit DNA synthesis. Neuron 15 6 , 1287–1298. doi:https://doi.org/10.1016/0896-6273(95)90008-x
75. Luikart, B.W., Nef, S., Virmani, T., Lush, M.E., Liu, Y., Kavalali, E.T., Parada, L.F., 2005. TrkB Has a Cell-Autonomous Role in the Establishment of Hippocampal Schaffer Collateral Synapses. J. Neurosci. 25 15 , 3774–3786. doi:https://doi.org/10.1523/JNEUROSCI.0041-05.2005
76. Luk, K.C., Kennedy, T.E., Sadikot, A.F., 2003. Glutamate Promotes Proliferation of Striatal Neuronal Progenitors by an NMDA Receptor-Mediated Mechanism. J Neurosci 23 6 , 2239–2250. doi:https://doi.org/10.1523/JNEUROSCI.23-06-02239.2003
77. Luk, K.C., Sadikot, A.F., 2004. Glutamate and regulation of proliferation in the developing mammalian telencephalon. Dev Neurosci 26 2–4 , 218–228. doi:https://doi.org/10.1159/000082139
78. Makinodan, M., Rosen, K.M., Ito, S., Corfas, G., 2012. A Critical Period for Social Experience–Dependent Oligodendrocyte Maturation and Myelination. Science 337 6100 , 1357–1360. doi:https://doi.org/10.1126/science.1220845
79. McKenzie, I.A., Ohayon, D., Li, H., de Faria, J.P., Emery, B., Tohyama, K., Richardson, W.D., 2014. Motor skill learning requires active central myelination. Science 346 6207 , 318–322. doi:https://doi.org/10.1126/science.1254960
80. Mesfin, F.B., Al-Dhahir, M.A., 2022. Gliomas, in: StatPearls. StatPearls Publishing, Treasure Island (FL).
81. Michotte, A., Neyns, B., Chaskis, C., Sadones, J., In 't Veld, P., 2004. Neuropathological and molecular aspects of low-grade and high-grade gliomas. Acta Neurol Belg 104 4 , 148–153.
82. Minichiello, L., 2009. TrkB signalling pathways in LTP and learning. Nat Rev Neurosci 10 12 , 850–860. doi:https://doi.org/10.1038/nrn2738
83. Mitew, S., Gobius, I., Fenlon, L.R., McDougall, S.J., Hawkes, D., Xing, Y.L., Bujalka, H., Gundlach, A.L., Richards, L.J., Kilpatrick, T.J., Merson, T.D., Emery, B., 2018. Pharmacogenetic stimulation of neuronal activity increases myelination in an axon-specific manner. Nat Commun 9, 306. doi:https://doi.org/10.1038/s41467-017-02719-2
84. Monje, M., Mitra, S.S., Freret, M.E., Raveh, T.B., Kim, J., Masek, M., Attema, J.L., Li, G., Haddix, T., Edwards, M.S.B., Fisher, P.G., Weissman, I.L., Rowitch, D.H., Vogel, H., Wong, A.J., Beachy, P.A., 2011. Hedgehog-responsive candidate cell of origin for diffuse intrinsic pontine glioma. Proc Natl Acad Sci U S A 108 11 , 4453–4458. doi:https://doi.org/10.1073/pnas.1101657108

85. Mount, C.W., Yalçın, B., Cunliffe-Koehler, K., Sundaresh, S., Monje, M., 2019. Monosynaptic tracing maps brain-wide afferent oligodendrocyte precursor cell connectivity. eLife 8, e49291. doi:https://doi.org/10.7554/eLife.49291
86. Moura, D.M.S., Brennan, E.J., Brock, R., Cocas, L.A., 2022. Neuron to Oligodendrocyte Precursor Cell Synapses: Protagonists in Oligodendrocyte Development and Myelination, and Targets for Therapeutics. Front Neurosci 15, 779125. doi:https://doi.org/10.3389/fnins.2021.779125
87. Nagaraja, S., Quezada, M.A., Gillespie, S.M., Arzt, M., Lennon, J.J., Woo, P.J., Hovestadt, V., Kambhampati, M., Filbin, M.G., Suva, M.L., Nazarian, J., Monje, M., 2019. Histone Variant and Cell Context Determine H3K27M Reprogramming of the Enhancer Landscape and Oncogenic State. Mol Cell 76 6 , 965–980.e12. doi:https://doi.org/10.1016/j.molcel.2019.08.030
88. Nagaraja, S., Vitanza, N.A., Woo, P.J., Taylor, K.R., Liu, F., Zhang, L., Li, M., Meng, W., Ponnuswami, A., Sun, W., Ma, J., Hulleman, E., Swigut, T., Wysocka, J., Tang, Y., Monje, M., 2017. Transcriptional Dependencies in Diffuse Intrinsic Pontine Glioma. Cancer Cell 31 5 , 635–652.e6. doi:https://doi.org/10.1016/j.ccell.2017.03.011
89. Nakata, H., Nakamura, S., 2007. Brain-derived neurotrophic factor regulates AMPA receptor trafficking to post-synaptic densities via IP3R and TRPC calcium signaling. FEBS Lett 581 10 , 2047–2054. doi:https://doi.org/10.1016/j.febslet.2007.04.041
90. Neftel, C., Laffy, J., Filbin, M.G., Hara, T., Shore, M.E., Rahme, G.J., Richman, A.R., Silverbush, D., Shaw, M.L., Hebert, C.M., Dewitt, J., Gritsch, S., Perez, E.M., Castro, L.N.G., Lan, X., Druck, N., Rodman, C., Dionne, D., Kaplan, A., Bertalan, M.S., Small, J., Pelton, K., Becker, S., Bonal, D., Nguyen, Q.-D., Servis, R.L., Fung, J.M., Mylvaganam, R., Mayr, L., Gojo, J., Haberler, C., Geyeregger, R., Czech, T., Slavc, I., Nahed, B.V., Curry, W.T., Carter, B.S., Wakimoto, H., Brastianos, P.K., Batchelor, Tracy, T., Stemmer-Rachamimov, A., Martinez-Lage, M., Frosch, M.P., Stamenkovic, I., Riggi, N., Rheinbay, E., Monje, M., Rozenblatt-Rosen, O., Cahill, D.P., Patel, A.P., Hunter, T., Verma, I.M., Ligon, K.L., Louis, D.N., Regev, A., Bernstein, B.E., Tirosh, I., Suvà, M.L., 2019. An integrative model of cellular states, plasticity and genetics for glioblastoma. Cell 178 4 , 835-849.e21. doi:https://doi.org/10.1016/j.cell.2019.06.024
91. Ono, K., Yasui, Y., Rutishauser, U., Miller, R.H., 1997. Focal Ventricular Origin and Migration of Oligodendrocyte Precursors into the Chick Optic Nerve. Neuron 19 2 , 283–292. doi:https://doi.org/10.1016/S0896-6273(00)80939-3
92. Osswald, M., Jung, E., Sahm, F., Solecki, G., Venkataramani, V., Blaes, J., Weil, S., Horstmann, H., Wiestler, B., Syed, M., Huang, L., Ratliff, M., Karimian Jazi, K., Kurz, F.T., Schmenger, T., Lemke, D., Gömmel, M., Pauli, M., Liao, Y., Häring, P., Pusch, S., Herl, V., Steinhäuser, C., Krunic, D., Jarahian, M., Miletic, H., Berghoff, A.S., Griesbeck, O., Kalamakis, G., Garaschuk, O., Preusser, M., Weiss, S., Liu, H., Heiland, S., Platten, M., Huber, P.E., Kuner, T., von Deimling, A., Wick, W., Winkler, F., 2015. Brain tumour cells interconnect to a functional and resistant network. Nature 528 7580 , 93–98. doi:https://doi.org/10.1038/nature16071
93. Ostrom, Q.T., Cioffi, G., Waite, K., Kruchko, C., Barnholtz-Sloan, J.S., 2021. CBTRUS Statistical Report: Primary Brain and Other Central Nervous System Tumors Diagnosed in the United States in 2014–2018. Neuro-Oncology 23 Supplement_3 , iii1–iii105. doi:https://doi.org/10.1093/neuonc/noab200
94. Pan, Y., Hysinger, J.D., Barron, T., Schindler, N.F., Cobb, O., Guo, X., Yalçın, B., Anastasaki, C., Mulinyawe, S.B., Ponnuswami, A., Scheaffer, S., Ma, Y., Chang, K.-C., Xia, X., Toonen, J.A., Lennon, J.J., Gibson, E.M., Huguenard, J.R., Liau, L.M., Goldberg, J.L., Monje, M., Gutmann, D.H., 2021. NF1 mutation drives neuronal activity-dependent initiation of optic glioma. Nature 594 7862 , 277–282. doi:https://doi.org/10.1038/s41586-021-03580-6
95. Pan, Y., Monje, M., 2022. Neuron-Glial Interactions in Health and Brain Cancer. Adv Biol (Weinh) e2200122. doi:https://doi.org/10.1002/adbi.202200122

96. Paredes, M.F., James, D., Gil-Perotin, S., Kim, H., Cotter, J.A., Ng, C., Sandoval, K., Rowitch, D.H., Xu, D., McQuillen, P.S., Garcia-Verdugo, J.-M., Huang, E.J., Alvarez-Buylla, A., 2016. Extensive migration of young neurons into the infant human frontal lobe. Science 354 6308 , aaf7073. doi:https://doi.org/10.1126/science.aaf7073
97. Paul, A., Chaker, Z., Doetsch, F., 2017. Hypothalamic regulation of regionally distinct adult neural stem cells and neurogenesis. Science 356 6345 , 1383–1386. doi:https://doi.org/10.1126/science.aal3839
98. Perea, G., Navarrete, M., Araque, A., 2009. Tripartite synapses: astrocytes process and control synaptic information. Trends in Neurosciences 32 8 , 421–431. doi:https://doi.org/10.1016/j.tins.2009.05.001
99. Potthoff, A.-L., Heiland, D.H., Evert, B.O., Almeida, F.R., Behringer, S.P., Dolf, A., Güresir, Á., Güresir, E., Joseph, K., Pietsch, T., Schuss, P., Herrlinger, U., Westhoff, M.-A., Vatter, H., Waha, A., Schneider, M., 2019. Inhibition of Gap Junctions Sensitizes Primary Glioblastoma Cells for Temozolomide. Cancers (Basel) 11 6 , E858. doi:https://doi.org/10.3390/cancers11060858
100. Sanai, N., Nguyen, T., Ihrie, R.A., Mirzadeh, Z., Tsai, H.-H., Wong, M., Gupta, N., Berger, M.S., Huang, E., Garcia-Verdugo, J.-M., Rowitch, D.H., Alvarez-Buylla, A., 2011. Corridors of Migrating Neurons in Human Brain and Their Decline during Infancy. Nature 478 7369 , 382–386. doi:https://doi.org/10.1038/nature10487
101. Sharma, P., Mesci, P., Carromeu, C., McClatchy, D.R., Schiapparelli, L., Yates, J.R., Muotri, A.R., Cline, H.T., 2019. Exosomes regulate neurogenesis and circuit assembly. Proc Natl Acad Sci U S A 116 32 , 16086–16094. doi:https://doi.org/10.1073/pnas.1902513116
102. Silberstein, S.D., Schoenen, J., Göbel, H., Diener, H.C., Elkind, A.H., Klapper, J.A., Howard, R.A., 2009. Tonabersat, a gap-junction modulator: efficacy and safety in two randomized, placebo-controlled, dose-ranging studies of acute migraine. Cephalalgia 29 Suppl 2, 17–27. doi:https://doi.org/10.1111/j.1468-2982.2009.01974.x
103. Singh, S.K., Clarke, I.D., Terasaki, M., Bonn, V.E., Hawkins, C., Squire, J., Dirks, P.B., 2003. Identification of a cancer stem cell in human brain tumors. Cancer Res 63 18 , 5821–5828.
104. Small, R.K., Riddle, P., Noble, M., 1987. Evidence for migration of oligodendrocyte – type-2 astrocyte progenitor cells into the developing rat optic nerve. Nature 328 6126 , 155–157. doi:https://doi.org/10.1038/328155a0
105. Solga, A.C., Gianino, S.M., Gutmann, D.H., 2014. NG2-cells are not the cell of origin for murine neurofibromatosis-1 (Nf1) optic glioma. Oncogene 33 3 , 289–299. doi:https://doi.org/10.1038/onc.2012.580
106. Solga, A.C., Toonen, J.A., Pan, Y., Cimino, P.J., Ma, Y., Castillon, G.A., Gianino, S.M., Ellisman, M.H., Lee, D.Y., Gutmann, D.H., 2017. The cell of origin dictates the temporal course of neurofibromatosis-1 (Nf1) low-grade glioma formation. Oncotarget 8 29 , 47206–47215. doi:https://doi.org/10.18632/oncotarget.17589
107. Song, J., Patel, R.V., Sharif, M., Ashokan, A., Michaelides, M., 2022. Chemogenetics as a neuromodulatory approach to treating neuropsychiatric diseases and disorders. Mol Ther 30 3 , 990–1005. doi:https://doi.org/10.1016/j.ymthe.2021.11.019
108. Song, J., Zhong, C., Bonaguidi, M.A., Sun, G.J., Hsu, D., Gu, Y., Meletis, K., Huang, Z.J., Ge, S., Enikolopov, G., Deisseroth, K., Luscher, B., Christian, K.M., Ming, G., Song, H., 2012. Neuronal circuitry mechanism regulating adult quiescent neural stem-cell fate decision. Nature 489 7414 , 150–154. doi:https://doi.org/10.1038/nature11306
109. Sreedharan, S., Maturi, N.P., Xie, Y., Sundström, A., Jarvius, M., Libard, S., Alafuzoff, I., Weishaupt, H., Fryknäs, M., Larsson, R., Swartling, F.J., Uhrbom, L., 2017. Mouse Models of Pediatric Supratentorial High-grade Glioma Reveal How Cell-of-Origin Influences Tumor Development and Phenotype. Cancer Research 77 3 , 802–812. doi:https://doi.org/10.1158/0008-5472.CAN-16-2482
110. Steadman, P.E., Xia, F., Ahmed, M., Mocle, A.J., Penning, A.R.A., Geraghty, A.C., Steenland, H.W., Monje, M., Josselyn, S.A., Frankland, P.W., 2020. Disruption of Oligodendrogenesis Impairs Memory Consolidation in Adult Mice. Neuron 105 1 , 150–164.e6. doi:https://doi.org/10.1016/j.neuron.2019.10.013

111. Steinhäser, C., Jabs, R., Kettenmann, H., 1994. Properties of GABA and glutamate responses in identified glial cells of the mouse hippocampal slice. Hippocampus 4 1 , 19–35. doi:https://doi.org/10.1002/hipo.450040105
112. Takahashi, T., 2021. Multiple Roles for Cholinergic Signaling from the Perspective of Stem Cell Function. Int J Mol Sci 22 2 , E666. doi:https://doi.org/10.3390/ijms22020666
113. Taylor, K.R., Barron, T., Zhang, H., Hui, A., Hartmann, G., Ni, L., Venkatesh, H.S., Du, P., Mancusi, R., Yalçin, B., Chau, I., Ponnuswami, A., Aziz-Bose, R., Monje, M., 2021. Glioma synapses recruit mechanisms of adaptive plasticity. doi:https://doi.org/10.1101/2021.11.04.467325
114. Tchoghandjian, A., Fernandez, C., Colin, C., El Ayachi, I., Voutsinos-Porche, B., Fina, F., Scavarda, D., Piercecchi-Marti, M.-D., Intagliata, D., Ouafik, L., Fraslon-Vanhulle, C., Figarella-Branger, D., 2009. Pilocytic astrocytoma of the optic pathway: a tumour deriving from radial glia cells with a specific gene signature. Brain 132 Pt 6 , 1523–1535. doi:https://doi.org/10.1093/brain/awp048
115. Tomita, Y., Shimazu, Y., Somasundaram, A., Tanaka, Y., Takata, N., Ishi, Y., Gadd, S., Hashizume, R., Angione, A., Pinero, G., Hambardzumyan, D., Brat, D.J., Hoeman, C.M., Becher, O.J., 2022. A novel mouse model of diffuse midline glioma initiated in neonatal oligodendrocyte progenitor cells highlights cell-of-origin dependent effects of H3K27M. Glia 70 9 , 1681–1698. doi:https://doi.org/10.1002/glia.24189
116. Trujillo, C.A., Schwindt, T.T., Martins, A.H., Alves, J.M., Mello, L.E., Ulrich, H., 2009. Novel perspectives of neural stem cell differentiation: from neurotransmitters to therapeutics. Cytometry A 75 1 , 38–53. doi:https://doi.org/10.1002/cyto.a.20666
117. Vélez-Fort, M., Maldonado, P.P., Butt, A.M., Audinat, E., Angulo, M.C., 2010. Postnatal Switch from Synaptic to Extrasynaptic Transmission between Interneurons and NG2 Cells. J Neurosci 30 20 , 6921–6929. doi:https://doi.org/10.1523/JNEUROSCI.0238-10.2010
118. Venkataramani, V., Tanev, D.I., Strahle, C., Studier-Fischer, A., Fankhauser, L., Kessler, T., Körber, C., Kardorff, M., Ratliff, M., Xie, R., Horstmann, H., Messer, M., Paik, S.P., Knabbe, J., Sahm, F., Kurz, F.T., Acikgöz, A.A., Herrmannsdörfer, F., Agarwal, A., Bergles, D.E., Chalmers, A., Miletic, H., Turcan, S., Mawrin, C., Hänggi, D., Liu, H.-K., Wick, W., Winkler, F., Kuner, T., 2019. Glutamatergic synaptic input to glioma cells drives brain tumour progression. Nature 573 7775 , 532–538. doi:https://doi.org/10.1038/s41586-019-1564-x
119. Venkataramani, V., Yang, Y., Schubert, M.C., Reyhan, E., Tetzlaff, S.K., Wißmann, N., Botz, M., Soyka, S.J., Beretta, C.A., Pramatarov, R.L., Fankhauser, L., Garofano, L., Freudenberg, A., Wagner, J., Tanev, D.I., Ratliff, M., Xie, R., Kessler, T., Hoffmann, D.C., Hai, L., Dörflinger, Y., Hoppe, S., Yabo, Y.A., Golebiewska, A., Niclou, S.P., Sahm, F., Lasorella, A., Slowik, M., Döring, L., Iavarone, A., Wick, W., Kuner, T., Winkler, F., 2022. Glioblastoma hijacks neuronal mechanisms for brain invasion. Cell 185 16 , 2899–2917.e31. doi:https://doi.org/10.1016/j.cell.2022.06.054
120. Venkatesh, H., Monje, M., 2017. Neuronal activity in ontogeny and oncology. Trends Cancer 3 2 , 89–112. doi:https://doi.org/10.1016/j.trecan.2016.12.008
121. Venkatesh, H.S., Johung, T.B., Caretti, V., Noll, A., Tang, Y., Nagaraja, S., Gibson, E.M., Mount, C.W., Polepalli, J., Mitra, S.S., Woo, P.J., Malenka, R.C., Vogel, H., Bredel, M., Mallick, P., Monje, M., 2015. Neuronal Activity Promotes Glioma Growth through Neuroligin-3 Secretion. Cell 161 4 , 803–816. doi:https://doi.org/10.1016/j.cell.2015.04.012
122. Venkatesh, H.S., Morishita, W., Geraghty, A.C., Silverbush, D., Gillespie, S.M., Arzt, M., Tam, L.T., Espenel, C., Ponnuswami, A., Ni, L., Woo, P.J., Taylor, K.R., Agarwal, A., Regev, A., Brang, D., Vogel, H., Hervey-Jumper, S., Bergles, D.E., Suvà, M.L., Malenka, R.C., Monje, M., 2019. Electrical and synaptic integration of glioma into neural circuits. Nature 573 7775 , 539–545. doi:https://doi.org/10.1038/s41586-019-1563-y
123. Venkatesh, H.S., Tam, L.T., Woo, P.J., Lennon, J., Nagaraja, S., Gillespie, S.M., Ni, J., Duveau, D.Y., Morris, P.J., Zhao, J.J., Thomas, C.J., Monje, M., 2017. Targeting neuronal activity-regulated neuroligin-3 dependency in high-grade glioma. Nature 549 7673 , 533–537. doi:https://doi.org/10.1038/nature24014

124. Webb, S.E., Moreau, M., Leclerc, C., Miller, A.L., 2005. Calcium transients and neural induction in vertebrates. Cell Calcium 37 5 , 375–385. doi:https://doi.org/10.1016/j.ceca.2005.01.005
125. Xiao, L., Ohayon, D., McKenzie, I.A., Sinclair-Wilson, A., Wright, J.L., Fudge, A.D., Emery, B., Li, H., Richardson, W.D., 2016. Rapid production of new oligodendrocytes is required in the earliest stages of motor skill learning. Nat Neurosci 19 9 , 1210–1217. doi:https://doi.org/10.1038/nn.4351
126. Younes, S.T., Herrington, B., 2020. In silico analysis identifies a putative cell-of-origin for BRAF fusion-positive cerebellar pilocytic astrocytoma. PLoS One 15 11 , e0242521. doi:https://doi.org/10.1371/journal.pone.0242521
127. Yu, K., Lin, C.-C.J., Hatcher, A., Lozzi, B., Kong, K., Huang-Hobbs, E., Cheng, Y.-T., Beechar, V.B., Zhu, W., Zhang, Y., Chen, F., Mills, G.B., Mohila, C.A., Creighton, C.J., Noebels, J.L., Scott, K.L., Deneen, B., 2020. PIK3CA variants selectively initiate brain hyperactivity during gliomagenesis. Nature 578 7793 , 166–171. doi:https://doi.org/10.1038/s41586-020-1952-2
128. Zeng, Q., Michael, I.P., Zhang, P., Saghafinia, S., Knott, G., Jiao, W., McCabe, B.D., Galván, J.A., Robinson, H.P.C., Zlobec, I., Ciriello, G., Hanahan, D., 2019. Synaptic proximity enables NMDAR signaling to promote brain metastasis. Nature 573 7775 , 526–531. doi:https://doi.org/10.1038/s41586-019-1576-6
129. Zhang, L., Zhang, S., Yao, J., Lowery, F.J., Zhang, Q., Huang, W.-C., Li, P., Li, M., Wang, X., Zhang, C., Wang, H., Ellis, K., Cheerathodi, M., McCarty, J.H., Palmieri, D., Saunus, J., Lakhani, S., Huang, S., Sahin, A.A., Aldape, K.D., Steeg, P.S., Yu, D., 2015. Microenvironment-induced PTEN loss by exosomal microRNA primes brain metastasis outgrowth. Nature 527 7576 , 100–104. doi:https://doi.org/10.1038/nature15376

Chapter 2
Neuron-Cancer Synaptic and Other Electrical Signaling

Humsa S. Venkatesh

Activity-regulated release of growth factors into the tumor microenvironment clearly represents part of the mechanism by which neuronal activity influences brain tumor growth [1–3] (i.e., NLGN3, BDNF, IGF-1 as discussed in previous chapters) but alone is insufficient to explain the magnitude of the effect the nervous system has on malignant disease progression. Recent studies have thus clarified direct electrochemical communication exists between neurons and cancer cells [4, 5]. This phenomenon is largely exemplified in the research of primary brain cancer pathophysiology, which will thus be used as a case study for this chapter.

Cancer progression mechanisms appear to emulate processes of normal cell development. Thus, mirroring normal oligodendrocyte precursor cells (OPCs), the putative cellular origins of glioma [6–10], both adult and pediatric gliomas, have been found to form bona fide synapses with neighboring neurons to receive micro-environmental cues. This is highly reminiscent of electrical communication at the axon-glial synapse that forms between neurons and OPCs in the developing brain [11, 12]. In back-to-back studies, authors found that in molecularly and clinically distinct types of glioma, a subpopulation of malignant cells express synapse-associated genes at high levels. Further, single-cell sequencing analysis of both patient glioma biopsy samples and patient-derived glioma xenograft tissue revealed the malignant cells that were most highly enriched for these synapse-related genes (glutamatergic receptors and synaptic structural proteins) represented the same sub-population of cells that most closely resembled OPCs, as assessed by lineage and stemness scores defined in previous studies [6]. These findings suggest that normal mechanisms of development may be hijacked for the purposes of malignant growth [4]. Using ultrastructural analyses with immuno- and correlative electron

H. S. Venkatesh (✉)
Department of Neurology, Brigham and Women's Hospital, Boston, MA, USA

Department of Neurology, Harvard Medical School, Boston, MA, USA
e-mail: hvenkatesh@bwh.harvard.edu

M. Amit, N. N. Scheff (eds.), *Cancer Neuroscience*,
https://doi.org/10.1007/978-3-031-32429-1_2

microscopy together with super-resolution light microscopy, synaptic structures were identified in glioma tissue that unambiguously characterized malignant gliomas as the postsynaptic cell type and suggested direct electrochemical communication between glioma cells and neurons in the tumor microenvironment [4, 5].

To determine whether these synaptic structures were truly electrophysiologically functional, pediatric gliomas were xenografted specifically into the CA1 region of the hippocampus, a very well-mapped circuit of the brain. Using whole-cell patch clamp recordings of the fluorescently (i.e., GFP) labeled tumor cells, electrophysiological responses to neuronal stimulation of CA3 Schaffer collateral afferent axons were recorded. A subpopulation of glioma cells exhibited evoked excitatory postsynaptic currents (EPSCs; <5 ms) [4]. Similar results were also found in adult glioma cultures [5]. In both cases, these EPSCs were found to be specifically mediated by calcium-permeable AMPA receptors and were abrogated by use of AMPA receptor inhibitors (both NBQX and NASPM), demonstrating glutamatergic input. Further, specifically in pediatric diffuse pontine glioma, malignant cells have now additionally been shown to receive synaptic signals from GABAergic neurons (Barron et al., Biorxviv 2022). These data, for the first time, indicate that malignant glioma cells possess the ability to functionally integrate into neural circuits to receive electrical input from neurons.

In addition to these direct synaptic currents, in a separate subpopulation of cells, neuronal activity evoked a second, non-synaptic electrophysiological response characterized by a prolonged (>1 s) depolarization. These longer duration currents were blocked by tetrodotoxin or barium and induced by potassium, indicating neuronal activity-dependent potassium flux reminiscent of astrocytic currents. The amplitude of these prolonged currents was reduced by gap junction inhibitors, supporting the concept that gap junction-mediated tumor interconnections can function to amplify evoked potassium currents in an electrically coupled network [4]. Further, using real-time two-photon calcium imaging, glioma cells were found to demonstrate depolarization and synchronous current flow through these malignant networks in response to neuronal stimulation. These observations suggest activity-regulated depolarizing currents propagate through cooperative tumor networks to influence glioma progression [4, 5]. Though neuronal activity was found to be the primary input to glioma cells that induced these calcium transients, recent work has additionally identified specific glioma "hub cells" that demonstrate autonomous rhythmic calcium activity mediated by potassium channel KCa3.1 [13]. This cell-intrinsic mechanism of depolarization defines an additional electrical cue that is critical to glioma progression. In the future, the mechanistic role of these calcium waves and associated activated downstream signaling pathways in glioma needs to be further studied to better understand depolarization-mediated growth.

Membrane depolarization of normal neural precursor cells regulates proliferation, differentiation, and survival in normal neurodevelopment [14–21]. Thus, understanding the functional consequence of this activity-mediated membrane depolarization in glioma cells is essential to understanding its therapeutic potential. Using in vivo optogenetic techniques to directly depolarize xenografted glioma cells, studies found that glioma depolarization robustly promoted proliferation [4], while pharmacologically or genetically blocking AMPA receptor mediated electrochemical signaling inhibited glioma xenograft growth and extended mouse survival [4, 5]. These studies highlight the clear therapeutic benefit of blocking neural signaling to malignant tissue and suggest a novel angle to treating these aggressive tumors. As further evidence for the functional significance of depolarization, in more recent studies, neuronal activity-mediated downstream calcium signals were found to promote invasion in adult gliomas [22]. Here, Venkataramani and colleagues found that neuronal activity induced downstream calcium signals in adult glioma cells enhanced the formation of tumor microtubes (TMs), ultralong, neurite-like membrane protrusions, that drive the invasion process [22–24]. Together, these studies indicate that the electrical signals relayed to malignant tumor networks can heavily influence the progression of brain tumors and represent a critical component of cancer pathophysiology that will need to be appreciated to effectively treat these cancers.

A plethora of adaptive plasticity mechanisms normally occur in the healthy brain as a means of shaping the form and function of distinct neuronal circuits [25–28]. Thus, it is possible that malignant neuron-cancer synapses can similarly be modulated (strengthened or weakened) based on external signals. Taylor and colleagues found that in addition to promoting growth, the activity-dependent secreted growth factor, brain-derived neurotrophic factor (BDNF), promotes the synaptic connections and strength of malignant synapses. By signaling through the receptor TrkB, BDNF promoted AMPA receptor trafficking to the glioma cell membrane, resulting in increased amplitude of glutamate-evoked currents in the malignant cells. BDNF-TrkB signaling further increased the number of neuron-glioma cell synapses in the tumor microenvironment. This study indicates that in addition to integrating into healthy neural circuits, gliomas are able to further reinforce these malignant synapses to drive progression. Further understanding of the complex mechanisms through which cancers can co-opt normal mechanisms of neural plasticity is needed in order to target and disrupt these dynamic neuron-cancer interactions.

Seizures induced by high levels of neuronal activity are often seen in patients presenting with glioma or brain metastases. Further, in preclinical models, gliomas have been shown to similarly create a hyperexcitable neuronal environment. This remodeling of the neural microenvironment is mediated through the secretion of various synaptogenic factors (glypican-3, thrombospondin-1), reduction of GABAergic inhibitory signaling, and glutamate release through the xc-cystine-glutamate transporter system [29–33]. Emphasizing this phenomenon, human intraoperative electrocorticography of awake patients just prior to resection also illustrated a clear increase in cortical excitability in the glioma-infiltrated brain even when compared to healthy tissue [4, 34]. Recent studies have now demonstrated that these glioma-induced changes to local neuronal hyperexcitability can further influence neural circuits involving cognition (Krishna et al., Nature 2023). Intracranial brain recordings during lexical retrieval language tasks found that gliomas remodel functional neural circuitry such that task-relevant neural responses activated tumor-infiltrated cortex. Furthermore, the increased level of functional connectivity negatively affected patient survival, suggesting that gliomas may contribute to network-level changes in patients impacting both cognition and growth (Krishna et al., Nature 2023). These studies emphasize the positive feedback mechanisms by which gliomas increase neuronal excitability and thus potentiate the mechanisms of activity-regulated glioma progression.

Brain metastases also represent a class of cancers in which these electrical signaling pathways may be critically vital to malignant progression. In metastatic breast cancer, Zeng et al. illustrated that malignant cells in the brain take advantage of neuronal signaling by sitting adjacent to neuronal synapses forming a "tripartite synapse." This allowed the uptake of glutamate, which then led to activation of NMDA receptor signaling in breast cancer cells that promoted metastatic outgrowth. Recent studies have further demonstrated that small cell lung cancers (SCLC) that have metastasized to the brain can similarly co-opt these neural mechanisms of plasticity to integrate into neural circuits. Elevated neuronal activity and concordant activity-dependent membrane depolarization was found to drive intracranial SCLC growth (Savchuk et al., Biorxiv 2023). Thus, there is a clear indication that metastatic brain cancer cells may be capable of similarly interacting with active neurons. Outside of the brain, ion channels have additionally been associated with driving the metastasis process. For example, genetic and pharmacologic modulation of potassium channel expression demonstrated that the alteration of resting membrane potential drives metastatic breast cancer progression [35]. Here, hyperpolarization led to cadherin-11 mediated MAPK signaling that was involved in cell migration and invasion. Further, reversing this change in membrane potential by inhibiting these potassium channels decreased overall metastasis in this triple negative breast cancer model [35]. Recent studies have more broadly suggested that a single sodium leak channel, NALCN, may regulate malignant cell dissemination and metastasis in a gastric, intestinal, and pancreatic adenocarcinomas [36]. Deletion of NALCN from these cancers did not alter tumor incidence but increased the number of circulating tumor cells and metastases. In the future, understanding whether direct synaptic and functional integration occurs in this subtype of cancers will be important to fully appreciate the electrical aspects of the metastatic process.

Outside of the brain, innervation has been shown to influence a number of different cancers across tissue types (as covered in other chapters). Yet, whether direct neuron-cancer electrical signaling extends to extracranial malignancies has yet to be determined. Moving forward, this will be an informative and necessary area of research. There already have been several studies suggesting that ion channels are critically important to the progression of a number of different cancers (reviewed here [37–40]). In this context, the dependence upon neural activity and the role of electrophysiological membrane depolarization will provide insight into novel cellular communication mechanisms that remain unexplored, in part due to the fact that technologies to visualize, modulate, and quantify electrical activity in the periphery remain limited. Still, as further interdisciplinary investigation into this field continues, it will be crucial to identify common mechanisms through which cancers co-opt the nervous system to fuel their own growth.

The neural regulation of cancers is a burgeoning field that highlights the nervous system's central role in facilitating tumor progression. The work described here

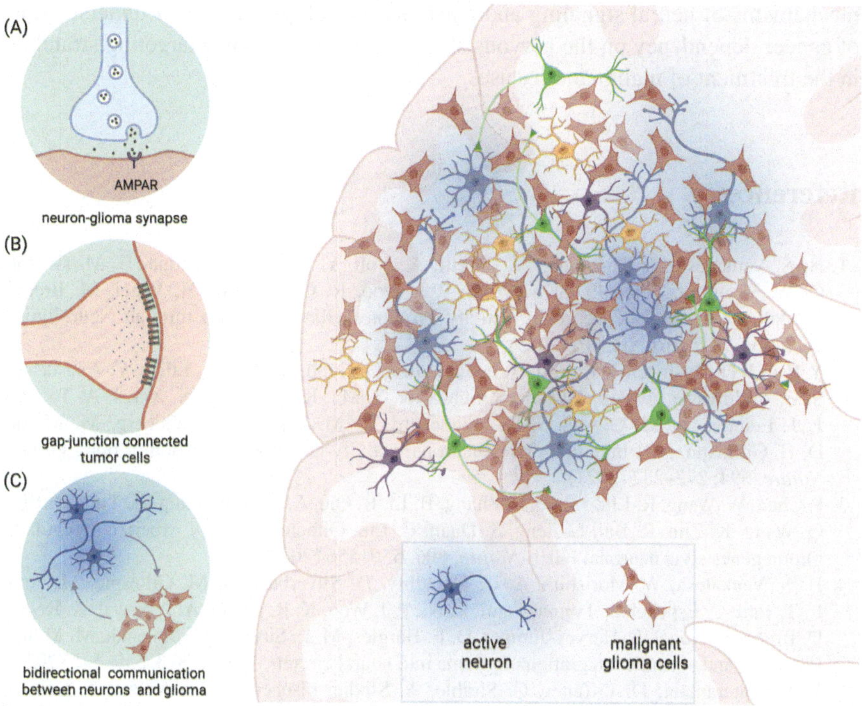

Fig. 2.1 Schematic of the glioma microenvironment as an example of the neural regulation of cancer (**a**) Synaptic communication between neurons and glioma cells induce membrane depolarization and downstream signaling in malignant tumor cells to drive growth (**b**) Neuronal activity driven currents in glioma are amplified through gap junction connected tumor cell networks (**c**) Bidirectional paracrine signaling between neurons and glioma cells includes the secretion of activity-dependent growth factors to induce a mitogenic effect on glioma, as well as cancer-derived factors that promote a hyperexcitable neuronal microenvironment

identifies activity-dependent mitogen secretion, synaptic neurotransmission, and gap junction-mediated electrical coupling as novel mechanisms controlling glioma development (Fig. 2.1). Future studies to elucidate through what mechanisms various extracranial cancers integrate electrical inputs will be critical to find targeted therapies that modulate the microenvironment. Destabilizing neuron-cancer interactions and preventing tumors from commandeering mechanisms of neural development and plasticity may provide a novel therapeutic angle to interrupt malignant growth.

As this field evolves, it is imperative that all axes of neuronal driven electrical communication with both neoplastic and nonneoplastic cells of the tumor microenvironment be thoroughly investigated. The clear parallels between cell development/regeneration and cancer growth imply that the role of the nervous system in normal contexts can provide unique insight into mechanisms of activity-mediated malignant progression. Taken together, the study of *cancer neuroscience* [41] emphasizes the underappreciated reliance of various malignancies on normal mechanisms of neural signaling and plasticity. Developing a deeper understanding of cancer dependency on the nervous system may clarify novel targeting strategies in the treatment of malignant diseases.

References

1. H. S. Venkatesh, T. B. Johung, V. Caretti, A. Noll, Y. Tang, S. Nagaraja, E. M. Gibson, C. W. Mount, J. Polepalli, S. S. Mitra, P. J. Woo, R. C. Malenka, H. Vogel, M. Bredel, P. Mallick, M. Monje, Neuronal Activity Promotes Glioma Growth through Neuroligin-3 Secretion. *Cell.* **161**, 803–16 (2015).
2. Y. Pan, J. D. Hysinger, T. Barron, N. F. Schindler, O. Cobb, X. Guo, B. Yalçın, C. Anastasaki, S. B. Mulinyawe, A. Ponnuswami, S. Scheaffer, Y. Ma, K.-C. Chang, X. Xia, J. A. Toonen, J. J. Lennon, E. M. Gibson, J. R. Huguenard, L. M. Liau, J. L. Goldberg, M. Monje, D. H. Gutmann, NF1 mutation drives neuronal activity-dependent initiation of optic glioma. *Nature.* **594**, 277–282 (2021).
3. P. Chen, W. Wang, R. Liu, J. Lyu, L. Zhang, B. Li, B. Qiu, A. Tian, W. Jiang, H. Ying, R. Jing, Q. Wang, K. Zhu, R. Bai, L. Zeng, S. Duan, C. Liu, Olfactory sensory experience regulates gliomagenesis via neuronal IGF1. *Nature.* **606**, 550–556 (2022).
4. H. S. Venkatesh, W. Morishita, A. C. Geraghty, D. Silverbush, S. M. Gillespie, M. Arzt, L. T. Tam, C. Espenel, A. Ponnuswami, L. Ni, P. J. Woo, K. R. Taylor, A. Agarwal, A. Regev, D. Brang, H. Vogel, S. Hervey-Jumper, D. E. Bergles, M. L. Suvà, R. C. Malenka, M. Monje, Electrical and synaptic integration of glioma into neural circuits. *Nature.* **573**, 539–545 (2019).
5. V. Venkataramani, D. I. Tanev, C. Strahle, A. Studier-Fischer, L. Fankhauser, T. Kessler, C. Körber, M. Kardorff, M. Ratliff, R. Xie, H. Horstmann, M. Messer, S. P. Paik, J. Knabbe, F. Sahm, F. T. Kurz, A. A. Acikgöz, F. Herrmannsdörfer, A. Agarwal, D. E. Bergles, A. Chalmers, H. Miletic, S. Turcan, C. Mawrin, D. Hänggi, H.-K. Liu, W. Wick, F. Winkler, T. Kuner, Glutamatergic synaptic input to glioma cells drives brain tumour progression. *Nature.* **573**, 532–538 (2019).
6. M. G. Filbin, I. Tirosh, V. Hovestadt, M. L. Shaw, L. E. Escalante, N. D. Mathewson, C. Neftel, N. Frank, K. Pelton, C. M. Hebert, C. Haberler, K. Yizhak, J. Gojo, K. Egervari, C. Mount, P. van Galen, D. M. Bonal, Q.-D. Nguyen, A. Beck, C. Sinai, T. Czech, C. Dorfer, L. Goumnerova, C. Lavarino, A. M. Carcaboso, J. Mora, R. Mylvaganam, C. C. Luo, A. Peyrl, M. Popović, A. Azizi, T. T. Batchelor, M. P. Frosch, M. Martinez-Lage, M. W. Kieran, P. Bandopadhayay, R. Beroukhim, G. Fritsch, G. Getz, O. Rozenblatt-Rosen, K. W. Wucherpfennig, D. N. Louis, M. Monje, I. Slavc, K. L. Ligon, T. R. Golub, A. Regev, B. E. Bernstein, M. L. Suvà, Developmental and oncogenic programs in H3K27M gliomas dissected by single-cell RNA-seq. *Science (80-.).* **360**, 331–335 (2018).
7. M. Monje, S. S. Mitra, M. E. Freret, T. B. Raveh, J. Kim, M. Masek, J. L. Attema, G. Li, T. Haddix, M. S. B. Edwards, P. G. Fisher, I. L. Weissman, D. H. Rowitch, H. Vogel, A. J. Wong, P. A. Beachy, Hedgehog-responsive candidate cell of origin for diffuse intrinsic pontine glioma. *Proc. Natl. Acad. Sci. U. S. A.* **108**, 4453–4458 (2011).
8. C. Liu, J. C. Sage, M. R. Miller, R. G. W. Verhaak, S. Hippenmeyer, H. Vogel, O. Foreman, R. T. Bronson, A. Nishiyama, L. Luo, H. Zong, Mosaic analysis with double markers reveals tumor cell of origin in glioma. *Cell.* **146**, 209–221 (2011).
9. S. Sugiarto, A. I. Persson, E. G. Munoz, M. Waldhuber, C. Lamagna, N. Andor, P. Hanecker, J. Ayers-Ringler, J. Phillips, J. Siu, D. A. Lim, S. Vandenberg, W. Stallcup, M. S. Berger, G. Bergers, W. A. Weiss, C. Petritsch, Asymmetry-defective oligodendrocyte progenitors are glioma precursors. *Cancer Cell.* **20**, 328–40 (2011).
10. A. I. Persson, C. Petritsch, F. J. Swartling, M. Itsara, F. J. Sim, R. Auvergne, D. D. Goldenberg, S. R. Vandenberg, K. N. Nguyen, S. Yakovenko, J. Ayers-Ringler, A. Nishiyama, W. B. Stallcup,

M. S. Berger, G. Bergers, T. R. McKnight, S. A. Goldman, W. A. Weiss, Non-stem cell origin for oligodendroglioma. *Cancer Cell*. **18**, 669–82 (2010).
11. D. E. Bergles, J. D. Roberts, P. Somogyi, C. E. Jahr, Glutamatergic synapses on oligodendrocyte precursor cells in the hippocampus. *Nature*. **405**, 187–91 (2000).
12. D. E. Bergles, R. Jabs, C. Steinhäuser, Neuron-glia synapses in the brain. *Brain Res. Rev.* **63**, 130–137 (2010).
13. D. Hausmann, D. C. Hoffmann, V. Venkataramani, E. Jung, S. Horschitz, S. K. Tetzlaff, A. Jabali, L. Hai, T. Kessler, D. D. Azoŕin, S. Weil, A. Kourtesakis, P. Sievers, A. Habel, M. O. Breckwoldt, M. A. Karreman, M. Ratliff, J. M. Messmer, Y. Yang, E. Reyhan, S. Wendler, C. Löb, C. Mayer, K. Figarella, M. Osswald, G. Solecki, F. Sahm, O. Garaschuk, T. Kuner, P. Koch, M. Schlesner, W. Wick, F. Winkler, Autonomous rhythmic activity in glioma networks drives brain tumour growth. *Nature*. **613**, 179–186 (2023).
14. J. J. LoTurco, D. F. Owens, M. J. Heath, M. B. Davis, A. R. Kriegstein, GABA and glutamate depolarize cortical progenitor cells and inhibit DNA synthesis. *Neuron*. **15**, 1287–98 (1995).
15. K. C. Luk, T. E. Kennedy, A. F. Sadikot, Glutamate promotes proliferation of striatal neuronal progenitors by an NMDA receptor-mediated mechanism. *J. Neurosci.* **23**, 2239–50 (2003).
16. R. S. Smith, C. J. Kenny, V. Ganesh, A. Jang, R. Borges-Monroy, J. N. Partlow, R. S. Hill, T. Shin, A. Y. Chen, R. N. Doan, A.-K. Anttonen, J. Ignatius, L. Medne, C. G. Bönnemann, J. L. Hecht, O. Salonen, A. J. Barkovich, A. Poduri, M. Wilke, M. C. Y. de Wit, G. M. S. Mancini, L. Sztriha, K. Im, D. Amrom, E. Andermann, R. Paetau, A.-E. Lehesjoki, C. A. Walsh, M. K. Lehtinen, Sodium Channel SCN3A (NaV1.3) Regulation of Human Cerebral Cortical Folding and Oral Motor Development. *Neuron*. **99**, 905–913.e7 (2018).
17. R. S. Smith, C. A. Walsh, Ion Channel Functions in Early Brain Development. *Trends Neurosci.* **43**, 103–114 (2020).
18. T. A. Weissman, P. A. Riquelme, L. Ivic, A. C. Flint, A. R. Kriegstein, Calcium waves propagate through radial glial cells and modulate proliferation in the developing neocortex. *Neuron*. **43**, 647–61 (2004).
19. I. Vitali, S. Fièvre, L. Telley, P. Oberst, S. Bariselli, L. Frangeul, N. Baumann, J. J. McMahon, E. Klingler, R. Bocchi, J. Z. Kiss, C. Bellone, D. L. Silver, D. Jabaudon, Progenitor Hyperpolarization Regulates the Sequential Generation of Neuronal Subtypes in the Developing Neocortex. *Cell*. **174**, 1264-1276.e15 (2018).
20. X. Gu, E. C. Olson, N. C. Spitzer, Spontaneous neuronal calcium spikes and waves during early differentiation. *J. Neurosci.* **14**, 6325–35 (1994).
21. E. Kougioumtzidou, T. Shimizu, N. B. Hamilton, K. Tohyama, R. Sprengel, H. Monyer, D. Attwell, W. D. Richardson, Signalling through AMPA receptors on oligodendrocyte precursors promotes myelination by enhancing oligodendrocyte survival. *Elife*. **6** (2017), https://doi.org/10.7554/eLife.28080.
22. V. Venkataramani, Y. Yang, M. C. Schubert, E. Reyhan, S. K. Tetzlaff, N. Wißmann, M. Botz, S. J. Soyka, C. A. Beretta, R. L. Pramatarov, L. Fankhauser, L. Garofano, A. Freudenberg, J. Wagner, D. I. Tanev, M. Ratliff, R. Xie, T. Kessler, D. C. Hoffmann, L. Hai, Y. Dörflinger, S. Hoppe, Y. A. Yabo, A. Golebiewska, S. P. Niclou, F. Sahm, A. Lasorella, M. Slowik, L. Döring, A. Iavarone, W. Wick, T. Kuner, F. Winkler, Glioblastoma hijacks neuronal mechanisms for brain invasion. *Cell*. **185**, 2899–2917.e31 (2022).
23. M. Osswald, E. Jung, F. Sahm, G. Solecki, V. Venkataramani, J. Blaes, S. Weil, H. Horstmann, B. Wiestler, M. Syed, L. Huang, M. Ratliff, K. Karimian Jazi, F. T. Kurz, T. Schmenger, D. Lemke, M. Gömmel, M. Pauli, Y. Liao, P. Häring, S. Pusch, V. Herl, C. Steinhäuser, D. Krunic, M. Jarahian, H. Miletic, A. S. Berghoff, O. Griesbeck, G. Kalamakis, O. Garaschuk, M. Preusser, S. Weiss, H. Liu, S. Heiland, M. Platten, P. E. Huber, T. Kuner, A. von Deimling, W. Wick, F. Winkler, Brain tumour cells interconnect to a functional and resistant network. *Nature*. **528**, 93–8 (2015).
24. E. Jung, M. Osswald, J. Blaes, B. Wiestler, F. Sahm, T. Schmenger, G. Solecki, K. Deumelandt, F. T. Kurz, R. Xie, S. Weil, O. Heil, C. Thomé, M. Gömmel, M. Syed, P. Häring, P. E. Huber, S. Heiland, M. Platten, A. von Deimling, W. Wick, F. Winkler, Tweety-Homolog 1 Drives Brain Colonization of Gliomas. *J. Neurosci.* **37** (2017) (available at http://www.jneurosci.org/content/37/29/6837).

25. T. V Bliss, T. Lomo, Plasticity in a monosynaptic cortical pathway. *J. Physiol.* **207**, 61P (1970).
26. A. Citri, R. C. Malenka, Synaptic Plasticity: Multiple Forms, Functions, and Mechanisms. *Neuropsychopharmacology*. **33**, 18–41 (2008).
27. R. C. Malenka, M. F. Bear, LTP and LTD: an embarrassment of riches. *Neuron*. **44**, 5–21 (2004).
28. H. Park, M. Poo, Neurotrophin regulation of neural circuit development and function. *Nat. Rev. Neurosci.* **14**, 7–23 (2013).
29. S. L. Campbell, S. C. Buckingham, H. Sontheimer, Human glioma cells induce hyperexcitability in cortical networks. *Epilepsia* (2012), https://doi.org/10.1111/j.1528-1167.2012.03557.x.
30. K. Yu, C.-C. J. Lin, A. Hatcher, B. Lozzi, K. Kong, E. Huang-Hobbs, Y.-T. Cheng, V. B. Beechar, W. Zhu, Y. Zhang, F. Chen, G. B. Mills, C. A. Mohila, C. J. Creighton, J. L. Noebels, K. L. Scott, B. Deneen, PIK3CA variants selectively initiate brain hyperactivity during gliomagenesis. *Nature*. **578**, 166–171 (2020).
31. C. C. John Lin, K. Yu, A. Hatcher, T. W. Huang, H. K. Lee, J. Carlson, M. C. Weston, F. Chen, Y. Zhang, W. Zhu, C. A. Mohila, N. Ahmed, A. J. Patel, B. R. Arenkiel, J. L. Noebels, C. J. Creighton, B. Deneen, Identification of diverse astrocyte populations and their malignant analogs. *Nat. Neurosci.* (2017), https://doi.org/10.1038/nn.4493.
32. S. L. Campbell, S. Robel, V. A. Cuddapah, S. Robert, S. C. Buckingham, K. T. Kahle, H. Sontheimer, GABAergic disinhibition and impaired KCC2 cotransporter activity underlie tumor-associated epilepsy. *Glia*. **63**, 23–36 (2015).
33. S. C. Buckingham, S. L. Campbell, B. R. Haas, V. Montana, S. Robel, T. Ogunrinu, H. Sontheimer, Glutamate release by primary brain tumors induces epileptic activity. *Nat. Med.* **17**, 1269–74 (2011).
34. A. A. Aabedi, B. Lipkin, J. Kaur, S. Kakaizada, C. Valdivia, S. Reihl, J. S. Young, A. T. Lee, S. Krishna, M. S. Berger, E. F. Chang, D. Brang, S. L. Hervey-Jumper, Functional alterations in cortical processing of speech in glioma-infiltrated cortex. *Proc. Natl. Acad. Sci.* **118**, e2108959118 (2021).
35. S. L. Payne, P. Ram, D. H. Srinivasan, T. T. Le, M. Levin, M. J. Oudin, Potassium channel-driven bioelectric signalling regulates metastasis in triple-negative breast cancer. *EBioMedicine*. **75**, 103767 (2022).
36. E. P. Rahrmann, D. Shorthouse, A. Jassim, L. P. Hu, M. Ortiz, B. Mahler-Araujo, P. Vogel, M. Paez-Ribes, A. Fatemi, G. J. Hannon, R. Iyer, J. A. Blundon, F. C. Lourenço, J. Kay, R. M. Nazarian, B. A. Hall, S. S. Zakharenko, D. J. Winton, L. Zhu, R. J. Gilbertson, The NALCN channel regulates metastasis and nonmalignant cell dissemination. *Nat. Genet.*, 1–12 (2022).
37. N. Prevarskaya, R. Skryma, Y. Shuba, Ion Channels in Cancer: Are Cancer Hallmarks Oncochannelopathies? *Physiol. Rev.* **98**, 559–621 (2018).
38. J. J. Fan, X. Huang, (Springer, Cham, 2020; https://link.springer.com/10.1007/112_2020_48), pp. 103–133.
39. A. Litan, S. A. Langhans, Cancer as a channelopathy: ion channels and pumps in tumor development and progression. *Front. Cell. Neurosci.* **9**, 86 (2015).
40. F. Lang, C. Stournaras, Ion channels in cancer: future perspectives and clinical potential. *Philos. Trans. R. Soc. Lond. B. Biol. Sci.* **369**, 20130108 (2014).
41. M. Monje, J. C. Borniger, N. J. D'Silva, B. Deneen, P. B. Dirks, F. Fattahi, P. S. Frenette, L. Garzia, D. H. Gutmann, D. Hanahan, S. L. Hervey-Jumper, H. Hondermarck, J. B. Hurov, A. Kepecs, S. M. Knox, A. C. Lloyd, C. Magnon, J. L. Saloman, R. A. Segal, E. K. Sloan, X. Sun, M. D. Taylor, K. J. Tracey, L. C. Trotman, D. A. Tuveson, T. C. Wang, R. A. White, F. Winkler, Roadmap for the Emerging Field of Cancer Neuroscience. *Cell*. **181**, 219–222 (2020).

Chapter 3
Mechanisms of Cancer-Induced Remodeling of the Central Nervous System

Saritha Krishna, Vardhaan Ambati, and Shawn L. Hervey-Jumper

Neuronal Regulation of Glioma

Neuronal Activity-Dependent Regulation of Glioma Growth and Progression

Unlike other organ systems, the nervous system is a major controlling and communicating system in the body which regulates stem and precursor cell behavior across a range of tissues. In the healthy brain, neuronal activity induces proliferation of oligodendrocyte precursor cells (OPCs) and pre-OPCs [27]. This activity-regulated OPC proliferation promotes new oligodendrocyte formation and plasticity of myelination in turn supporting neural circuit functions underlying learning and memory. Intriguingly, parallel mechanisms of neuronal activity-dependent growth of brain cancers, which molecularly resemble OPCs, have been reported in preclinical models, wherein normal myelin plasticity mechanisms are either dysregulated or hijacked to promote malignant glioma cell proliferation [15]. Using patient-derived orthotopic xenograft model, the profound influence of neuronal activity in promoting the growth and spread of tumor cells in diffuse gliomas and glioblastoma was demonstrated [46]. Optogenetic activation of cortical projection neurons in mice xenografted with human high-grade glioma cells increased glioma cell proliferation and promoted tumor growth and progression in a circuit-specific manner [46]. Further investigation into the mechanisms that mediate this neuronal activity-regulated glioma growth by the above group and several other researchers

S. Krishna · V. Ambati · S. L. Hervey-Jumper (✉)
Department of Neurological Surgery, University of California, San Francisco, San Francisco, CA, USA
e-mail: Shawn.Hervey-Jumper@ucsf.edu

M. Amit, N. N. Scheff (eds.), *Cancer Neuroscience*,
https://doi.org/10.1007/978-3-031-32429-1_3

demonstrated different modes of both direct/synaptic and indirect/non-synaptic paracrine and autocrine interactions between neurons and glioma cells in the tumor microenvironment.

Molecular Mechanisms of Neuronal Regulation of Glioma Growth

Activity-Regulated Secretion of Paracrine Factors

One of the initial reports of activity-regulated paracrine factors mediating glioma proliferation discovered that optogenetic stimulation of acute cortical slices resulted in the activity-dependent secretion of mitogenic proteins in the conditioned media, which in turn increased the proliferation of patient-derived tumor cells from different high-grade glioma subtypes, including adult and pediatric glioblastoma, anaplastic oligodendroglioma, and H3K27M+ diffuse midline gliomas [44]. Among several secreted factors, synaptic protein neuroligin-3 (NLGN3) was found to be the primary factor that showed the most robust effect on glioma proliferation, along with lesser contributions from other soluble factors such as brain-derived neurotrophic factor (BDNF) and 78 kDa glucose-regulated protein (GRP78) [44]. They further found that secreted NLGN3 promotes glioma proliferation through phosphatidylinositol 3-kinase (PI3K) pathway and also induces feedforward NLGN3 expression and associated upregulation of synapse-associated genes in glioma cells. It was also demonstrated that a disintegrin and metalloproteinase domain-containing protein 10 (ADAM10) is released into the synaptic cleft upon neuronal activity which subsequently functions to cleave and release NLGN3 in a neuronal activity-dependent manner. Blocking access of glioma cells to neuronally derived NLGN3 and/or pharmacological interference with ADAM10 via the use of ADAM10 inhibitors significantly suppressed glioma growth and progression in patient-derived orthotopic xenograft animal models, suggesting a new therapeutic strategy for the treatment of glioma [45]. Interestingly, a recent study showed higher levels of tumor-derived NLGN3 which enhances ADAM10 expression and NLGN3 cleavage to facilitate its secretion from glioma cells suggesting that similar to the neuronal counterpart, glioma-derived NLGN3 engages in a positive feedback loop to promote glioma proliferation [10]. Taken together, the above findings indicate that besides blocking neuronal activity-induced NLGN3 signaling, targeting glioma-derived NLGN3 could prove as an effective strategy for combinatorial treatment of malignant glioma.

Neurotrophins

Among neurotrophins, BDNF deserves special attention because of its established role in promoting myelination, synaptic connectivity, and synaptic strength in the normal healthy brain. In health, BDNF regulates synaptic plasticity via signaling through its membrane receptor tropomyosin receptor kinase B (TrkB) to recruit calcium signaling pathways and promote α-amino-3-hydroxy-5-methyl-4-isoxazole propionic acid (AMPA) receptor trafficking to the postsynaptic neuronal membrane [29]. BDNF-TrkB signaling was found to be crucial in regulating the number of malignant synapses formed between neurons and glioma cells [32]. Blockade of neuronal activity-regulated BDNF secretion using the pan-Trk inhibitor entrectinib was shown to significantly decrease high-grade glioma cell proliferation and growth both in vitro and in vivo [32]. Examination of malignant glioma tissues and glioma-derived brain tumor initiating cells identified the presence of BDNF receptors TrkB and tropomyosin receptor kinase C (TrkC) as well as other neurotrophins such as nerve growth factor (NGF) and neurotrophin 3 (NT3) [23]. These findings highlight the potential of glioma cells to release neurotrophins in an autocrine fashion to sustain their own growth and survival [49].

Neurotransmitters

Similar to neurotrophins, many studies have begun to recognize the role of neurotransmitters in the tumor microenvironment and highlight the possibility of neuronal activity in regulating glioma cell behavior via neurotransmitter release. One of the earliest cases linking glioma cell response to neurotransmission came from a group who reported lower incidence of glioma in patients treated with tricyclic antidepressants which act primarily via reuptake inhibition of serotonin and norepinephrine [47]. This association was further validated experimentally when low-grade glioma-bearing mice treated with tricyclic antidepressant, imipramine, showed reduced glioma cell proliferation and prolonged survival [37]. This effect was mechanistically attributed to an induction in autophagy and apoptosis, leading to programmed death of tumor cells [37]. Dopamine, another predominant neurotransmitter in the brain, also regulates autophagy and apoptosis via dopamine receptor D4. Pharmacologic blockade of dopamine receptor D4 expressed on glioblastoma cells effectively inhibited glioma cell proliferation through inhibition of autophagy and downstream induction of apoptosis [13].

γ-Aminobutyric acid (GABA), an inhibitory neurotransmitter in the normal adult brain, evokes an activity-dependent Cl^- influx through the $GABA_A$ receptor, thereby hyperpolarizing the neuronal cell membrane. Given the role of GABA in regulating the growth of many cell types such as neural stem cells, glial precursor cells, and oligodendrocytes [48], there is growing interest in understanding the functional relevance of GABAergic signaling in glioma growth and progression. The putative impact of GABAergic signaling via activation of $GABA_A$ receptor studied in low-grade gliomas demonstrated a reduction in glioma cell proliferation; the mechanism of action appears to involve excess intracellular Cl^- concentration mediated by Na^+-K^+–$2Cl^-$ cotransporter (NKCC1) and pathological reversal of the chloride gradient, promoting Cl^- efflux upon GABA binding and glioma cell depolarization [21]. Importantly, several studies reported the absence or downregulation of functional $GABA_A$ receptor in high-grade gliomas and further indicated this as a mechanism by which malignant gliomas downregulate GABAergic neurotransmission and counteract the inhibitory effect of GABA on glioma cell proliferation to favor tumor growth [17].

Besides acting as a major player in regulating glioma cell proliferation, recent findings of the critical role of NKCC1 in driving tumor-associated neuronal hyperexcitability in concert with the cotransporter potassium chloride cotransporter (KCC2) has received increasing attention. Specifically, malignant gliomas such as glioblastoma induces an increase in neuronal NKCC1 expression which results in elevated intracellular Cl^- concentration and reversal of the chloride gradient, promoting Cl^- efflux upon GABA binding and cell depolarization [17]. This depolarizing effect of GABA further results in peritumoral neuronal hyperexcitability and epileptiform activities found in the peritumoral cortex of glioma patients. Prior studies which investigated the contribution of GABAergic and glutamatergic mechanisms to neuronal hyperexcitability and epilepsy in glioma patients focused primarily on the peritumoral brain tissue into which the glioma infiltrates [8, 31]. However, it is noteworthy that the high rate and morbidity associated with tumor-associated epilepsy persists, despite targeting the peritumoral microenvironment. Hence, the contribution of NKCC1-mediated GABAergic mechanisms underlying tumor-intrinsic network activity and associated epileptiform activity in high-grade glioma patients is an area in need of investigation.

Aside from dysregulation of GABAergic signaling, substantial changes in the peritumoral tissue with extracellular glutamate concentration reaching excitotoxic levels are known to contribute to increased tumor progression and tumor-associated epilepsy [17]. Glutamate is a major excitatory neurotransmitter in the brain signals through metabotropic glutamate receptors (mGluR) and NMDA receptors resulting in increased calcium entry as well as cell proliferation, migration, survival, and cell differentiation of neural progenitor cells. Aberrant activation of neuronal glutamate receptors, particularly NMDA receptors, can lead to sustained calcium influx resulting in excitotoxic cell death. Parallel mechanisms of NMDA receptor-mediated neuronal death have been implicated in malignant glioma wherein glioma cells secrete exuberant levels of glutamate to the extracellular space resulting in peritumoral neuronal death observed in in vitro and in vivo studies [38]. This non-synaptically secreted glutamate from glioma cells also serves as a trophic factor promoting glioblastoma cell survival, growth, and migration [11], suggesting this neurotransmitter could promote tumor growth in an autocrine and paracrine fashion. Accordingly, previous studies reporting slower tumor growth and longer survival of mice bearing glutamate-secreting gliomas as well as recent in vitro reports of decreased tumor cell survival, migration, and subsequently increased radiosensitivity in glioblastoma cell lines following treatment with NMDA receptor antagonists MK801, memantine, or ifenprodil underscore the clinical potential of targeting glutamate receptors in glioblastoma treatment [28, 40].

Glioma cells also express a subclass of ionotropic AMPA receptor, which is calcium permeable. Accordingly, several in vitro studies which targeted AMPA glutamate receptors or AMPA receptor-mediated calcium signaling demonstrated a decrease in cell proliferation and migration of human astrocytoma and glioblastoma cell lines and primary glioma culture [18, 35]; the mechanism of action appears to involve attenuation of the downstream pro-oncogenic Akt and ERK/MAPK signaling pathways [34]. Noncompetitive AMPA receptor antagonist, perampanel, is widely used for reducing seizure frequency in patients with high-grade glioma and focal epilepsy [22]. Interestingly, the anti-tumorigenic potential of this drug was recently demonstrated in glioma mouse model and patient-derived glioblastoma cell lines [22, 42, 46], suggesting perampanel may be more beneficial in terms of antitumor efficacy than other anticonvulsant drugs in the treatment of malignant glioma.

Neuronal Activity-Evoked Direct Electrochemical Signaling

In addition to the abovementioned paracrine factors, active neurons in the tumor microenvironment also drive glioma growth and progression by creating direct synaptic contact with glioma cells. Two recent studies examined primary glioma tissue samples and patient-derived xenografts from both high-grade (adult glioblastoma and pediatric diffuse intrinsic pontine glioma) and low-grade glioma (adult IDH-mutant diffuse and anaplastic astrocytomas) cells and demonstrated clear synaptic structures between presynaptic neurons and postsynaptic glioma cells [42, 46]. Importantly, further evaluation of these neuron-to-glioma synapses revealed all classical features of a glutamatergic chemical synapse, mediated by calcium-permeable AMPA receptors. Functional and electrophysiological characterization of these glutamatergic synapses showed two types of currents in glioma cells, namely, fast excitatory postsynaptic currents and prolonged slow inward currents, that were occurring either spontaneously or evoked by neuronal action potentials. Further experiments suggested that these activity-induced glioma excitatory postsynaptic currents are depolarizing with direct optogenetic depolarization of glioma xenografts increasing tumor cell proliferation in vivo, while pharmacological or genetic interference with synaptic transmission shows a tumor growth inhibitory effect [42, 46]. Thus, electrochemical signaling mediated by functional glutamatergic neuron-to-glioma synapses represents another neuronal activity-regulated mechanism of glioma growth and progression.

Neuronal Activity-Dependent Regulation of Glioma Initiation

While we previously discussed mechanisms of neuronal activity promoting glioma growth and progression in models of high-grade glioma, new research suggests a key role of neuronal activity in the initiation and growth of low-grade glioma [3, 33]. This phenomenon was demonstrated using a genetic mouse model with mutations in the neurofibromatosis 1 tumor suppressor gene, *Nf1*, in which mice consistently develop tumors of the optic nerve and chiasm at a predictable postnatal age [33]. The optic nerve contains axons of retinal ganglion cells, whose activity is influenced by environmental light conditions and visual experience. It was discovered that rearing NF1 mutant mice in complete darkness from 6 weeks of age or 9–16 weeks of age, representing time points before tumor initiation or during tumor development and growth, either prevented the formation of tumors or resulted in tumors that were much smaller compared to littermate control mice with normal visual experience. Furthermore, the finding that optogenetic stimulation of the optic nerve increased nerve volume and tumor content of the optic pathway validated the equally important role of neuronal activity in promoting the growth of low-grade glioma similar to high-grade gliomas discussed in the above sessions. The neuronal activity-regulated paracrine factor NLGN3, previously identified to be associated with growth of high-grade gliomas, was found to influence optic pathway glioma initiation and growth, as NLGN3 knockout prevented optic pathway glioma formation in mice. Critically, the aberrantly increased shedding of NLGN3 due to hyperexcitability of the *Nf1*-mutant optic nerve was attenuated by ADAM10 inhibitor, decreasing tumor formation and/or growth of *Nf1*-mutant mice [33]. Overall, these findings indicate neuronal hyperexcitability and subsequent activity-dependent release of paracrine factors as a major driver of optic glioma formation in the setting of NF1 predisposition syndrome [3].

Similar to the finding that optic nerve activity plays in optic pathway low-grade glioma initiation and growth, recent study shows that manipulating the activity of receptor neurons involved in olfaction can directly regulate the incidence of high-grade gliomas of the olfactory bulb [9]. Mechanistically, an activity-dependent secretion of insulin-like growth factor 1 (IGF-1) signaling was found to be associated with increased olfactory neuronal activity and gliomagenesis; targeting IGF-1 significantly suppressed tumor formation [9].

Neuronal Activity-Dependent Regulation of Glioma Invasion

In addition to the strong influence on tumor initiation, maintenance, and progression, neuronal activity has been recently found to induce glioma invasion of healthy brain tissue. Using a series of key experiments, recent work identified a subpopulation of glioblastoma cells that resemble neuronal precursor cells [43]. They found that these neuron-like glioblastoma cells lack connection to other tumor cells and exploit neuronal synaptic inputs to hijack mechanisms that are physiologically intended for neuronal cell migration during development to invade and colonize the brain. Critically, suppression of neuronal activity using deep anesthesia resulted in decreased tumor microtube branching and turnover along with decreased glioblastoma invasion speed suggesting that calcium transients might play a key role in neuronal activity-driven glioblastoma invasion. Furthermore, optogenetic activation of neurons in vivo resulted in increased calcium events, confirming a link between neuronal activity and complex calcium dynamics. Additionally, blockade of calcium currents using the calcium chelator BAPTA-AM significantly reduced microtube formation and glioblastoma invasion, connecting neuronal activity to glioblastoma invasion via calcium events [43]. These findings extend the field's understanding of glioblastoma invasion and yield potential avenues for glioblastoma treatment including targeting direct synaptic input to unconnected glioblastoma cell population.

Neuronally Driven Glioma-Glioma Connections

Apart from the fast-depolarizing excitatory postsynaptic current observed in a small population of glioma cells, neuronal activity was found to induce non-synaptic long-lasting potassium-evoked current in a subset of glioma cells in coculture with neurons and brain slices of tumor-xenograft mice [30]. The subpopulation of glioma cells that respond to neuronal activity with slow, inward, potassium-evoked currents were also demonstrated to form long processes called tumor microtubes and engage in gap junction-coupled interconnected tumor network. This connexin 43 gap junction-mediated coupling of glioma cells further helps to propagate electrochemical signals and calcium currents to spread through the entire glioma network, promote tumor cell proliferation and migration, as well as confer resistance to standard radio- and chemotherapy [30]. Moreover, the molecule with a key role in the generation of the abovementioned tumor microtubes and driving subsequent processes including invasion, proliferation, cell-to-cell interaction, and resistance to radiotherapy was found to be neuronal growth-associated protein 43 (GAP43) which could also be a possible molecular pathway target in glioma treatment to tackle tumor microtube formation and function [30]. Accordingly, approaches to disrupt intercellular connectivity of glioma cells by targeting tumor microtubes or gap junctions by several groups showed promising effects on controlling growth and invasive behavior of glioma cells. For instance, following treatment with a gap junction blocker, meclofenamate, resulted in a significant reduction in the intercellular tumor microtube network and a concomitant increase in chemotherapy (i.e., temozolomide) sensitivity of glioblastoma cells injected into human neocortical slice cultures [36]. These results are in accordance with previous in vivo preclinical studies demonstrating a decrease in frequency and amplitude of glioma network currents and an associated reduction in glioma growth and expansion using gap junction blockers such as carbenoxolone, meclofenamate, and tonabersat [12, 30, 42]. Overall, these findings demonstrate a therapeutic potential of gap junction inhibitors in human glioma therapy.

Glioma Regulation of Neuronal Excitability

In the previous sections, we have outlined various direct and indirect mechanisms by which excitatory neuronal activity promotes glioma growth and progression. On the other hand, emerging work also suggests that neuron-to-glioma communication is bidirectional with glioma progression leading to neuronal hyperactivity and excitability. Different mechanisms have been proposed by which glioma cells make neighboring neurons hyperexcitable resulting in abnormal, excessive neuronal firing. As mentioned above, several studies have demonstrated significantly higher levels of glutamate secretion from glioma cells, which in turn has been shown to positively regulate tumor cell proliferation and growth [35, 38]. Notably, this defective glutamatergic signaling was also attributed to the peritumoral hyperexcitability observed in in vitro brain slices from glioma-implanted mice [7] and seizure development in glioma-bearing mice [6]. Further experiments showed decreased GABAergic inhibition and significant loss of inhibitory interneurons in the peritumoral cortex, which also contributed to glioma-induced neuronal hyperexcitability [8]. Another mechanism by which glioma cells establish a hyperexcitable peritumoral neuronal environment is through a subpopulation of astrocytic-like glioma cells that secrete synaptogenic factors, such as glypican-3 (GPC3) [19]. Importantly, this subpopulation of tumor cells with enhanced synaptogenic ability evolved during the course of tumor progression and correlated with onset of seizure activity in the tumor-bearing mice [19]. Further investigations into the mechanisms of tumor-induced neuronal hyperexcitability discovered glioma-specific point mutations in the *Pik3ca* oncogene in genetic mouse models of glioblastoma PMID: 31996845. Gliomas driven by *Pik3ca* variants selectively secreted GPC3 and promoted neuronal hyperexcitability and seizures in tumor-bearing mice [50]. Overall, these findings indicate that drugs targeting glioma-derived paracrine factors and glioma-induced malignant synaptogenesis may confer additional therapeutic benefit to glioma patients.

Besides secreting synaptogenic factors, glioma cells have also been demonstrated to actively communicate with neurons via the release of extracellular vesicles, including exosomes. The addition of exosomes derived from U87 glioblastoma cell line and low-grade glioma patients to mature neurons disrupted synchrony and increased excitability of mature neurons [39]. It was further observed that there was an upregulation of the neuronal cytoskeletal protein, Arp2/3, on exosome-treated neurons, which in turn correlated with the increased frequency of neuronal spontaneous synaptic responses [39]. This finding is in agreement with a previous study which showed extracellular vesicle-mediated acceleration of synaptogenesis and increase in synaptic activity of neurons from mice implanted with glioma cells [14]. Unexpectedly, inhibition of exosome release from glioma cells slowed down tumor growth in vivo, suggesting that extracellular vesicle-mediated interaction between glioma cells and neurons alters the tumor microenvironment to foster glioma growth. Using in vivo wide-field optical mapping, alterations in both neuronal activity and functional hemodynamics during tumor progression of awake mice implanted with glioma cells were monitored [26]. Besides causing significant alterations in neuronal synchrony, tumor growth and progression resulted in profound disruption of neurovascular coupling in the tumor-burdened region, further exacerbating seizure-related brain damage in glioma-bearing mice [26]. Taken together, these studies suggest that besides exerting a direct influence on neurons, glioma cells also closely interact with other cell types, such as endothelial cells and pericytes in the tumor microenvironment, to hijack vascular responses and aggravate changes in glioma-induced neuronal activity.

Apart from the complex interplay occurring with active neurons, neoplastic glioma cells also interact with infiltrating immune cells in the tumor microenvironment to create a tumor-promoting neuroinflammatory niche that acts in favor of tumor cell proliferation, invasion, and survival [2]. Unlike other neuroinflammatory brain pathologies like multiple sclerosis and encephalitis where the neuro-immune axis is in a hyperactive state, in the context of glioma, malignant cells secrete a variety of cytokines, chemokines, and growth factors into the microenvironment, which actively shifts the neuro-immune axis to an immunosuppressive phenotype [25]. However, as the tumor grows and evolves, competent immune cells, namely, microglia, natural killer cells, as well as reactive astrocytes, release pro-inflammatory mediators, such as transforming growth factor-beta (TGF-β) and tumor necrosis factor-alpha (TNF-α), further exacerbating the inflammatory milieu of the peritumoral microenvironment [2]. Importantly, this immune-mediated neuroinflammation has been identified as a key cellular mechanism contributing to the cortical network excitability progression in the context of tumor-related epilepsy [16]. Specifically, reactive microgliosis accompanied by higher levels of microglial infiltration in the tumor and peritumoral regions has been found to correlate positively with cortical hyperexcitability and tumor progression in preclinical animal models [16]. Apart from the tumor-associated inflammatory responses, neuroinflammation as a consequence of aggressive chemo- and radiotherapies has also been demonstrated to alter neuronal integrity and contribute to the functional network connectivity disruption and cognitive decline observed in glioma patients [4, 5, 24, 41]. Given this strong evidence of cross talk between cancer and the neuro-immune system, modulation of the neuro-immune interactions offers a promising approach to regulate cancer growth and progression as well as to address and modulate cognitive side effects developing after chemo- and radiotherapies. As a result, cancer immunotherapies aimed to attenuate the neurotoxic cross talk while enhancing the endogenous immune response and restore immunity in the tumor microenvironment have become an emerging therapeutic option for gliomas.

Glioma-Induced Neuron-Neuron Connections and Network Remodeling

Our collective understanding of nervous system control of glioblastoma initiation and progression as well as the influence of glioblastoma on the nervous system function is based largely on in vitro and in vivo preclinical glioma models. Human disease has remained difficult to interrogate largely due to limited access, and the usability of glioblastoma-infiltrated brain has given rise to additional questions regarding the mechanistic functional significance of glioma cell integration into brain electrical network and its impact on neuronal network dynamics, cognition, and survival in glioma patients. Clinical evidence of glioma-induced neuronal hyperexcitability has also been confirmed in human patients. Intraoperative electrocorticography recordings taken before surgical resection in awake, resting adult glioblastoma patients demonstrated significantly increased high gamma frequency power in the tumor-infiltrated brain compared to more normal healthy-appearing cortex, demonstrating hyperexcitability of the glioma-infiltrated cortex in humans

[46]. Based on the clinical evidence of gliomas actively promoting neuronal excitability, a recent work studied the extent to which glioma-infiltrated cortex can meaningfully participate in cognitive processing [1]. Using subdural electrocorticography, the glioma-infiltrated cortex was found to participate in coordinated neural responses during speech production in a manner similar to normal-appearing cortex but recruited a diffuse spatial network. Importantly, neuronal activation within glioma-infiltrated language areas suffered from a loss of information storage capacity when challenged on nuanced aspects of speech processing, such as vocalization of mono- versus polysyllabic words. The result in task performance was that the glioma-injured cortex may retain the ability to participate in basic cognitive tasks but yet lose computational ability for demanding aspects of task performance [1]. These findings highlight the retained, albeit reduced, ability of the glioma-infiltrated cortex to engage in cognitive processing and underscore the functional consequences of glioma-to-neuron network connectivity in brain tumor patients.

A recent study showed further evidence in support of the impact that glioma-brain interactions have on cognition and survival [20]. They showed that beyond the promotion of glioma proliferation and invasion, neuronal activity within the glioblastoma-infiltrated cortex participated in task-related functional circuit activation of language network in the human brain. Further experiments showed that the functionally connected regions within the tumor are enriched for a glioblastoma subpopulation that selectively secretes synaptogenic factors, including thrombospondin-1 (THBS1). Critically, glioblastoma patients with greater tumor network integration and increased connectivity with the brain exhibited worse language and survival outcomes, underscoring the pathophysiological significance of glioma integration into neural circuits [20]. Together, these findings are indicative of a key role of THBS1 signaling in glioma-induced synaptic remodeling, and therefore drugs targeting THBS1 such as gabapentin and pregabalin represent a promising strategy for regulating the impact of glioma on neural circuit dynamics and associated behavioral phenotypes.

Therapeutic Implications and Future Work

In summary, the works described in this chapter illustrate active cross talk between cancer cells and neurons and describe both neuronal- and glioma-derived mechanisms that manipulate the tumor microenvironment to accelerate glioma growth and progression (Fig. 3.1). Various approaches that could be exploited to uncouple the bidirectional signaling between neuronal populations and cancer cells by i) modulating or suppressing tumor-promoting neuronal signaling to glioma cells, ii) disrupting the connections tumor cells make between each other by targeting gap junctions, and iii) interrupting the integration of brain cancer into functional circuits by inhibiting tumor-induced synaptogenesis all allow for new developments in the diagnosis and treatment of brain cancer. Additionally, several FDA-approved drugs used for other central nervous system disorders, such as memantine, perampanel, and gabapentin, could be repurposed to target glioma-associated signaling

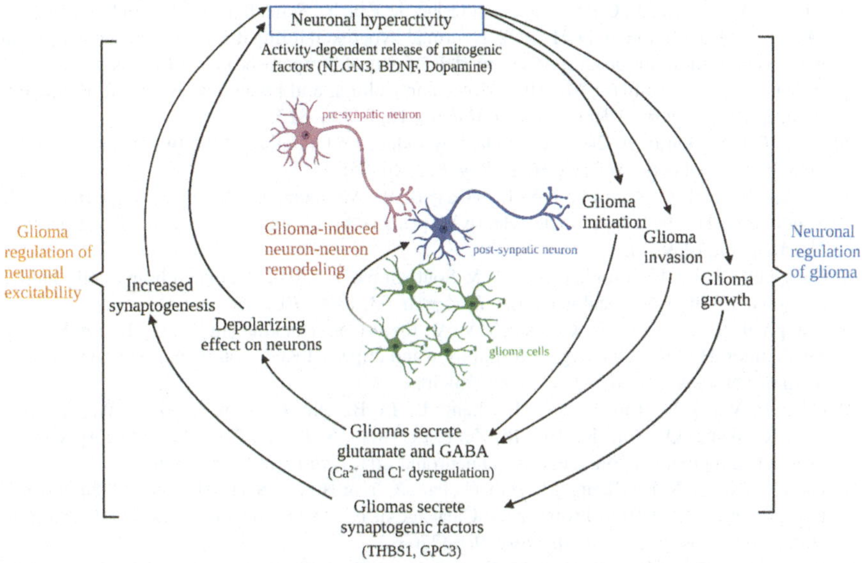

Fig. 3.1 Schematic representation of the known mechanisms by which gliomas and neurons communicate within the tumor microenvironment. Increased neuronal activity and excitability regulates the initiation, growth, and invasion of glioma through activity-dependent release of trophic factors such as neuroligin-3 (NLGN3), brain-derived neurotrophic factor (BDNF), dopamine, etc. Glioma cells in turn depolarize neurons by secreting neurotransmitters such as glutamate and GABA and increase the synaptic activity of neurons via synaptogenic proteins, such as thrombospondin-1 (THBS1) and Glypican-3 (GPC3). This synaptic integration of glioma cells into neuronal circuits induces synaptic remodeling and influences neuronal network dynamics. Images are created with BioRender.com

and expedite therapeutic drug development for malignant gliomas. While many of these drugs are undergoing preclinical testing, others are in phase I and II trials of clinical development. Although much remains to be discovered, increasing knowledge of the complex interactions between neurons and glioma cells along with the influence of neuronal and tumor regulation of other cell types in the tumor microenvironment provide new opportunities for improving therapeutic outcomes of glioma patients.

References

1. Aabedi, A. A., Lipkin, B., Kaur, J., Kakaizada, S., Valdivia, C., Reihl, S., Young, J. S., Lee, A. T., Krishna, S., Berger, M. S., Chang, E. F., Brang, D. & Hervey-Jumper, S. L. 2021. Functional alterations in cortical processing of speech in glioma-infiltrated cortex. *Proc Natl Acad Sci U S A,* 118.
2. Al-Kharboosh, R., Refaey, K., Lara-Velazquez, M., Grewal, S. S., Imitola, J. & Quinones-Hinojosa, A. 2020. Inflammatory Mediators in Glioma Microenvironment Play a Dual Role in Gliomagenesis and Mesenchymal Stem Cell Homing: Implication for Cellular Therapy. *Mayo Clin Proc Innov Qual Outcomes,* 4**,** 443–459.

3. Anastasaki, C., Mo, J., Chen, J. K., Chatterjee, J., Pan, Y., Scheaffer, S. M., Cobb, O., Monje, M., Le, L. Q. & Gutmann, D. H. 2022. Neuronal hyperexcitability drives central and peripheral nervous system tumor progression in models of neurofibromatosis-1. *Nat Commun,* 13, 2785.
4. Balentova, S. & Adamkov, M. 2015. Molecular, Cellular and Functional Effects of Radiation-Induced Brain Injury: A Review. *Int J Mol Sci,* 16, 27796–815.
5. Berg, T. J. & Pietras, A. 2022. Radiotherapy-induced remodeling of the tumor microenvironment by stromal cells. *Semin Cancer Biol,* 86, 846–856.
6. Buckingham, S. C., Campbell, S. L., Haas, B. R., Montana, V., Robel, S., Ogunrinu, T. & Sontheimer, H. 2011. Glutamate release by primary brain tumors induces epileptic activity. *Nat Med,* 17, 1269–74.
7. Campbell, S. L., Buckingham, S. C. & Sontheimer, H. 2012. Human glioma cells induce hyperexcitability in cortical networks. *Epilepsia,* 53, 1360–70.
8. Campbell, S. L., Robel, S., Cuddapah, V. A., Robert, S., Buckingham, S. C., Kahle, K. T. & Sontheimer, H. 2015. Gabaergic disinhibition and impaired Kcc2 cotransporter activity underlie tumor-associated epilepsy. *Glia,* 63, 23–36.
9. Chen, P., Wang, W., Liu, R., Lyu, J., Zhang, L., Li, B., Qiu, B., Tian, A., Jiang, W., Ying, H., Jing, R., Wang, Q., Zhu, K., Bai, R., Zeng, L., Duan, S. & Liu, C. 2022. Olfactory sensory experience regulates gliomagenesis via neuronal Igf1. *Nature,* 606, 550–556.
10. Dang, N. N., Li, X. B., Zhang, M., Han, C., Fan, X. Y. & Huang, S. H. 2021. Nlgn3 Upregulates Expression of Adam10 to Promote the Cleavage of Nlgn3 via Activating the Lyn Pathway in Human Gliomas. *Front Cell Dev Biol,* 9, 662763.
11. De Groot, J. & Sontheimer, H. 2011. Glutamate and the biology of gliomas. *Glia,* 59, 1181–9.
12. De Meulenaere, V., Bonte, E., Verhoeven, J., Kalala Okito, J. P., Pieters, L., Vral, A., De Wever, O., Leybaert, L., Goethals, I., Vanhove, C., Descamps, B. & Deblaere, K. 2019. Adjuvant therapeutic potential of tonabersat in the standard treatment of glioblastoma: A preclinical F98 glioblastoma rat model study. *PloS One,* 14, e0224130.
13. Dolma, S., Selvadurai, H. J., Lan, X., Lee, L., Kushida, M., Voisin, V., Whetstone, H., So, M., Aviv, T., Park, N., Zhu, X., Xu, C., Head, R., Rowland, K. J., Bernstein, M., Clarke, I. D., Bader, G., Harrington, L., Brumell, J. H., Tyers, M. & Dirks, P. B. 2016. Inhibition of Dopamine Receptor D4 Impedes Autophagic Flux, Proliferation, and Survival of Glioblastoma Stem Cells. *Cancer Cell,* 29, 859–873.
14. Gao, X., Zhang, Z., Mashimo, T., Shen, B., Nyagilo, J., Wang, H., Wang, Y., Liu, Z., Mulgaonkar, A., Hu, X. L., Piccirillo, S. G. M., Eskiocak, U., Dave, D. P., Qin, S., Yang, Y., Sun, X., Fu, Y. X., Zong, H., Sun, W., Bachoo, R. M. & Ge, W. P. 2020. Gliomas Interact with Non-glioma Brain Cells via Extracellular Vesicles. *Cell Rep,* 30, 2489–2500 e5.
15. Gibson, E. M., Geraghty, A. C. & Monje, M. 2018. Bad wrap: Myelin and myelin plasticity in health and disease. *Dev Neurobiol,* 78, 123–135.
16. Hatcher, A., Yu, K., Meyer, J., Aiba, I., Deneen, B. & Noebels, J. L. 2020. Pathogenesis of peritumoral hyperexcitability in an immunocompetent Crispr-based glioblastoma model. *J Clin Invest,* 130, 2286–2300.
17. Hills, K. E., Kostarelos, K. & Wykes, R. C. 2022. Converging Mechanisms of Epileptogenesis and Their Insight in Glioblastoma. *Front Mol Neurosci,* 15, 903115.
18. Ishiuchi, S., Yoshida, Y., Sugawara, K., Aihara, M., Ohtani, T., Watanabe, T., Saito, N., Tsuzuki, K., Okado, H., Miwa, A., Nakazato, Y. & Ozawa, S. 2007. Ca2+-permeable Ampa receptors regulate growth of human glioblastoma via Akt activation. *J Neurosci,* 27, 7987–8001.
19. John Lin, C. C., Yu, K., Hatcher, A., Huang, T. W., Lee, H. K., Carlson, J., Weston, M. C., Chen, F., Zhang, Y., Zhu, W., Mohila, C. A., Ahmed, N., Patel, A. J., Arenkiel, B. R., Noebels, J. L., Creighton, C. J. & Deneen, B. 2017. Identification of diverse astrocyte populations and their malignant analogs. *Nat Neurosci,* 20, 396–405.
20. Krishna, S., Choudhury, A., Keough, M. B., Seo, K., Ni, L., Kakaizada, S., Lee, A., Aabedi, A., Popova, G., Lipkin, B., Cao, C., Nava Gonzales, C., Sudharshan, R., Egladyous, A., Almeida, N., Zhang, Y., Molinaro, A. M., Venkatesh, H. S., Daniel, A. G. S., Shamardani, K., Hyer, J., Chang, E. F., Findlay, A., Phillips, J. J., Nagarajan, S., Raleigh, D. R., Brang, D., Monje, M., & Hervey-Jumper, S. L. 2023. Glioblastoma remodelling of human neural circuits decreases survival. *Nature, 617*, 599–607.

21. Labrakakis, C., Patt, S., Hartmann, J. & Kettenmann, H. 1998. Functional Gaba(A) receptors on human glioma cells. *Eur J Neurosci,* 10, 231–8.
22. Lange, F., Hornschemeyer, J. & Kirschstein, T. 2021. Glutamatergic Mechanisms in Glioblastoma and Tumor-Associated Epilepsy. *Cells,* 10.
23. Lawn, S., Krishna, N., Pisklakova, A., Qu, X., Fenstermacher, D. A., Fournier, M., Vrionis, F. D., Tran, N., Chan, J. A., Kenchappa, R. S. & Forsyth, P. A. 2015. Neurotrophin signaling via TrkB and TrkC receptors promotes the growth of brain tumor-initiating cells. *J Biol Chem,* 290, 3814–24.
24. Mitchell, T. J., Seitzman, B. A., Ballard, N., Petersen, S. E., Shimony, J. S. & Leuthardt, E. C. 2020. Human Brain Functional Network Organization Is Disrupted After Whole-Brain Radiation Therapy. *Brain Connect,* 10, 29–38.
25. Mitchell, D., Shireman, J., Sierra Potchanant, E. A., Lara-Velazquez, M. & Dey, M. 2021. Neuroinflammation in Autoimmune Disease and Primary Brain Tumors: The Quest for Striking the Right Balance. *Front Cell Neurosci,* 15, 716947.
26. Montgomery, M. K., Kim, S. H., Dovas, A., Zhao, H. T., Goldberg, A. R., Xu, W., Yagielski, A. J., Cambareri, M. K., Patel, K. B., Mela, A., Humala, N., Thibodeaux, D. N., Shaik, M. A., Ma, Y., Grinband, J., Chow, D. S., Schevon, C., Canoll, P. & Hillman, E. M. C. 2020. Glioma-Induced Alterations in Neuronal Activity and Neurovascular Coupling during Disease Progression. *Cell Rep,* 31, 107500.
27. Moura, D. M. S., Brennan, E. J., Brock, R. & Cocas, L. A. 2021. Neuron to Oligodendrocyte Precursor Cell Synapses: Protagonists in Oligodendrocyte Development and Myelination, and Targets for Therapeutics. *Front Neurosci,* 15, 779125.
28. Muller-Langle, A., Lutz, H., Hehlgans, S., Rodel, F., Rau, K. & Laube, B. 2019. Nmda Receptor-Mediated Signaling Pathways Enhance Radiation Resistance, Survival and Migration in Glioblastoma Cells-A Potential Target for Adjuvant Radiotherapy. *Cancers (Basel),* 11.
29. Numakawa, T. & Odaka, H. 2021. Brain-Derived Neurotrophic Factor Signaling in the Pathophysiology of Alzheimer's Disease: Beneficial Effects of Flavonoids for Neuroprotection. *Int J Mol Sci,* 22.
30. Osswald, M., Jung, E., Sahm, F., Solecki, G., Venkataramani, V., Blaes, J., Weil, S., Horstmann, H., Wiestler, B., Syed, M., Huang, L., Ratliff, M., Karimian Jazi, K., Kurz, F. T., Schmenger, T., Lemke, D., Gommel, M., Pauli, M., Liao, Y., Haring, P., Pusch, S., Herl, V., Steinhauser, C., Krunic, D., Jarahian, M., Miletic, H., Berghoff, A. S., Griesbeck, O., Kalamakis, G., Garaschuk, O., Preusser, M., Weiss, S., Liu, H., Heiland, S., Platten, M., Huber, P. E., Kuner, T., Von Deimling, A., Wick, W. & Winkler, F. 2015. Brain tumour cells interconnect to a functional and resistant network. *Nature,* 528, 93–8.
31. Pallud, J., Le Van Quyen, M., Bielle, F., Pellegrino, C., Varlet, P., Cresto, N., Baulac, M., Duyckaerts, C., Kourdougli, N., Chazal, G., Devaux, B., Rivera, C., Miles, R., Capelle, L. & Huberfeld, G. 2014. Cortical Gabaergic excitation contributes to epileptic activities around human glioma. *Sci Transl Med,* 6, 244ra89.
32. Pan, Y. & Monje, M. 2022. Neuron-Glial Interactions in Health and Brain Cancer. *Adv Biol (Weinh),* 6, e2200122.
33. Pan, Y., Hysinger, J. D., Barron, T., Schindler, N. F., Cobb, O., Guo, X., Yalcin, B., Anastasaki, C., Mulinyawe, S. B., Ponnuswami, A., Scheaffer, S., Ma, Y., Chang, K. C., Xia, X., Toonen, J. A., Lennon, J. J., Gibson, E. M., Huguenard, J. R., Liau, L. M., Goldberg, J. L., Monje, M. & Gutmann, D. H. 2021. Nf1 mutation drives neuronal activity-dependent initiation of optic glioma. *Nature,* 594, 277–282.
34. Pei, Z., Lee, K. C., Khan, A., Erisnor, G. & Wang, H. Y. 2020. Pathway analysis of glutamate-mediated, calcium-related signaling in glioma progression. *Biochem Pharmacol,* 176, 113814.
35. Rzeski, W., Ikonomidou, C. & Turski, L. 2002. Glutamate antagonists limit tumor growth. *Biochem Pharmacol,* 64, 1195–200.
36. Schneider, M., Vollmer, L., Potthoff, A. L., Ravi, V. M., Evert, B. O., Rahman, M. A., Sarowar, S., Kueckelhaus, J., Will, P., Zurhorst, D., Joseph, K., Maier, J. P., Neidert, N., D'errico, P., Meyer-Luehmann, M., Hofmann, U. G., Dolf, A., Salomoni, P., Guresir, E., Enger, P. O., Chekenya, M., Pietsch, T., Schuss, P., Schnell, O., Westhoff, M. A., Beck, J., Vatter, H., Waha,

A., Herrlinger, U. & Heiland, D. H. 2021. Meclofenamate causes loss of cellular tethering and decoupling of functional networks in glioblastoma. *Neuro Oncol,* 23, 1885–1897.
37. Shchors, K., Massaras, A. & Hanahan, D. 2015. Dual Targeting of the Autophagic Regulatory Circuitry in Gliomas with Repurposed Drugs Elicits Cell-Lethal Autophagy and Therapeutic Benefit. *Cancer Cell,* 28, 456–471.
38. Sontheimer, H. 2008. A role for glutamate in growth and invasion of primary brain tumors. *J Neurochem,* 105, 287–95.
39. Spelat, R., Jihua, N., Sanchez Trivino, C. A., Pifferi, S., Pozzi, D., Manzati, M., Mortal, S., Schiavo, I., Spada, F., Zanchetta, M. E., Ius, T., Manini, I., Rolle, I. G., Parisse, P., Millan, A. P., Bianconi, G., Cesca, F., Giugliano, M., Menini, A., Cesselli, D., Skrap, M. & Torre, V. 2022. The dual action of glioma-derived exosomes on neuronal activity: synchronization and disruption of synchrony. *Cell Death Dis,* 13, 705.
40. Takano, T., Lin, J. H., Arcuino, G., Gao, Q., Yang, J. & Nedergaard, M. 2001. Glutamate release promotes growth of malignant gliomas. *Nat Med,* 7, 1010–5.
41. Tannock, I. F., Ahles, T. A., Ganz, P. A. & Van Dam, F. S. 2004. Cognitive impairment associated with chemotherapy for cancer: report of a workshop. *J Clin Oncol,* 22, 2233–9.
42. Venkataramani, V., Tanev, D. I., Strahle, C., Studier-Fischer, A., Fankhauser, L., Kessler, T., Korber, C., Kardorff, M., Ratliff, M., Xie, R., Horstmann, H., Messer, M., Paik, S. P., Knabbe, J., Sahm, F., Kurz, F. T., Acikgoz, A. A., Herrmannsdorfer, F., Agarwal, A., Bergles, D. E., Chalmers, A., Miletic, H., Turcan, S., Mawrin, C., Hanggi, D., Liu, H. K., Wick, W., Winkler, F. & Kuner, T. 2019. Glutamatergic synaptic input to glioma cells drives brain tumour progression. *Nature,* 573, 532–538.
43. Venkataramani, V., Yang, Y., Schubert, M. C., Reyhan, E., Tetzlaff, S. K., Wissmann, N., Botz, M., Soyka, S. J., Beretta, C. A., Pramatarov, R. L., Fankhauser, L., Garofano, L., Freudenberg, A., Wagner, J., Tanev, D. I., Ratliff, M., Xie, R., Kessler, T., Hoffmann, D. C., Hai, L., Dorflinger, Y., Hoppe, S., Yabo, Y. A., Golebiewska, A., Niclou, S. P., Sahm, F., Lasorella, A., Slowik, M., Doring, L., Iavarone, A., Wick, W., Kuner, T. & Winkler, F. 2022. Glioblastoma hijacks neuronal mechanisms for brain invasion. *Cell,* 185, 2899–2917 e31.
44. Venkatesh, H. S., Johung, T. B., Caretti, V., Noll, A., Tang, Y., Nagaraja, S., Gibson, E. M., Mount, C. W., Polepalli, J., Mitra, S. S., Woo, P. J., Malenka, R. C., Vogel, H., Bredel, M., Mallick, P. & Monje, M. 2015. Neuronal Activity Promotes Glioma Growth through Neuroligin-3 Secretion. *Cell,* 161, 803–16.
45. Venkatesh, H. S., Tam, L. T., Woo, P. J., Lennon, J., Nagaraja, S., Gillespie, S. M., Ni, J., Duveau, D. Y., Morris, P. J., Zhao, J. J., Thomas, C. J. & Monje, M. 2017. Targeting neuronal activity-regulated neuroligin-3 dependency in high-grade glioma. *Nature,* 549, 533–537.
46. Venkatesh, H. S., Morishita, W., Geraghty, A. C., Silverbush, D., Gillespie, S. M., Arzt, M., Tam, L. T., Espenel, C., Ponnuswami, A., Ni, L., Woo, P. J., Taylor, K. R., Agarwal, A., Regev, A., Brang, D., Vogel, H., Hervey-Jumper, S., Bergles, D. E., Suva, M. L., Malenka, R. C. & Monje, M. 2019. Electrical and synaptic integration of glioma into neural circuits. *Nature,* 573, 539–545.
47. Walker, A. J., Card, T., Bates, T. E. & Muir, K. 2011. Tricyclic antidepressants and the incidence of certain cancers: a study using the Gprd. *Br J Cancer,* 104, 193–7.
48. Wu, C. & Sun, D. 2015. Gaba receptors in brain development, function, and injury. *Metab Brain Dis,* 30, 367–79.
49. Xiong, J., Zhou, L., Yang, M., Lim, Y., Zhu, Y. H., Fu, D. L., Li, Z. W., Zhong, J. H., Xiao, Z. C. & Zhou, X. F. 2013. Probdnf and its receptors are upregulated in glioma and inhibit the growth of glioma cells in vitro. *Neuro Oncol,* 15, 990–1007.
50. Yu, K., Lin, C. J., Hatcher, A., Lozzi, B., Kong, K., Huang-Hobbs, E., Cheng, Y. T., Beechar, V. B., Zhu, W., Zhang, Y., Chen, F., Mills, G. B., Mohila, C. A., Creighton, C. J., Noebels, J. L., Scott, K. L. & Deneen, B. 2020. Pik3ca variants selectively initiate brain hyperactivity during gliomagenesis. *Nature,* 578, 166–171.

Part II
Cancer and the Peripheral Nervous System

Chapter 4
Neural Influence on Cancer Invasion and Metastasis

Ligia B. Schmitd, Cindy Perez-Pacheco, and Nisha J. D'Silva

Introduction

Cancer-related mortality is mostly caused by metastasis, a process by which cancer cells spread from primary tumors to invade and colonize distant sites. To metastasize, cancer cells acquire a phenotype that enables dissemination, survival, and proliferation outside the primary tumor microenvironment [1]. The mechanisms that induce metastasis of cancer cells are not fully understood. Recent studies show that nerves are important players in the tumor microenvironment, highlighting the dynamic communication between cancer cells and nerves [2–5]. Tumor-associated nerves manipulate cancer and non-cancer cells in the tumor microenvironment to provide a milieu favorable to tumor progression. Interestingly, nerve-related parameters such as high nerve density [3, 6, 7], perineural invasion, and nerve-tumor distance [8] have been associated with poor survival, recurrence, and metastasis in cancer patients. High nerve density correlates with increased cancer cell proliferation and with expression of proteins relevant for cancer cell survival [7].

Perineural invasion, observed in primary tumors at multiple sites, is the most well-known pro-tumoral phenotype associated with nerves. Thus, most of the literature

---

Authors Ligia B. Schmitd and Cindy Perez-Pacheco have equally contributed to this chapter.

L. B. Schmitd · C. Perez-Pacheco
Department of Periodontics and Oral Medicine, University of Michigan School of Dentistry, Ann Arbor, MI, USA

N. J. D'Silva (✉)
Department of Periodontics and Oral Medicine, University of Michigan School of Dentistry, Ann Arbor, MI, USA

Pathology, University of Michigan Medical School, Ann Arbor, MI, USA
e-mail: njdsilva@umich.edu

M. Amit, N. N. Scheff (eds.), *Cancer Neuroscience*,
https://doi.org/10.1007/978-3-031-32429-1_4

exploring the role of nerves in cancer progression focuses on perineural invasion. Not all tumor types have significant perineural invasion. For example, melanomas have low nerve density and usually do not exhibit perineural invasion [9]. However, melanoma growth can be modulated by activation of Schwann cells derived from damaged nerves [10]. Understanding the mechanistic underpinnings of nerve-tumor interactions will reveal potential treatment targets in the tumor microenvironment.

Peripheral nerves regulate multiple functions that vary with the type of innervation. These include sensory (afferent) and motor (efferent) branches. Motor nerves are divided into somatic and autonomic with the latter subdivided into sympathetic and parasympathetic. The tissue-specific distribution of nerves and variety of functions may trigger different molecular mechanisms contributing to tumor progression in different organs. In this chapter, we will explore how different cells within nerves advance tumor invasion and metastasis.

Nerve Structure

Peripheral nerves have a unique structural organization that encompasses multiple cells and layers. Peripheral nerves contain neuronal axons and glial cells named Schwann cells that are myelinating or nonmyelinating. Myelinating Schwann cells surround large caliber axons to create the myelin sheath, a protective envelope that facilitates transmission of electrical impulses through the axon. Nonmyelinating Schwann cells support axons at the nerve trunk and terminals [11]. These are subdivided into two distinct types: Remak cells surround a bundle of small caliber axons [11], whereas terminal Schwann cells are present at the synaptic interface of the neuromuscular junction and mechanosensory lamellar corpuscles [12]. Myelinating Schwann cells can become nonmyelinating in diseases where myelin is lost. Conversely, nonmyelinating Schwann cells retain the potential to myelinate, highlighting the plasticity of these cells.

Recently, using single cell sequencing of healthy and diseased peripheral nerves to characterize nerves based on gene expression, new populations of cells were discovered [13, 14]. The predominant cell types in healthy peripheral nerves are fibroblasts and nonmyelinating Schwann cells, followed by vascular endothelial cells and myelinating Schwann cells (see Table 4.1 for cells types). Nerves are wrapped by three sheaths of mesenchymal tissue: epineurium, perineurium, and endoneurium. Nerve fascicles, surrounded by perineurium, also contain endoneurial mesenchymal cells in close proximity with axons and Schwann cells, including mesenchymal precursor cells that contribute to regeneration [15, 16]. Figure 4.1 shows the nerve structure and cell types. Of note, the cell profile of different types of nerves has not been described; for example, it is unknown if the composition of sensory nerves differs from that of sympathetic nerves.

Cell composition of diseased nerves was studied in autoimmune neuritis, which exhibits an increase in lymphocytes [14]. In injured nerves, a mesenchymal population with progenitor characteristics is present. These cells have transcriptional similarities with pre-adipocytes and osteochondro-progenitors [14], highlighting the fascinating regenerative capacity that nerves exert in the surrounding tissues.

Table 4.1 Cell types in peripheral nerves from healthy mice (brachial plexus and sciatic nerve)

Cell type	Prevalence[a]
Fibroblast	+ + + + + +
Nonmyelinating Schwann cells	+ + + + +
Vascular endothelial cells	+ + + +
Myelinating Schwann cells	+ + + +
T cells	+ + +
Vascular smooth muscle cells	+ + +
Pericytes	+ +
Macrophages	+ +
B cells	+ +
Myeloid lineage cells	+
Lymphatic endothelial cells	+
Other endothelial cells	+

[a]Adapted from Wolbert et al. [13]

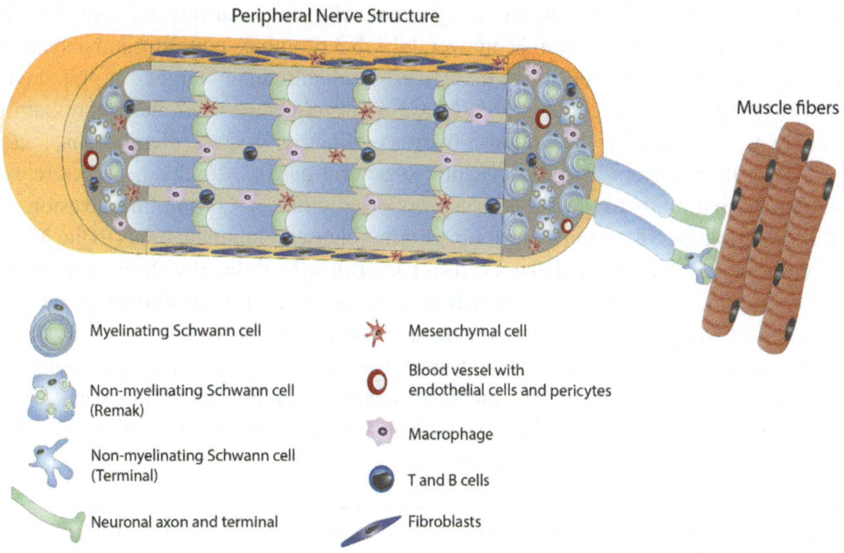

Fig. 4.1 Schematic illustration of a peripheral nerve: Neuronal axons are supported by both myelinating and non-myelinating Schwann cells. Nerves contain other cell types such as fibroblasts, macrophages and immune cells. Endothelial cells and pericytes line the blood vessels within the nerve. Another type of non-myelinating cell, terminal Schwann cell, supports the neuronal synapsis at the neuromuscular junction. Mesenchymal cells are present in all three nerve layers and are crucial for the regenerative capacity of the nerve

In injury, Schwann cells can dedifferentiate to acquire mesenchymal characteristics and invade the nerve stump to drive regeneration [11].

Cancer has many similarities to a wound, including promotion of a nerve injury phenotype [10]. Here we summarize the literature on how cells within nerves influence cancer invasion and metastasis.

Influence of Cells Within Nerves on Tumor Invasion and Metastasis

Neuronal Cells

Direct Effect on Tumor Cells

Direct communication between neurons and primary cancer cells was observed in glioma, a central nervous system tumor; neurons and human glioma cells form synapses within the tumor microenvironment [17]. Neurons release neuroligin-3 (NLGN3), a synaptic protein that is required for glioma progression [18]. Interestingly, in patients with high-grade glioma, the intensity of *NLGN3* expression is associated with poor overall survival [18]. NLGN3 induces expression of several synaptic genes in glioma cells [17]. Coculture of glioma cells and neurons without NLGN3 suppresses synapse formation. Moreover, NLGN3 promotes neuron-glioma synapses through phosphatidylinositol-3-kinase (PI3K)-mammalian target of rapamycin (mTOR) signaling. In glioma cells, NLGN3 subsequently promotes expression of glutamate, especially α-amino-3-hydroxy-5-methyl-4-isoxazolepropionic acid (AMPA), its receptors, and postsynaptic structural genes. Moreover, nerves release glutamate that binds to glutamate receptors on glioma cells promoting invasion and glutamate release. In an autocrine loop, glioma-released glutamate binds to the AMPA receptor on glioma cells and also exerts a paracrine effect, thereby stimulating neuronal cells [18]. These data suggest that neurons are relevan t for glioma growth and progression via NLGN3 secretion and synapsis neurotransmission.

In the peripheral nervous system, tumor-associated nerves secrete neurotransmitters, chemokines, neuropeptides, and neurotrophins that play important roles in cancer invasion and metastasis. The mechanisms of nerve-tumor interaction described in this section are summarized in Fig. 4.2.

Fig. 4.2 (continued) IFNγ on CD4 and CD8 enhancing tumor growth and metastasis. In macrophages, norepinephrine increases macrophage infiltration into the tumor and differentiation into the M2 phenotype, promoting tumor cell proliferation and invasion. Parasympathetic nerves release acetylcholine that downregulates IFNγ in CD8 T cells and impairs tumor infiltration. In addition, acetylcholine enhances the abundance of Th2 phenotype in CD4 T cells, promoting an immune-suppressive tumor microenvironment. Pink and grey arrows indicate the effect of norepinephrine and acetylcholine on T cells, respectively. Abbreviations: *ADRβ2* adrenergic receptor beta 2, *CCL2* C-C motif chemokine ligand 2, *CCR2* C-C motif chemokine receptor 2, *ChRM1* type 1 muscarinic cholinergic receptor, *ChRM3* type 3 muscarinic cholinergic receptor, *CXCL12* C-X-C motif chemokine l ligand 12, *CXCR4* C-X-C motif chemokine receptor 4, *EMT* epithelial-mesenchymal transition, *FOXP3* forkhead box protein P3, *GalR2* galanin receptor 2, *IFN-γ* interferon gamma, *nAChR* acetylchoine nicotinic receptor, *NK1R* neurokinin receptor 1, *PD1* programmed cell death protein 1

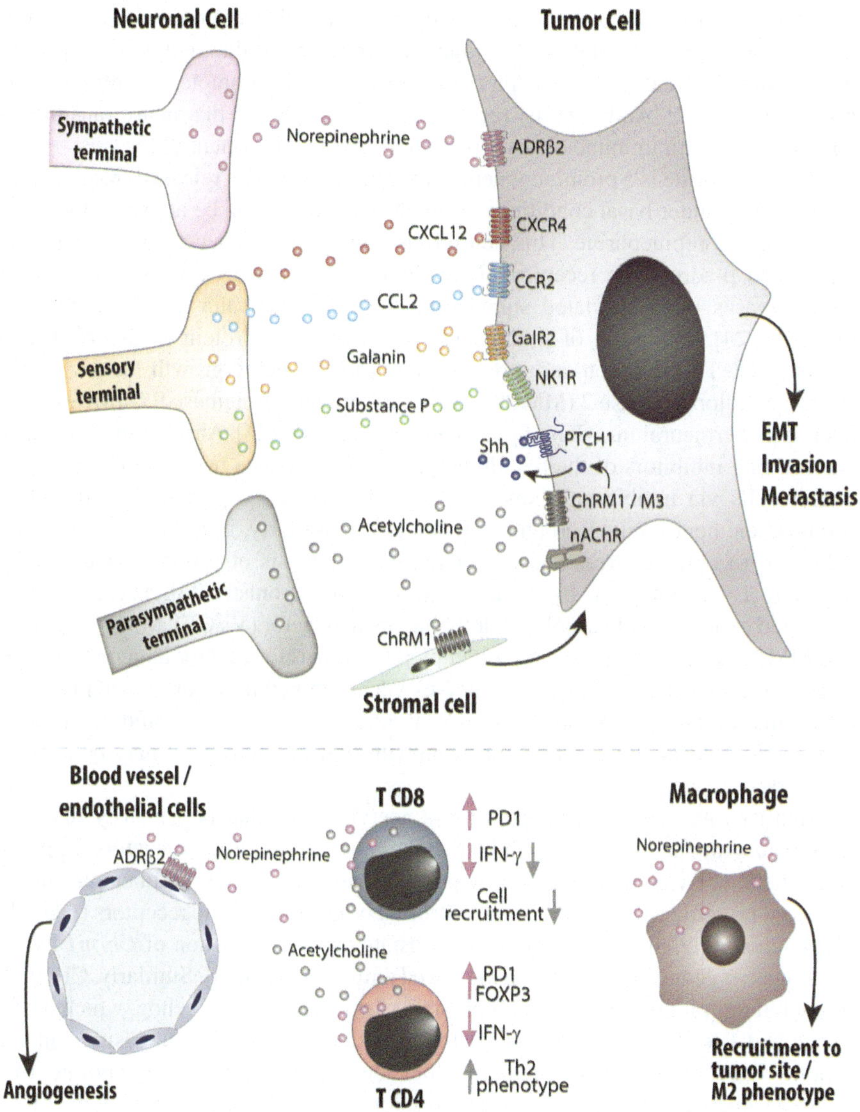

Fig. 4.2 Effect of different types of neurons on tumor invasion and metastasis: Many molecular interactions driven by neuronal cells influence tumor behavior to promote invasion. Sympathetic nerves secrete the catecholamine norepinephrine, which binds to β-adrenergic receptors in tumor cells to promote invasion and metastasis. Sensory nerves release chemokines such as CCL2 and CXCL12, and neuropeptides Galanin and Substance P that bind to their receptors to stimulate tumor cell migration and perineural invasion. Similarly, parasympathetic cholinergic fibers release acetylcholine that promotes invasion, migration, and metastasis of cancer cells via stromal ChRM1 signals, and through ChRM1/M3 and nAChR present in cancer cells. Nerve-secreted neurotransmitters also impact non cancer cells present in the tumor microenvironment. Norepinephrine binds to β-adrenergic receptors present on endothelial cells to promote angiogenesis. Furthermore, norepinephrine increases PD-1 on CD4 and CD8 T cells, increases FOXP3 on CD4, and decreases

Neurotransmitters are chemicals that carry information between neurons and/or nonneuronal cells to modify the function/behavior of the recipient cell [19]. Neurotransmitters also bind receptors on cancer cells to promote proliferation and migration [20–22]. Additionally, recent studies have shown that neurotransmitters communicate with immune cells in the tumor microenvironment [23].

Norepinephrine is a prominent neurotransmitter, primarily released from sympathetic nerves under basal conditions. Sympathetic nerves can be activated by stress and secrete norepinephrine. This neurotransmitter can be detected by cancer cells via α- and β-adrenergic receptors. Norepinephrine enhances cancer cell invasion and increases phosphorylated signal transducer and activator of transcription 3 (STAT3) [24]. Inhibition of β-adrenergic receptors and protein kinase A (PKA) reverses STAT3 activation and reduces expression of nerve growth factor (NGF), matrix metalloproteinase 2 (MMP2), and matrix metalloproteinase 9 (MMP9). This attenuates perineural invasion of pancreatic cancer cells [24]. Another in vitro study showed that inhibitors of the β2-adrenergic receptor reduce invasion of pancreatic cancer cells via inhibition of nuclear factor κB, activator protein 1, and cAMP response element binding protein (CREB) [25]. Similarly, in gastric cancer, mice subjected to chronic stress showed increased metastasis of primary cancer cells. This effect was enhanced by β-adrenergic receptor agonists [26]. These agonists promoted expression of vascular endothelial growth factor (VEGF), MMP2, MMP7, and MMP9, an effect that was inhibited by β2-adrenergic receptor antagonists [26]. These data suggest that sympathetic nerves via norepinephrine participate in cancer dissemination and metastasis. Moreover, β-adrenergic receptors could be suitable markers for cancer progression, and administration of antagonists may be a treatment option.

Another important neurotransmitter is acetylcholine that is primarily released from parasympathetic nerves to induce nicotinic and muscarinic receptors. In prostate cancer, acetylcholine released by parasympathetic cholinergic fibers promotes invasion, migration, and metastasis via stromal type 1 muscarinic receptors (*Chrm1*) [3]. Interestingly, in a mouse model of prostate cancer, inhibition of *Chrm1* by an antagonist or genetic manipulation improved animal survival [3]. Similarly, CHRM1 promoted migration and invasion of prostate cancer cells via hedgehog, which binds to patched 1 to allow activation of GLI family zinc finger 1 [27]. In gastric cancer, genetic knockout or pharmacologic blockade of muscarinic acetylcholine M3 (*Chrm3*) receptor suppresses tumor progression [28]. In vitro, activation of the nicotinic acetylcholine receptor via the p38 MAPK signaling pathway stimulated MMP2 and MMP9 and induced invasion of colon cancer cells [29]. In poorly differentiated non-small cell lung carcinoma, acetylcholine promoted invasion via α5 and α7 nicotinic acetylcholine receptors [30]. Taken together, acetylcholine receptors may be potential targets for some cancers.

Neuropeptides are neuromodulators in the central and peripheral nervous system. Neurons secrete neuropeptides that bind to G protein-coupled receptors on adjacent cells. In in vitro and in vivo models in which oral cancer cells overexpressing galanin receptor 2 (HNSCC-GALR2) were cocultured or implanted with dorsal root ganglia, nerve-released galanin stimulated neuron-cancer interaction [2].

HNSCC-GALR2 were more invasive toward dorsal root ganglia, while downregulation of GALR2 inhibited this effect [2]. Furthermore, Scanlon et al. elucidated that galanin induces prostaglandin E2 secretion via nuclear factor of activated T cell (NFATC2) mediated transcription of cyclooxygenase-2 (COX-2). Downregulation of NFATC2 inhibited cancer cell migration toward dorsal root ganglia. Substance P, a neuropeptide expressed by sensory neurons, stimulates pancreatic cancer cell migration [31]. A coculture model of human pancreatic cancer cells, neonatal dorsal root ganglia and substance P, showed migration of these cancer cells in a peak-shaped cluster approaching dorsal root ganglia. Inhibition of neurokinin 1 (NK-1R), substance P receptor, abrogated this effect. In addition, substance P induced the expression of MMP2 in pancreatic cancer cells, and NK-1R antagonists inhibit these effects [31]. These studies show that nerves contribute to an aggressive phenotype in cancer cells via release of neuropeptides.

Chemokines are small secreted proteins that regulate cell migration, proliferation, and survival through formation of a concentration gradient [32]. Together with their receptors, chemokines are important mediators in perineural invasion. A proteomic profiler chemokine array used to screen chemokines released by dorsal root ganglia identified the chemokine C-C motif ligand 2 (CCL2) [33]. CCL2 modulates inflammatory processes regulating chemotaxis of myeloid and lymphoid cells. In prostate cancer cells, expression of the CCL2 receptor (C-C motif chemokine receptor 2, CCR2) correlated with expression of MAPK and Akt pathways and migration toward CCL2 and dorsal root ganglia. Coculture of cancer cells and dorsal root ganglia showed that CCR2 facilitates perineural invasion, which is inhibited in cocultures with dorsal root ganglia from CCL2 knockout mice. Furthermore, in a murine sciatic nerve perineural invasion model, prostate cancer cells without CCR2 show diminished invasion of nerves [33]. Similarly, C-X-C motif ligand 12 (CXCL12) released from nerves stimulated perineural invasion and chemotactic migration of C-X-C motif receptor 4 (CXCR4)-positive pancreatic cancer cells [34]. This evidence suggests that nerve-released chemokines modulate chemotaxis of cancer cells into nerves, a route of cancer metastasis.

Effects on the Tumor Microenvironment

The tumor microenvironment contains nerves, fibroblasts, immune cells, blood and lymphatic vessels, extracellular matrix, and soluble molecules that work together to support tumor progression. For example, in vitro and in vivo experiments showed that adrenergic signals have an important role in tumor angiogenesis. Sympathetic nerve fibers induce β-adrenergic receptors on endothelial cells in the tumor microenvironment to enhance angiogenesis and prostate cancer growth [35]. In β2-adrenergic receptor-deficient mice, oxidative phosphorylation of endothelial cells increased and altered cytoskeletal organization and migration and inhibited angiogenesis [35]. This evidence suggests that suppression of the adrenergic receptor via β-blockers could interfere with tumoral vascular support, thereby impairing tumor progression.

The tumor-associated immune cell population can be divided into three types: tumor-antagonizing, tumor-promoting, and controversial immune cells [36]. Tumor-antagonizing immune cells are the effector T cells (CD8 cytotoxic T cells and effector CD4 T cells), natural killer cells (NK), dendritic cells (DCs), M1-polarized macrophages, and N1-polarized neutrophils. Tumor-promoting immune cells consist of regulatory T cells (Treg) and myeloid-derived suppressor cells (MDSCs) [36]. B cells belong to the controversial group of immune cells because the behavior/function in the tumor microenvironment is not completely clear. T lymphocytes expressing CD4 are cytokine producers, also called T helper (Th) cells, and are subdivided into Th1 and Th2 [37]. Th1 and Th2 cells trigger different immune response pathways. Th1 cells drive the "cellular immunity" type 1 pathway to fight viruses and other intracellular pathogens and eliminate tumor cells. Th2 cells direct the "humoral immunity" type 2 pathway to fight extracellular organisms [38].

The impact of nerves on tumor-associated immune cells may vary in a nerve-type dependent manner. In a spontaneous breast cancer model in mice, genetic neurostimulation of sympathetic nerves increased nerve Ca^{2+} activity, tumor norepinephrine content, tumor growth, and lung metastasis. In contrast, denervation of sympathetic nerves by virus injection inhibited these phenotypes. In these denervated tumors, expression of programmed cell death protein 1 (PD-1) diminished on CD4 and CD8 tumor-infiltrating lymphocytes. Forkhead box protein P3 (FOXP3) also decreased on CD4 tumor-infiltrating lymphocytes, while expression of interferon gamma (IFN-γ) increased on CD4 and CD8 tumor-infiltrating lymphocytes. Parasympathetic nerves decreased PD-1 and PD-L1 expression and decelerated breast cancer progression, whereas denervation accelerated cancer progression [4]. Therefore, we can infer that parasympathetic nerves modulate the T cell response within the tumor microenvironment to have an immune-protective effect, while sympathetic nerves enhance tumor growth and metastasis via immunosuppression.

However, the impact of parasympathetic fibers on the tumor microenvironment may be dependent on cancer site. Parasympathetic cholinergic nerves secrete acetylcholine that acts on immune cells to modulate their functions. In an orthotopic pancreatic ductal adenocarcinoma model, cholinergic signaling impaired recruitment of CD8 T cells [39]. In addition, acetylcholine downregulated IFN-γ production by CD8 T cells and drove differentiation and abundance of Th2 phenotype in CD4 cells. Bilateral vagotomy prevented perineural invasion in tumor-bearing mice, and an increase in T cells and higher Th1/Th2 ratios were observed. This study suggests that acetylcholine signaling in perineural invasion enhances tumor growth via suppression of T cell function in pancreatic ductal adenocarcinoma [39].

Stress-induced norepinephrine released by sympathetic fibers induced metastasis of breast cancer via β-adrenergic receptors [40]. Adrenergic signaling increased infiltration of CD11b F4/80 macrophages. This leads to increased expression of prometastatic genes including *Cox2*, *Mmp9*, *Vegf*, and vascular cell adhesion molecule 1 (*Vcam1*) as well as macrophage chemoattractant and growth factor colony stimulating factor 1 (*Csf1*) in the primary tumor. To elucidate the effect of sympathetic nerves on macrophages, bone marrow-derived macrophages were treated with norepinephrine leading to an increase in M2-phenotype-related gene expression, arginase 1 (*Arg1*), and transforming growth factor-beta (*Tgfβ*), while expression of a M1-phenotype-related gene (nitric oxide synthase 2 – *Nos2*) decreased [40]. Propranolol, a β-adrenergic receptor antagonist, or CSF-1 receptor kinase inhibitor, decreased macrophage infiltration and metastasis in a stress-induced breast cancer model [40].

The immune checkpoint PD-1 and its ligand PD-L1, which inhibit the adaptive immune system, are targets for immunotherapy. PD-L1 binds to PD-1 to prevent T cells from eliminating cancer cells, thereby facilitating tumor progression [36]. Immunotherapy targeting the PD1/PD-L1 axis can effectively block pro-tumoral activity. Immune checkpoint inhibitors bind to both PD-1 and PD-L1 and block ligand-receptor binding. This leaves T cells free to neutralize cancer cells. PD-L1 protein expression has been detected in tumor-associated nerves in human prostate cancer, and the PD-L1-positive nerve density in tumor stroma is associated with reduced CD8 tumor-infiltrating lymphocytes and with biochemical recurrence [41]. This may indicate that combined immunotherapies should be considered in prostate cancer.

Additionally, using a pain model, Chen et al. [42] showed that nonmalignant tissues such as the brain, dorsal root ganglia, and skin produce endogenous PD-L1 protein in mice. Furthermore, PD-L1 inhibited acute and chronic pain. This group also demonstrated that murine melanomas present high levels of PD-L1 and that tumor onset is not associated with spontaneous pain. Interestingly, PD-L1 inhibition in melanoma-induced tumor reduced the withdrawal pain threshold and induced spontaneous pain. In addition, PD-L1 inhibition in mouse melanoma increased excitability in nerve fibers. Taken together, this data suggest that melanoma-related PD-L1 inactivates intratumoral sensory nerves via PD-1 to mask tumor-related pain.

Schwann Cells

Direct Effects on Tumor Cells

Schwann cells are associated with epithelial-to-mesenchymal transition, invasion, migration, and metastasis of cancer cells [43]. Since pancreatic tumors have high nerve density [6] and perineural invasion [44], several mechanisms of Schwann cell-tumor interaction promoting invasion and metastasis are described and summarized in Fig. 4.3.

Schwann cells can leave the nerve environment to colonize tumors [45]. For example, pancreatic tumor cells attract Schwann cells during disease onset via secretion of the chemokine C-X-C motif ligand 12 (CXCL-12) [46]. Additionally, macrophages interact with cancer-associated fibroblasts to enhance Schwann cell migration [47, 48].

Tumor cells adhere to Schwann cells via interaction between Mucin 1 (MUC1) present in tumor cells and myelin-associated glycoprotein (MAG) in Schwann cells. Both proteins are detected histologically at sites of perineural invasion, suggesting

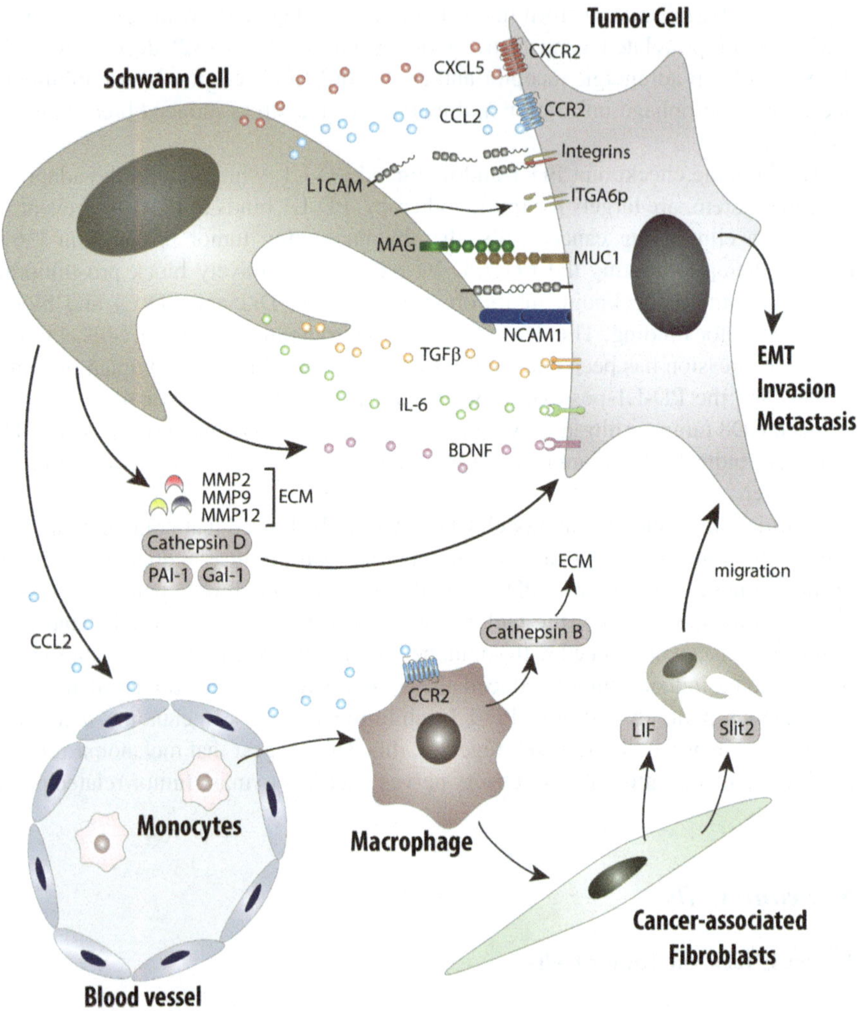

Fig. 4.3 Effect of Schwann cells on invasion and metastasis: Schwann cells bind to and recruit tumor cells to the nerve via adhesion molecules MAG, NCAM1, and L1CAM. Secreted L1CAM also acts as a chemoattractant on tumor cells. Furthermore, many secreted molecules from Schwann cells induce invasion, epithelial-to-mesenchymal transition, and metastasis of cancer cells. Chemokines CXCL5 and CCL2 induce a more mesenchymal/ migratory phenotype in tumor cells. Additionally, CCL2 recruits macrophages to the tumor microenvironment. Macrophages help degrade extracellular matrix, facilitate invasion, and interact with cancer-associated fibroblasts, further stimulating Schwann cell migration to the tumor. TGBß1 triggers epithelial to mesenchymal transition in tumor cells via SMAD2; IL-6 acts through STAT3 to induce a similar outcome. Other Schwann cell-derived factors such as Cathepsin D, PAI-1 and Gal-1 encourage tumor cell migration. Unknown factors lead to cleavage of integrin alpha 6 molecules in tumor cells, to generate a variant that confers increased motility (ITGA6p). Schwann cells also secrete gelatinases MMP2, MMP9 and MMP12 to degrade extracellular matrix, thereby facilitating tumor cell migration and invasion. The neurotrophic factor BDNF, also derived from Schwann cells, induces migration of tumor cells via the TrkB receptor. Abbreviations: *BDNF* brain-derived

participation in invasion of tumor cells into nerves [49]. Furthermore, adhesion proteins expressed in Schwann cells and pancreatic tumor cells such as neural cell adhesion molecule 1 (NCAM1) [50] and L1 cell adhesion molecule (L1CAM) [51] have been implicated in mechanisms of perineural invasion, as discussed below.

In vitro studies show that Schwann cells extend protrusions to contact cancer cells [50]. NCAM1 accumulates in these filopodia contributing to recruitment of tumor cells toward Schwann cells. One study showed that cancer migration toward Schwann cells is dependent on cell-cell contact since exposure of cancer cells to conditioned medium from Schwann cells did not increase tumor attraction [50]. Interestingly, migration of cancer cells toward fibroblasts was much lower, suggesting a dominant effect of Schwann cells on cancer cell invasion. Additionally, tumor cells injected into the sciatic nerve in NCAM1 knockout mice were less migratory than in wild-type mice. However, since cells were injected into the nerve, it is difficult to determine the extent to which NCAM1 contributes to initiation of perineural invasion in vivo. Also, participation of other nerve cells in this process cannot be excluded, since NCAM1 was knocked out in all mouse cells.

Contradicting the dependence on Schwann cell-cancer cell contact for perineural invasion [50], secreted L1CAM from Schwann cells increases tumor migration independent of contact [51]. L1CAM shed from Schwann cells is a chemoattractant to pancreatic cancer cells and triggers increased expression of gelatinases MMP2 and MMP9, facilitating the breakdown of extracellular matrix, which is necessary for invasion. This may be a mechanism that initiates cancer invasion toward nerves. Pancreatic cancer cells also express L1CAM. When this adhesion molecule is silenced in both Schwann cells and cancer cells, migration is reduced to a minimum, supporting that L1CAM-mediated cell-cell interactions may be important for perineural invasion [51].

Currently, sufficient evidence supports paracrine signaling between Schwann cells and pancreatic cancer, independent of cell contact. In vitro, TGFß1 secreted by Schwann cells modulates invasion of pancreatic cancer cells [52]. Conditioned medium from human Schwann cells derived from plexiform neurofibromas or healthy nerve (primary cells) contains high levels of TGFß1 and increases migration, invasion, and tumorigenesis via SMAD2 signaling in cancer cells [52]. Additionally, pancreatic cancer cells induce expression of interleukin 6 (IL-6) in Schwann cells in vitro, which stimulates epithelial-to-mesenchymal transition and invasion of tumor cells via IL-6 receptor-STAT3 signaling [53]. However, Schwann cell-derived factors are poorly characterized. Secretome analysis of Schwann cell-derived exosomes, usually in a context unrelated to cancer, revealed molecules that

Fig 4.3 (continued) neurotrophic factor, *CXCL5* C-X-C motif chemokine ligand 5, *CXCR2* C-X-C motif chemokine receptor 2, *CCL2* C-C motif chemokine ligand 2, *CCR2* C-C motif chemokine receptor 2 CCL, *ECM* extracellular matrix, *EMT* epithelial-mesenchymal transition, *Gal-1* Galectin-1, *IL-6* interleukin 6, *IL-6R* interleukin 6 receptor, *ITGA6* integrin subunit alpha 6, *L1CAM* L1 cell adhesion molecule, *LIF* LIF interleukin 6 family cytokine, *MAG* myelin-associated glycoprotein, *MMP* matrix metallopeptidase, *MUC1* mucin 1, cell surface-associated, *NCAM1* neural cell adhesion molecule 1, *PAI-1* plasminogen activator inhibitor-1, *Slit2* Slit guidance ligand 2, *TGFß* transforming growth factor beta

induce nerve regeneration and inhibit inflammation [54]. Proteomic analysis of primary human Schwann cells from a healthy donor showed candidate proteins (MMP2, cathepsin D, plasminogen activator inhibitor-1, galectin-1) that increase pancreatic cancer invasion in vitro [55], but the mechanisms are unexplored. To add a layer of complexity to these candidate paracrine mechanisms, secretomes of specific types of Schwann cells are not available.

An in vitro study in pancreatic and prostate cancer dissected the contribution of different types of Schwann cells to invasion [56]. Conditioned medium from myelinating Schwann cells increases cancer cell invasion via cleavage of integrin alpha 6 (ITGA6) receptor to produce a more motile variant (ITGA6p). ITGA6p is linked to an aggressive/invasive phenotype in different cancers [57]. Nonmyelinating Schwann cells have an opposite effect on integrin cleavage and tumor invasion, highlighting the dynamic interactions between cancer and Schwann cells in different phases of differentiation [56].

Perineural invasion, which occurs in a large proportion of head and neck cancers, is associated with poor clinical outcomes [8, 58]. Human oral squamous cell carcinoma cells migrate toward Schwann cells in vitro [59]. Addition of brain-derived neurotrophic factor (BDNF) increases migration leading to intermingling between the two cell types in a migration assay, while blocking BDNF receptor tropomyosin receptor kinase B (TrkB) abrogates the effect [59]. Since both cell types express BDNF and TrkB, it is possible that a feed-forward loop occurs during perineural invasion. However, this hypothesis needs further mechanistic investigation. Similar findings were noted in adenoid cystic carcinoma, a salivary gland cancer, which is a type of head and neck cancer that frequently exhibits perineural invasion [60]. Coculture of adenoid cystic carcinoma cell line SACC-LM and Schwann cells increased epithelial-to-mesenchymal transition; tumor cells also acquire a Schwann cell-like phenotype verified by expression of markers S100A4 and GFAP. This change is due to Schwann cell-derived BDNF since inhibition of TrkB dampens the effect [60]. Expression of S100A4 and GFAP in adenoid cystic carcinomas in clinical specimens is associated with the presence of perineural invasion [60].

A recent study in oral squamous cell carcinoma showed that dietary palmitic acid induces expression of CD36 and increases metastasis in mice. These CD36-expressing tumor cells stimulate Schwann cell colonization in the tumor bulk via secretion of the neuropeptide galanin [43]. Downstream of galanin signaling, Schwann cells secrete a specialized pro-regenerative extracellular matrix supporting nerve growth. Palmitic acid creates a pro-metastatic memory in these tumors that may be due to Schwann cells "preparing" the environment for tumor invasion [43]. Galanin is a nerve-regeneration neuropeptide that was previously implicated in perineural invasion in oral squamous cell carcinoma via its effects in both tumor and neuronal cells, as discussed earlier in this chapter [2]. However, the effects of galanin in tumor progression may not be limited to Schwann cells and neurons, since this neuropeptide is also expressed by immune cells [61].

The paracrine signaling between Schwann cells and tumor cells to increase invasiveness has also been described in lung and cervical cancers. Schwann cells increase growth, migration, and invasion of cervical cancer cells in vitro via secretion of a chemokine, CCL2 [62]. Schwann cells cocultured with tumor show higher expression of CCL2 compared to Schwann cells alone. Additionally, increased secretion of MMP2, MMP9, and MMP12 suggests that Schwann cells degrade extracellular matrix to migrate perhaps toward cancer sites and create an extracellular environment more prone to cancer migration. Schwann cell-derived CCL2 binds to CCR2 in tumor cells, inducing an epithelial-to-mesenchymal phenotype that is reduced via silencing of CCL2 [62]. Differing from findings in pancreatic cancer [51], expression of MMP2 and MMP9 is unchanged in cervical tumor cells [62], highlighting that the mechanisms described here may be cancer type-specific. Chemokine-induced epithelial-to-mesenchymal transition was also described in lung cancer [63]. Schwann cell-derived C-X-C motif ligand 5 (CXCL5) induces a more migratory phenotype in tumor cells via increase of epithelial-to-mesenchymal transition-associated proteins twist and snail [63].

Overall, in vitro and in vivo mechanistic studies highlight that Schwann cells in different cancer types interact directly with tumor cells, participating in tumor invasion via both cell contact and paracrine signaling. Because of the regenerative capacity and great plasticity of Schwann cells, other currently unknown mechanisms may be implicated in perineural invasion, as well as tumor progression and distant metastasis.

Effects on Tumor Microenvironment

Schwann cells can interact with macrophages in the tumor microenvironment to promote tumor invasion. CCL2, a chemokine released by Schwann cells, recruits monocytes to the site of perineural invasion. Once differentiated into macrophages, these cells release cathepsin B to help digest the extracellular matrix in the nerve, facilitating cancer invasion. When tumor cells are injected into the sciatic nerve to induce perineural invasion, an increased population of macrophages and enhanced tumor spread throughout the nerve is observed in wild-type compared to CCL2 knockout mice [64]. Although CCL2 expression from Schwann cells in invaded nerves is evident, this study did not address whether the mechanism is also relevant for initiation of perineural invasion.

Endoneurial Macrophages

Peripheral nerve macrophages can be either tissue-resident or recruited from circulating monocytes that migrate into nerves. Recruited macrophages are found in the perineurial space between nerve fascicles and epineurium, while tissue-resident endoneurial macrophages are in the endoneurium, in close association with Schwann cells and axons [65].

In human pancreatic cancer samples, there is an accumulation of macrophages around nerves [64, 66]. Macrophages are commonly observed in perineural invasion-positive cancer specimens compared to benign tissue, especially located around nerves with cancer [67] suggesting that macrophages contribute to perineural invasion. Higher infiltration of macrophages is also associated with poor survival [67].

Macrophages and cancer-associated fibroblasts in the tissue microenvironment modulate Schwann cell behavior, increasing migration into the tumor [47, 48]. When cancer-associated fibroblasts are cocultured with macrophages, fibroblasts secrete high levels of leukemia inhibitor factor (LIF), which belongs to the IL-6 cytokine family [47], and slit guidance ligand 2 (SLIT2), an axon guidance factor [48]. These factors induce Schwann cell migration toward pancreatic cancer cells [47, 48]. SLIT2 activates the ß-catenin pathway in Schwann cells to enhance motility [48], while LIF activates JAK/STAT3/AKT signaling. Higher LIF expression in tumors correlates with increased nerve density [47], indicating that macrophages may drive nerve-tumor interactions via recruitment of Schwann cells. In turn, Schwann cells attract cancer cells toward nerves, as discussed earlier in this section. Similarly, SLIT2 expression is associated with increased perineural invasion [48], but the exact mechanisms of invasion are not fully explored. These studies evaluated macrophages in tumor stroma, rather than endoneurial macrophages, which are not well characterized; there are no specific markers to establish a nerve origin.

There is little evidence for the participation of endoneurial macrophages in tumor progression. Endoneurial macrophages dissociated from dorsal root ganglia enhance invasion of pancreatic cancer cells in vitro, to a greater extent when activated with conditioned medium from cancer [66]. These activated endoneurial macrophages have increased glial cell-derived neurotrophic factor (GDNF) expression, which enhances ERK phosphorylation in tumor cells via RET (rearranged during transfection) tyrosine kinase receptor activity. Tumors injected into the sciatic nerve in CCR2 global knockout mice have less macrophage infiltration and are less migratory along the nerve, compared to wild-type mice [66]. CCR2 is an important receptor activated during macrophage recruitment; however, it is not specific for endoneurial macrophages; these are likely a small population of cells [13]. It is reasonable that tumor-associated or newly recruited macrophages in tumor stroma contribute to the effect as much or more than endoneurial macrophages. As previously discussed, Schwann cells can recruit monocytes from circulation to the nerve vicinity to promote perineural invasion [64]. Therefore, the contribution of endoneurial macrophages to tumor invasion warrants further exploration.

Summary

Here we summarize the mechanisms underlying tumor invasion and metastasis that are driven by nerve activity. Neuronal cells, Schwann cells, and endoneurial macrophages interact with cancer cells and non-cancer cells in the tumor microenvironment to promote tumor cell invasion and migration and an epithelial-to-mesenchymal phenotype. The study of interaction/communication between neuronal and other cells, with the tumor microenvironment, is an emerging and important field for understanding cancer progression. Elucidation of these mechanisms may reveal new targets or therapeutic approaches to improve patient survival.

References

1. Lambert AW, Pattabiraman DR, Weinberg RA. Emerging Biological Principles of Metastasis. Cell. 2017;168(4):670–91.
2. Scanlon CS, Banerjee R, Inglehart RC, Liu M, Russo N, Hariharan A, et al. Galanin modulates the neural niche to favour perineural invasion in head and neck cancer. Nat Commun. 2015;6:6885.
3. Magnon C, Hall SJ, Lin J, Xue X, Gerber L, Freedland SJ, et al. Autonomic nerve development contributes to prostate cancer progression. Science. 2013;341(6142):1236361.
4. Kamiya A, Hayama Y, Kato S, Shimomura A, Shimomura T, Irie K, et al. Genetic manipulation of autonomic nerve fiber innervation and activity and its effect on breast cancer progression. Nat Neurosci. 2019;22(8):1289–305.
5. Amit M, Takahashi H, Dragomir MP, Lindemann A, Gleber-Netto FO, Pickering CR, et al. Loss of p53 drives neuron reprogramming in head and neck cancer. Nature. 2020;578(7795):449–54.
6. Schmitd LB, Perez-Pacheco C, D'Silva NJ. Nerve density in cancer: Less is better. FASEB Bioadv. 2021;3(10):773–86.
7. Olar A, He D, Florentin D, Ding Y, Wheeler T, Ayala G. Biological correlates of prostate cancer perineural invasion diameter. Hum Pathol. 2014;45(7):1365–9.
8. Schmitd LB, Beesley LJ, Russo N, Bellile EL, Inglehart RC, Liu M, et al. Redefining Perineural Invasion: Integration of Biology With Clinical Outcome. Neoplasia. 2018;20(7):657–67.
9. Varey AHR, Goumas C, Hong AM, Mann GJ, Fogarty GB, Stretch JR, et al. Neurotropic melanoma: an analysis of the clinicopathological features, management strategies and survival outcomes for 671 patients treated at a tertiary referral center. Mod Pathol. 2017;30(11):1538–50.
10. Shurin GV, Kruglov O, Ding F, Lin Y, Hao X, Keskinov AA, et al. Melanoma-Induced Reprogramming of Schwann Cell Signaling Aids Tumor Growth. Cancer Res. 2019;79(10):2736–47.
11. Clements MP, Byrne E, Camarillo Guerrero LF, Cattin AL, Zakka L, Ashraf A, et al. The Wound Microenvironment Reprograms Schwann Cells to Invasive Mesenchymal-like Cells to Drive Peripheral Nerve Regeneration. Neuron. 2017;96(1):98–114.e7.
12. Koike T, Ebara S, Tanaka S, Kase M, Hirahara Y, Hayashi S, et al. Distribution, fine structure, and three-dimensional innervation of lamellar corpuscles in rat plantar skin. Cell Tissue Res. 2021;386(3):477–90.
13. Wolbert J, Li X, Heming M, Mausberg AK, Akkermann D, Frydrychowicz C, et al. Redefining the heterogeneity of peripheral nerve cells in health and autoimmunity. Proc Natl Acad Sci U S A. 2020;117(17):9466–76.
14. Carr MJ, Toma JS, Johnston APW, Steadman PE, Yuzwa SA, Mahmud N, et al. Mesenchymal Precursor Cells in Adult Nerves Contribute to Mammalian Tissue Repair and Regeneration. Cell Stem Cell. 2019;24(2):240–56.e9.
15. Joseph NM, Mukouyama YS, Mosher JT, Jaegle M, Crone SA, Dormand EL, et al. Neural crest stem cells undergo multilineage differentiation in developing peripheral nerves to generate endoneurial fibroblasts in addition to Schwann cells. Development. 2004;131(22):5599–612.
16. Richard L, Védrenne N, Vallat JM, Funalot B. Characterization of Endoneurial Fibroblast-like Cells from Human and Rat Peripheral Nerves. J Histochem Cytochem. 2014;62(6):424–35.

17. Venkatesh HS, Morishita W, Geraghty AC, Silverbush D, Gillespie SM, Arzt M, et al. Electrical and synaptic integration of glioma into neural circuits. Nature. 2019;573(7775):539–45.
18. Venkatesh HS, Johung TB, Caretti V, Noll A, Tang Y, Nagaraja S, et al. Neuronal Activity Promotes Glioma Growth through Neuroligin-3 Secretion. Cell. 2015;161(4):803–16.
19. Hyman SE. Neurotransmitters. Curr Biol. 2005;15(5):R154–8.
20. Renz BW, Takahashi R, Tanaka T, Macchini M, Hayakawa Y, Dantes Z, et al. β2 Adrenergic-Neurotrophin Feedforward Loop Promotes Pancreatic Cancer. Cancer Cell. 2018;33(1):75–90.e7.
21. Sarrouilhe D, Clarhaut J, Defamie N, Mesnil M. Serotonin and cancer: what is the link? Curr Mol Med. 2015;15(1):62–77.
22. Lang K, Bastian P. Neurotransmitter effects on tumor cells and leukocytes. Prog Exp Tumor Res. 2007;39:99–121.
23. Binnewies M, Roberts EW, Kersten K, Chan V, Fearon DF, Merad M, et al. Understanding the tumor immune microenvironment (TIME) for effective therapy. Nat Med. 2018;24(5):541–50.
24. Guo K, Ma Q, Li J, Wang Z, Shan T, Li W, et al. Interaction of the sympathetic nerve with pancreatic cancer cells promotes perineural invasion through the activation of STAT3 signaling. Mol Cancer Ther. 2013;12(3):264–73.
25. Zhang D, Ma QY, Hu HT, Zhang M. β2-adrenergic antagonists suppress pancreatic cancer cell invasion by inhibiting CREB, NFκB and AP-1. Cancer Biol Ther. 2010;10(1):19–29.
26. Zhang X, Zhang Y, He Z, Yin K, Li B, Zhang L, et al. Chronic stress promotes gastric cancer progression and metastasis: an essential role for ADRB2. Cell Death Dis. 2019;10(11):788.
27. Yin QQ, Xu LH, Zhang M, Xu C. Muscarinic acetylcholine receptor M1 mediates prostate cancer cell migration and invasion through hedgehog signaling. Asian J Androl. 2018;20(6):608–14.
28. Zhao CM, Hayakawa Y, Kodama Y, Muthupalani S, Westphalen CB, Andersen GT, et al. Denervation suppresses gastric tumorigenesis. Sci Transl Med. 2014;6(250):250ra115.
29. Xiang T, Fei R, Wang Z, Shen Z, Qian J, Chen W. Nicotine enhances invasion and metastasis of human colorectal cancer cells through the nicotinic acetylcholine receptor downstream p38 MAPK signaling pathway. Oncol Rep. 2016;35(1):205–10.
30. Medjber K, Freidja ML, Grelet S, Lorenzato M, Maouche K, Nawrocki-Raby B, et al. Role of nicotinic acetylcholine receptors in cell proliferation and tumour invasion in broncho-pulmonary carcinomas. Lung Cancer. 2015;87(3):258–64.
31. Li X, Ma G, Ma Q, Li W, Liu J, Han L, et al. Neurotransmitter substance P mediates pancreatic cancer perineural invasion via NK-1R in cancer cells. Mol Cancer Res. 2013;11(3):294–302.
32. Jiang BC, Liu T, Gao YJ. Chemokines in chronic pain: cellular and molecular mechanisms and therapeutic potential. Pharmacol Ther. 2020;212:107581.
33. He S, He S, Chen CH, Deborde S, Bakst RL, Chernichenko N, et al. The chemokine (CCL2-CCR2) signaling axis mediates perineural invasion. Mol Cancer Res. 2015;13(2):380–90.
34. Xu Q, Wang Z, Chen X, Duan W, Lei J, Zong L, et al. Stromal-derived factor-1α/CXCL12-CXCR4 chemotactic pathway promotes perineural invasion in pancreatic cancer. Oncotarget. 2015;6(7):4717–32.
35. Zahalka AH, Arnal-Estapé A, Maryanovich M, Nakahara F, Cruz CD, Finley LWS, et al. Adrenergic nerves activate an angio-metabolic switch in prostate cancer. Science. 2017;358(6361):321–6.
36. Lei X, Lei Y, Li JK, Du WX, Li RG, Yang J, et al. Immune cells within the tumor microenvironment: Biological functions and roles in cancer immunotherapy. Cancer Lett. 2020;470:126–33.
37. Bretscher P. On Analyzing How the Th1/Th2 Phenotype of an Immune Response Is Determined: Classical Observations Must Not Be Ignored. Front Immunol. 2019;10:1234.
38. Kidd P. Th1/Th2 balance: the hypothesis, its limitations, and implications for health and disease. Altern Med Rev. 2003;8(3):223–46.
39. Yang MW, Tao LY, Jiang YS, Yang JY, Huo YM, Liu DJ, et al. Perineural Invasion Reprograms the Immune Microenvironment through Cholinergic Signaling in Pancreatic Ductal Adenocarcinoma. Cancer Res. 2020;80(10):1991–2003.

40. Sloan EK, Priceman SJ, Cox BF, Yu S, Pimentel MA, Tangkanangnukul V, et al. The sympathetic nervous system induces a metastatic switch in primary breast cancer. Cancer Res. 2010;70(18):7042–52.
41. Mo RJ, Han ZD, Liang YK, Ye JH, Wu SL, Lin SX, et al. Expression of PD-L1 in tumor-associated nerves correlates with reduced CD8(+) tumor-associated lymphocytes and poor prognosis in prostate cancer. Int J Cancer. 2019;144(12):3099–110.
42. Chen G, Kim YH, Li H, Luo H, Liu DL, Zhang ZJ, et al. PD-L1 inhibits acute and chronic pain by suppressing nociceptive neuron activity via PD-1. Nat Neurosci. 2017;20(7):917–26.
43. Pascual G, Domínguez D, Elosúa-Bayes M, Beckedorff F, Laudanna C, Bigas C, et al. Dietary palmitic acid promotes a prometastatic memory via Schwann cells. Nature. 2021;599(7885):485–90.
44. Liebig C, Ayala G, Wilks JA, Berger DH, Albo D. Perineural invasion in cancer: a review of the literature. Cancer. 2009;115(15):3379–91.
45. Demir IE, Boldis A, Pfitzinger PL, Teller S, Brunner E, Klose N, et al. Investigation of Schwann cells at neoplastic cell sites before the onset of cancer invasion. J Natl Cancer Inst. 2014;106(8).
46. Demir IE, Kujundzic K, Pfitzinger PL, Saricaoglu Ö C, Teller S, Kehl T, et al. Early pancreatic cancer lesions suppress pain through CXCL12-mediated chemoattraction of Schwann cells. Proc Natl Acad Sci U S A. 2017;114(1):E85–e94.
47. Bressy C, Lac S, Nigri J, Leca J, Roques J, Lavaut MN, et al. LIF Drives Neural Remodeling in Pancreatic Cancer and Offers a New Candidate Biomarker. Cancer Res. 2018;78(4):909–21.
48. Secq V, Leca J, Bressy C, Guillaumond F, Skrobuk P, Nigri J, et al. Stromal SLIT2 impacts on pancreatic cancer-associated neural remodeling. Cell Death Dis. 2015;6(1):e1592.
49. Swanson BJ, McDermott KM, Singh PK, Eggers JP, Crocker PR, Hollingsworth MA. MUC1 is a counter-receptor for myelin-associated glycoprotein (Siglec-4a) and their interaction contributes to adhesion in pancreatic cancer perineural invasion. Cancer Res. 2007;67(21):10222–9.
50. Deborde S, Omelchenko T, Lyubchik A, Zhou Y, He S, McNamara WF, et al. Schwann cells induce cancer cell dispersion and invasion. J Clin Invest. 2016;126(4):1538–54.
51. Na'ara S, Amit M, Gil Z. L1CAM induces perineural invasion of pancreas cancer cells by upregulation of metalloproteinase expression. Oncogene. 2019;38(4):596–608.
52. Roger E, Martel S, Bertrand-Chapel A, Depollier A, Chuvin N, Pommier RM, et al. Schwann cells support oncogenic potential of pancreatic cancer cells through TGFβ signaling. Cell Death Dis. 2019;10(12):886.
53. Su D, Guo X, Huang L, Ye H, Li Z, Lin L, et al. Tumor-neuroglia interaction promotes pancreatic cancer metastasis. Theranostics. 2020;10(11):5029–47.
54. Wei Z, Fan B, Ding H, Liu Y, Tang H, Pan D, et al. Proteomics analysis of Schwann cell-derived exosomes: a novel therapeutic strategy for central nervous system injury. Mol Cell Biochem. 2019;457(1–2):51–9.
55. Ferdoushi A, Li X, Griffin N, Faulkner S, Jamaluddin MFB, Gao F, et al. Schwann Cell Stimulation of Pancreatic Cancer Cells: A Proteomic Analysis. Front Oncol. 2020;10:1601.
56. Sroka IC, Chopra H, Das L, Gard JM, Nagle RB, Cress AE. Schwann Cells Increase Prostate and Pancreatic Tumor Cell Invasion Using Laminin Binding A6 Integrin. J Cell Biochem. 2016;117(2):491–9.
57. Kacsinta AD, Rubenstein CS, Sroka IC, Pawar S, Gard JM, Nagle RB, et al. Intracellular modifiers of integrin alpha 6p production in aggressive prostate and breast cancer cell lines. Biochem Biophys Res Commun. 2014;454(2):335–40.
58. Schmitd LB, Scanlon CS, D'Silva NJ. Perineural Invasion in Head and Neck Cancer. J Dent Res. 2018;97(7):742–50.
59. Ein L, Bracho O, Mei C, Patel J, Boyle T, Monje P, et al. Inhibition of tropomyosine receptor kinase B on the migration of human Schwann cell and dispersion of oral tongue squamous cell carcinoma in vitro. Head Neck. 2019;41(12):4069–75.

60. Shan C, Wei J, Hou R, Wu B, Yang Z, Wang L, et al. Schwann cells promote EMT and the Schwann-like differentiation of salivary adenoid cystic carcinoma cells via the BDNF/TrkB axis. Oncol Rep. 2016;35(1):427–35.
61. Costa de Medeiros M. LM, Banerjee R., Bellile E., D'Silva NJ., Rossa Jr. C. *(Preprint) Galanin mediates tumor-induced immunosuppression in head and neck squamous cell carcinoma*. Cellular Oncology. August 2021.
62. Huang T, Fan Q, Wang Y, Cui Y, Wang Z, Yang L, et al. Schwann Cell-Derived CCL2 Promotes the Perineural Invasion of Cervical Cancer. Front Oncol. 2020;10:19.
63. Zhou Y, Shurin GV, Zhong H, Bunimovich YL, Han B, Shurin MR. Schwann Cells Augment Cell Spreading and Metastasis of Lung Cancer. Cancer Res. 2018;78(20):5927–39.
64. Bakst RL, Xiong H, Chen CH, Deborde S, Lyubchik A, Zhou Y, et al. Inflammatory Monocytes Promote Perineural Invasion via CCL2-Mediated Recruitment and Cathepsin B Expression. Cancer Res. 2017;77(22):6400–14.
65. Kieseier BC, Hartung HP, Wiendl H. Immune circuitry in the peripheral nervous system. Curr Opin Neurol. 2006;19(5):437–45.
66. Cavel O, Shomron O, Shabtay A, Vital J, Trejo-Leider L, Weizman N, et al. Endoneurial macrophages induce perineural invasion of pancreatic cancer cells by secretion of GDNF and activation of RET tyrosine kinase receptor. Cancer Res. 2012;72(22):5733–43.
67. Zeng L, Guo Y, Liang J, Chen S, Peng P, Zhang Q, et al. Perineural Invasion and TAMs in Pancreatic Ductal Adenocarcinomas: Review of the Original Pathology Reports Using Immunohistochemical Enhancement and Relationships with Clinicopathological Features. J Cancer. 2014;5(9):754–60.

Chapter 5
Cancer Induced Remodeling of the Peripheral Nervous System

Anthony C. Restaino and Paola D. Vermeer

Remodeling of a tumor's surrounding environment has been recognized as a hallmark of cancer for decades [1]. For example, foundational studies in the mid-1900s identified unique immune cell populations surrounding metastatic tumors as critical components of the tumor microenvironment (TME) [2]. Additional studies later identified cancer cell lines that had increased tumorigenicity, or tumor forming potential, in vivo compared to in vitro, establishing a precedent for the interaction between the cancer cells and their surrounding environment in the promotion of tumor formation [3]. Today, a greater molecular understanding of the recruitment and function of immune cells within the TME has led to one of the most important therapeutic advancements of our time, immunotherapy. Immunotherapy is the utilization and stimulation of a patient's own immune system for the treatment of cancer and has resulted in improved treatment of various aggressive cancers including melanoma, small cell lung cancer, and head and neck squamous cell carcinoma [4–8]. As our technological progress enables a deeper molecular understanding of the TME, we gain further insights into its cellular relationships. With this ever-growing understanding, the complexity of the TME is slowly unraveling, and, with it, we gain a deeper molecular comprehension of tumor biology, disease progression, and important components we can target for therapeutic intervention. This field of study has opened the door for advancements in patient treatment, ushering in a new wave of targeted therapies and personalized medicine for cancer patients.

Most recently, the identification of functional neurons within the TME has transformed the field of cancer biology to one that must now become more interdisciplinary than ever before. This exciting time not only enables, but rather demands, cancer

A. C. Restaino
Cancer Biology and Immunotherapies Group, Sanford Research, Sioux Falls, SD, USA

P. D. Vermeer (✉)
University of South Dakota Sanford School of Medicine, Vermillion, SD, USA
e-mail: Paola.Vermeer@sanfordhealth.org

M. Amit, N. N. Scheff (eds.), *Cancer Neuroscience*,
https://doi.org/10.1007/978-3-031-32429-1_5

biologists, immunologists, and neuroscientists to work elbow-to-elbow to understand how these components of the TME influence one another and ultimately drive disease progression. The first foundational evidence of peripheral neurons participating in cancer initiation and disease progression demonstrated that nerve density increases as prostate cancer advances [9]. Moreover, surgical or chemical sympathectomy, or genetic ablation of adrenergic nerves, reduces cancer initiation and progression in a model of prostate cancer [10]. Together, these studies suggest that intra-tumoral nerves actively participate in disease. Additional studies in various cancer models quickly contributed to this emerging field, providing further support for an active role of the nervous system in cancer [11–15]. Cancer-mediated mechanisms driving the remodeling of the peripheral nervous system were quickly identified and include a role for growth factors, chemokines, and genetic material. As neuroscience techniques have, and continue to be, repurposed for cancer innervation experiments, additional insights into the relationship between cancer and intratumoral nerves will develop. Here, we will discuss the current knowledge of the composition of neural tissue in the tumor microenvironment, how tumors promote and guide their own innervation, and future directions in the investigation of the relationship between the peripheral nervous system and cancer.

Tumor Profiling Promotes Differences in Innervation and Peripheral Nerve Remodeling

Although various tumors have demonstrated the potential to mediate and promote their own innervation, the presence and function of nerves in the tumor environment is not clear-cut. Studies in solid and hematological cancers indicate that the type and density of innervation and remodeling depends on the cancer type and genetic profile of the tumor itself. Moreover, there is immense heterogeneity in the type of peripheral nerves that have been identified; several studies have detected adrenergic, cholinergic, and sensory innervation of solid tumors [11–17]. For example, studies focused on prostate cancer demonstrate that a vast majority of innervating fibers are sympathetic in nature [9, 10]. Sympathetic nerves are a component of the autonomic nervous system, are endogenously present in almost all tissues, and mainly function through the release of catecholamines, primarily norepinephrine. This is a characteristic seen across many cancer models, including breast, pancreatic, and colon cancers. Sympathetic innervation of these cancers primarily promotes tumor progression and metastasis [9, 10, 12–14]. In models of pancreatic ductal adenocarcinoma, sympathetic neurons promoting cancer metastasis and proliferation by signaling through β-adrenergic receptors expressed on tumor cells [12, 18]. Similarly, sympathetic denervation of breast and tongue cancers decreases tumor growth and invasiveness of associated cancers [13, 14].

Another commonly identified nerve type found within the TME includes nociceptive, or sensory, nerves. Sensory nerves function to regulate and respond to environmental changes in tissues and organs [19, 20]. The presence of sensory nerves has not been as widely reported as sympathetic nerves but has been identified in melanoma, head and neck squamous cell carcinoma, cervical and gastric carcinoma, pancreatic ductal adenocarcinoma, and breast cancer patient samples [21–25]. However, unlike sympathetic innervation, the function of sensory neurons in cancer is debated; while various models suggest that increased sensory innervation is protective against cancer growth and progression, others demonstrate correlations with poor survival [23, 26]. While the extent and limitations of sensory nerves on tumor progression has yet to be fully investigated, the impact of cancer on nociceptive nerves and cancer-associated pain has been investigated in greater detail. Cancer-associated pain is a common symptom of various solid tumors, and has been associated with cancer progression [27]. Previously, this symptom was thought to result primarily from inflammation or physical compression of surrounding tissue; however, mounting evidence suggests that intra-tumoral nociceptive neurons also contribute to cancer-associated pain [28–30]. Further studies investigating the interaction between tumor cells and components of the nervous system, including Schwann cells and other glial cells, demonstrate their participation in facilitating nociceptive behaviors and pain responses [31, 32].

The least prevalent nerve type identified in solid tumors is cholinergic. These nerves function primarily within the central nervous system but are also found in peripheral systems and function in aspects of arousal, attention, and memory [33]. Both cancer promoting and suppressive functions have been ascribed to cholinergic tumor innervation. For instance, in pancreatic ductal carcinoma, cholinergic fibers suppress tumor formation via signaling through muscarinic receptors; treatment of tumors with muscarinic agonists slowed tumor progression [34]. In thyroid and gastric carcinoma, tumor growth is promoted by intra-tumoral cholinergic signaling, indicating a cancer-specific response to innervation [16, 17].

While most studies in solid tumors suggest that intra-tumoral nerves promote tumorigenesis and disease progression, evidence in hematological cancers is different. Here, a leading hypothesis suggests that tumor-induced neuronal remodeling of the bone matrix results in neurodegeneration of existing fibers that is associated with disease progression [35, 36]. Foundational studies conducted by the Frenette Laboratory demonstrate that upon development of hematological malignancies, sympathetic innervation of the bone marrow and the hematopoietic stem cell niche becomes ablated. Consistent with this, additional studies indicate that the ablation of nerves in the bone marrow environment is a conserved process in the development of hematological cancers [35, 36]. This contradictory nature of nerves in solid and hematological cancers indicates a potential innate difference in types of tumors and their relationship with innervation. One potential component that can explain this difference is the function of nerves in the initiating tissue. The relationship between nerves and the hematopoietic stem cell (HSC) niche, the origin of hematological cancers, is a structured one. Both nociceptive and sympathetic nerves facilitate various components of HSC functions and behaviors. Through the secretion of calcitonin gene-related peptide (CGRP) and other neuropeptides, nociceptive neurons facilitate and promote mobilization and differentiation of HSCs [37, 38].

Blockade of these neuropeptides, or ablation of nociceptive nerves in the bone marrow, results in loss of HSC mobilization. Additionally, adrenergic nerves maintain HSC integrity, and ablation of preexisting adrenergic nerves accelerates aging of the HSC population. This ablation is rescued following pharmacological intervention [39]. On the other hand, nerves play an important role in the organization, repair, and regeneration of epithelial tissue [40–42]. While these tissues are sparsely innervated in homeostatic conditions, their nerve density increases during wound healing and repair processes [34–36]. Thus, the presence of nerves in various carcinomas and cancers of epithelial origin is not surprising as these cancers co-opt and aberrantly promote these processes [9–18, 43]. Additionally, some epithelial cancers preferentially arise in specific niches maintained and supported by innervated tissues. One such example of this is basal skin cancer. These tumors arise in specific hair follicle niches innervated by sensory nerves; importantly, nerve density increases during the development of the cancer. Furthermore, chemical ablation of these nerves reduces the incidence and development of cancer in these niches [44]. However, even within solid tumors, there are differences in innervation and response to neuropeptide signaling [45]. The differences in neuronal contributions to the development and progression of these cancers indicates an innate relationship between the two and suggest that developing cancers mediate changes by remodeling the peripheral nervous system to their benefit.

Although evidence supports the notion that the prevalence of tumor innervation and the types of nerves innervating tumors might depend on the environment and tissue they develop in, studies in neurological cancers, including gliomas and meningiomas, provide additional evidence that the genetic tumor background also influences remodeling potential. In a recent study investigating the effect of glioblastomas with *Pik3ca* mutations, researchers identified that particular variants of *Pik3ca* result in hyperexcitability and synapse remodeling in in vivo mouse models [46]. Furthermore, these specific mutant variants increase secreted glypican family proteins, directly promoting changes in neuronal hyperexcitability. While conducted in a central nervous system cancer, these results indicate that even mutational variations can contribute to dynamic changes in remodeling of the nervous system. Additionally, a recent study in head and neck squamous cell carcinoma indicated that loss of p53 expression promoted increased neurite outgrowth in an in vitro trigeminal nerve model [47]. This study identified several p53 loss-of-function mutations that drive changes in microRNAs (miRNA) packaged within secreted extracellular vesicles that promote increased neurite outgrowth. Together, these studies indicate that changes in oncogenic drivers, and even particular variants, influence secreted components that participate in nerve remodeling.

Taken together, these early studies demonstrate that cancers and the nervous system undergo bidirectionally mediated remodeling which, ultimately, influences disease progression. Below, we focus on the current understanding of neural remodeling in peripheral cancers.

Cancer-Mediated Remodeling Mechanisms

Remodeling by Cancer-Secreted Proteins and Cytokines

Early studies investigating tumor innervation focused on cancer-secreted growth factors, cytokines, and chemokines that remodel the peripheral nervous system. Neurotrophins, a family of proteins involved in survival, development, and growth of neurons, have been a particular focus of this avenue of research [48].

Nerve growth factor (NGF), an established and heavily studied neurotrophin, has routinely been identified as a secreted factor in numerous in vitro cancer models [49–52]. Moreover, positive staining of prostate cancer patient samples for ProNGF, an NGF precursor, correlates with patient Gleason score, a clinical scale for describing prostate cancer differentiation and stage, and, as such, holds potential as a biomarker for tumor severity [50]. The identification of this NGF precursor highlights not only the local generation of this neurotrophin but also its potential function in remodeling tissue adjacent nerves. Similar correlations between NGF staining and cancer severity have been demonstrated in thyroid, pancreatic, and breast cancers, indicating a conserved relationship between NGF production and cancer severity [49, 51, 52]. Tumor secreted NGF has also been demonstrated to facilitate cancer pain [53]. Treatment with anti-NGF antibody prophylactically or following tumor implantation reduces pain-associated behaviors in mice, demonstrating that anti-NGF therapy can not only be used as a cancer therapeutic but also to potentially mitigate cancer-associated pain [53].

NGF is not the only cancer-secreted neuronal factors identified that can remodel peripheral neurons within the TME. EphrinB1 is a transmembrane ligand that functions in processes of axon guidance, axonal transport, cytoskeleton rearrangement, and neuronal migration. In a human papillomavirus induced (HPV+) model of head and neck squamous cell carcinoma (HNSCC), EphrinB1 is packaged within secreted exosomes, a small extracellular vesicle (sEV), which promote increased axonogenesis in a PC12 cell in vitro and tumors in vivo [22]. Importantly, these exosomes lack NGF, and brain derived neurotrophic factor (BDNF), suggesting tumor innervation was not dependent on these neurotrophic factors. Moreover, inhibition of exosome release from tumor cells (induced genetically or pharmacologically) decreased tumor innervation as well as tumor growth [22]. These data suggest that tumor-released sEVs contribute to recruitment of intra-tumoral nerves to the TME and support a contribution of these nerves to disease progression [22]. Together, these findings suggest that tumors release various factors (sEVs and soluble factors) that induce remodeling of peripheral nerves which promote to disease progression.

Cancer-secreted factors involved in peripheral nerve remodeling have not been limited to neurotrophins; cytokines also participate in this remodeling. The identification of various cytokine and chemokine signaling axes provides additional evidence that the remodeling of the peripheral nervous system by metastatic disease likely involves multiple complex signals.

For instance, the CXCL12-CXCR4 signaling axis is a cytokine signaling pathway implicated in various aspects of cancer biology. In a murine prostate cancer model, in vivo treatment of tumors with CXCL12 increases innervation near the tumor site [54]. Here, signaling of CXCL12 via tumor expressed CXCR4 increases NGF expression, which promotes increased innervation of the tumor space [54]. In a pancreatic ductal carcinoma murine model, expression of CXCL10 and CCL12 from dorsal root ganglia promotes pancreatic cancer cell migration toward infiltrative nerves, while inhibition of CXCL10 and CCL12 receptors on tumor cells decreases cell migration and reduces pain behaviors in mice as evidenced by reduced sensory nerve hypersensitivity and nerve hypertrophy [55]. These data indicate that a responsive signal from the pancreatic carcinoma cells promotes pain, nerve sensitivity, and nerve hypertrophy. Cytokines and chemokines are prominent signaling molecules, and although they don't robustly or directly contribute to nerve remodeling, they could be responsible for initiating changes in cancer cell secretions that are a contributing factor to peripheral nerve remodeling.

Indirect effects of cytokine signaling have also been identified as potential mediators of TME remodeling and cancer progression. Endoneurial macrophages are associated with neural structures from pancreatic adenocarcinoma (PDA) patients suffering from perineural invasion (PNI) [56–58]. Further analysis demonstrated that endoneurial macrophages were stimulated by cancer cells to secrete glial-derived neurotropic factor (GDNF) [56–58]. GDNF promotes PNI through binding to the tyrosine kinase receptor, RET, on PDA cells; pharmacological or genetic inhibition of GDNF-RET interaction inhibits PNI [59].

Remodeling by Transfer of Genetic Material

Although a large research focus in the field of tumor innervation includes tumor secreted proteins and vesicles, a contribution of cancer-derived genetic material, namely, miRNA, to tumor innervation is emerging. As described previously, sEVs and other extracellular vesicles released from cancer cells promote neurite outgrowth and neurite remodeling [21, 22, 47]. Extracellular vesicles are small, membrane-bound vesicles released by cells; these vesicles contain proteins, lipids, transmembrane receptors, mRNA, DNA, and miRNA. miRNAs have an established and defined role in several cellular processes, including neurodevelopment, synaptic plasticity, nerve remodeling, and nerve injury repair [60–64]. miRNA are small segments of single stranded RNA, typically 18–22 nucleotides in length, that suppress translation of mRNA through complexing with the Dicer enzyme. In this way, miRNAs are capable of rapidly altering cellular function and programming by inhibiting protein formation. miRNAs have recently become an area of focus in cancer research, with seminal papers identifying them as driving forces for cancer metastasis, progression, and TME remodeling [65–67]. In fact, early papers focused on the potential of utilizing miRNA as potential biomarkers for cancer progression and prognosis [68–70]. miRNAs have subsequently been identified as potentiating aspects of tumor progression, including metastasis, via direct regulation of Smad7, the TME, as well as mediating cellular reprogramming of fibroblasts and cancer cells [65, 67]. Additionally, various studies have identified the function of miRNAs as regulating neurodevelopment and nerve injury repair [60–64]. miRNAs have been identified within various compartments of neurons, including growth cones, axons, and dendrites [71–73]. The localization and expression of miRNAs changes in response to cellular alterations, such as nerve injury and nerve regeneration [62–64]. Given the length of their axons, mature neurons are complex and large cells, which makes reprogramming and altering cellular composition difficult at distal aspects of the cell, specifically the axon. Neurons overcome this hurdle by transporting miRNAs to the axon, a process that involves well-studied proteins such as Fragile X messenger ribonucleoprotein (FMRP) [74]. In this way, protein expression can be exquisitely regulated in time and space. Utilizing these established components of the nervous system allows for more expedited restructuring and reprogramming. These studies have demonstrated a conserved function of miRNA in nerve development and repair. More recently, studies focused on a contribution of miRNAs on nerve remodeling in the context of cancer and the tumor microenvironment have emerged. In an influential paper published by Amit et al., the miRNA cargo from HNSCC sEVs was identified as promoting axonogenesis and neurite outgrowth in an in vitro trigeminal nerve model. The identified miRNAs, including miR-21-5p and miR-324, promote trigeminal nerve axonogenesis, by targeting pathways involved in neuritogenesis [47, 75]. For example, a target of the miR-21 miRNA is the tumor suppressor transcript, PTEN. In nerves, overexpression of PTEN interferes with the development of neurons and neuritogenesis by inhibiting the PI3K/AKT signaling pathway [76]. Additionally, miR-34a was identified as an inhibitor of neuritogenesis, and its expression decreased following loss of function of the oncogene p53. Interestingly, miR-34a may regulate cell fate and differentiation in neural tissue [47, 77, 78]. Taken together, these studies suggest that cancer-derived miRNAs co-opt and reprogram the function of recipient cells. A difficulty associated with miRNA, as opposed to chemokines or neurotrophic proteins, is that individual miRNA can have multiple targets. Careful consideration, validation, and identification of affected pathways are vital for defining a concise mechanism of action taken by identified miRNA. Having a strong understanding of the endogenous miRNAs that regulate neuronal functions is essential for understanding these processes.

Remodeling by Cellular Stress and Dysfunction

As neural remodeling also occurs in response to injury and cellular stress, additional work also has investigated the effects of cellular stress on tumor innervation and remodeling. The tumor microenvironment is hypoxic and acidic; while this is advantageous for developing and growing tumors, it promotes a stressful environment for normal adjacent tissue. Using an in vitro model of endoplasmic reticulum

(ER) stress, through the administration of the pharmacologic agents tunicamycin or thapsigargin, demonstrates the molecular consequences of this stress on peripheral nerve remodeling [79]. Here, ER stress promotes downstream expression of proBDNF, a precursor for BDNF, which results in axonogenesis [74]. Interestingly, treatment with the common chemotherapeutic, 5-fluorouracil, further increases ER stress, proBDNF secretion, and, ultimately, axonogenesis [79, 80]. In fact, ER stress increases following chemotherapeutic treatment and functions in the protection of cancer cells [80]. These results suggest that changes in the TME, either a natural result of tumor progression or due to therapeutic intervention, can alter cancer cellular functions (transcription of specific genes) and peripheral nerve remodeling.

ER stress is not alone in mediating nerve remodeling in cancer. A study in pancreatic ductal adenocarcinoma shows that hyperglycemia promotes nerve remodeling and, further, induces perineural invasion [70], a process whereby tumor cells invade through nerve layers and are detected within neural structures upon histological staining and which is associated with poor prognosis. Here, hyperglycemia increases secretion of NGF from pancreatic cancer cells, resulting in enhanced axonogenesis. Additionally, hyperglycemia reduces migration of Schwann cells, promoting a nerve regeneration response, ultimately potentiating tumor innervation [81]. Additional evidence using in vivo cancer models indicates that metabolic changes are not restricted to cancer cells but also occur within infiltrating neurons; these changes alter nerve function and remodeling.

In an in vivo murine model of breast cancer, alterations in hepatic glucose metabolism and sleep behavior were identified but were not a result of inflammation [82, 83]. Further analysis indicated that immunofluorescence of the transcription factor c-Fos, a marker for neuronal activity, was increased in the hypocretin/orexin neurons and that metabolic abnormalities and sleep disturbances were reduced following treatment with hypocretin/orexin neuron antagonists and sympathetic denervation [72]. Evidence for environmental or cellular stresses as well as changes in cellular metabolism as driving forces of tumor innervation and nerve remodeling provides a unique area of research to investigate how progressive changes in the tumor environment drive tumor innervation.

Recent Advancements and Future Directions

As discussed above, the relationship between the peripheral nervous system and cancer is a rapidly expanding area of research. To date, studies have identified various cancer-secreted neurotrophic factors and chemokines that promote the innervation of the tumor environment. These secreted factors promote functional changes in locoregional nerves resulting in the axonogenesis and neuritogenesis that culminate in their infiltration of the tumor microenvironment. Additionally, the reprogramming of these nerves can be stimulated by miRNAs packaged in sEVs, changes in cellular metabolic states and available metabolites, and cellular stress. Once recruited to the tumor space, these nerves remain functional, but experience changes in sensitivity, hyperactivity, and aberrant nerve signaling.

Another area that has been gaining additional attention in this field is the indirect influence of cancer on nerves through interacting and manipulating neuronal support cells, such as Schwann cells and other glial cells, and immune cells. As mentioned above, alterations in the TME can impair functions of glial cells triggering nerve repair functions and cancer-associated pain [31, 32]. Additionally, cancer-secreted factors (in a mouse model of oral cancer) can stimulate Schwann cell activation, leading to increased secretion of NGF and tumor necrosis factor-alpha (TNF-α) and inducing nociceptive behaviors in cancer-burdened mice [32]. Furthermore, metabolic changes instigated by the development of cancer, namely, the production of a hyperglycemic environment, facilitate increased NGF production from Schwann cells and promote innervation of the TME [81]. Additionally, endoneurial macrophages have demonstrated a supportive function in promoting perineural invasion in PDA [56–59]. Together, these studies illustrate not only the breadth of the field of cancer neuroscience but also the complexity of this relationship that is still being uncovered.

To date, studies in this emerging field provide a strong foundation for an active role of cancer cells in manipulating and altering the peripheral nervous system for its benefit, yet many questions and directions still need to be addressed and explored. These questions focus on the influence of intra-tumoral nerves on cancer pain, the impact of neuronal support cells on cancer progression, and neuro-immune interactions, remain largely unanswered.

A greater understanding of the relationship between the peripheral nervous system, cancer, and the surrounding TME requires strengthening and expanding the toolbox of techniques and procedures available to cancer biologists, by utilizing and co-opting existing techniques from the field of neuroscience. Imaging studies in human samples and murine tumor models have relied primarily on immunohistochemistry and immunofluorescent staining, techniques which provide limited insights into neuronal type due to lack of specific protein markers. While these techniques allow for easy and rapid visualization of intra-tumoral nerves, they limit the ability to fully visualize the infiltration of these structures. Perfecting the techniques of CLARITY and tissue clearing to better appreciate the full expanse of the infiltration of intra-tumoral nerves will enhance our understanding of the neural composition of peripheral malignancies. Additionally, nerve tracing experiments have assisted in identifying the origin of intra-tumoral nerves, but utilizing these techniques to better understand the complexity and depth of these connections with existing neural structures will allow for a better understanding of the reach of influence of cancer in the nervous system [84]. Finally, while the presence of nerves within the tumor microenvironment is established, studies investigating their electrophysiological function are few and far between. Repurposing existing neuroscience techniques, including patch-clamp electrophysiology as well as multielectrode arrays, will allow for better understanding of how tumors alter the activity and signaling of intra-tumoral nerves and potentially provide insights into how processes such as cancer pain are initiated and promoted.

The field of cancer neuroscience is an exciting avenue of cancer research, providing opportunities to better understand basic tumor biology, identify new potential therapies for patients, and address various aspects of cancer manifestations including cancer pain and perineural invasion. Increasing our understanding of how cancer manipulates and regulates the peripheral nervous system will only strengthen our ability to identify the functions these cellular components contribute to cancer initiation, progression, therapy resistance, and metastasis.

References

1. Hanahan D, Weinberg RA. Hallmarks of cancer: the next generation. *Cell*. 2011;144(5):646–674.
2. Klein G, Sjogren HO, Klein E, Hellstrom KE. Demonstration of resistance against methylcholanthrene-induced sarcomas in the primary autochthonous host. *Cancer Res*. 1960;20:1561–1572.
3. Halachmi E, Witz IP. Differential tumorigenicity of 3T3 cells transformed in vitro with polyoma virus and in vivo selection for high tumorigenicity. *Cancer Res*. 1989;49(9):2383–2389.
4. Huang AC, Zappasodi R. A decade of checkpoint blockade immunotherapy in melanoma: understanding the molecular basis for immune sensitivity and resistance. *Nat Immunol*. 2022;23(5):660–670.
5. Ralli M, Botticelli A, Visconti IC, et al. Immunotherapy in the Treatment of Metastatic Melanoma: Current Knowledge and Future Directions. J Immunol Res. 2020;2020:9235638.
6. Reck M, Remon J, Hellmann MD. First-Line Immunotherapy for Non-Small-Cell Lung Cancer [published correction appears in J Clin Oncol. 2022 Apr 10;40(11):1265]. J Clin Oncol. 2022;40(6):586–597.
7. von Witzleben A, Wang C, Laban S, Savelyeva N, Ottensmeier CH. HNSCC: Tumour Antigens and Their Targeting by Immunotherapy. Cells. 2020;9(9):2103.
8. Moskovitz J, Moy J, Ferris RL. Immunotherapy for Head and Neck Squamous Cell Carcinoma. Curr Oncol Rep. 2018;20(2):22.
9. Ayala GE, Dai H, Powell M, et al. Cancer-related axonogenesis and neurogenesis in prostate cancer. *Clin Cancer Res*. 2008;14(23):7593–7603.
10. Magnon C, Hall SJ, Lin J, et al. Autonomic nerve development contributes to prostate cancer progression. *Science*. 2013;341(6142):1236361.
11. Saloman JL, Albers KM, Li D, et al. Ablation of sensory neurons in a genetic model of pancreatic ductal adenocarcinoma slows initiation and progression of cancer. *Proc Natl Acad Sci U S A*.
12. Ferdoushi A, Griffin N, Marsland M, et al. Tumor innervation and clinical outcome in pancreatic cancer. *Sci Rep*. 2021;11(1):7390.
13. Raju B, Haug SR, Ibrahim SO, Heyeraas KJ. Sympathectomy decreases size and invasiveness of tongue cancer in rats. *Neuroscience*. 2007;149(3):715–725.
14. Kappos EA, Engels PE, Tremp M, et al. Denervation leads to volume regression in breast cancer. *J Plast Reconstr Aesthet Surg*. 2018;71(6):833–839.
15. Huang D, Su S, Cui X, et al. Nerve fibers in breast cancer tissues indicate aggressive tumor progression. *Medicine (Baltimore)*. 2014;93(27):e172.
16. Wang Z, Liu W, Wang C, Li Y, Ai Z. Acetylcholine promotes the self-renewal and immune escape of CD133+ thyroid cancer cells through activation of CD133-Akt pathway. *Cancer Lett*. 2020;471:116–124.
17. Hayakawa Y, Sakitani K, Konishi M, et al. Nerve Growth Factor Promotes Gastric Tumorigenesis through Aberrant Cholinergic Signaling. *Cancer Cell*. 2017;31(1):21–34.
18. Wakiya T, Ishido K, Yoshizawa T, Kanda T, Hakamada K. Roles of the nervous system in pancreatic cancer. *Ann Gastroenterol Surg*. 2021;5(5):623–633. Published 2021 Mar 29.
19. Willis W. D., Jr (2009). The role of TRPV1 receptors in pain evoked by noxious thermal and chemical stimuli. *Experimental brain research*, *196*(1), 5–11.

20. Cao X, Kajino-Sakamoto R, Doss A, Aballay A. Distinct Roles of Sensory Neurons in Mediating Pathogen Avoidance and Neuropeptide-Dependent Immune Regulation. *Cell Rep.* 2017;21(6):1442–1451.
21. Lucido CT, Wynja E, Madeo M, et al. Innervation of cervical carcinoma is mediated by cancer-derived exosomes. *Gynecol Oncol.* 2019;154(1):228–235.
22. Madeo M, Colbert PL, Vermeer DW, et al. Cancer exosomes induce tumor innervation. *Nat Commun.* 2018;9(1):4284.
23. Keskinov AA, Tapias V, Watkins SC, Ma Y, Shurin MR, Shurin GV. Impact of the Sensory Neurons on Melanoma Growth In Vivo. *PLoS One.* 2016;11(5):e0156095.
24. Saloman JL, Albers KM, Li D, et al. Ablation of sensory neurons in a genetic model of pancreatic ductal adenocarcinoma slows initiation and progression of cancer. *Proc Natl Acad Sci U S A.* 2016;113(11):3078–3083.
25. Erin N. Role of sensory neurons, neuroimmune pathways, and transient receptor potential vanilloid 1 (TRPV1) channels in a murine model of breast cancer metastasis. *Cancer Immunol Immunother.* 2020;69(2):307–314.
26. Prazeres, P., Leonel, C., Silva, W. N., Rocha, B., Santos, G., Costa, A. C., Picoli, C. C., Sena, I., Gonçalves, W. A., Vieira, M. S., Costa, P., Campos, L., Lopes, M., Costa, M. R., Resende, R. R., Cunha, T. M., Mintz, A., & Birbrair, A. (2020). Ablation of sensory nerves favours melanoma progression. *Journal of cellular and molecular medicine, 24*(17), 9574–9589.
27. Quinten C, Coens C, Mauer M, et al. Baseline quality of life as a prognostic indicator of survival: a meta-analysis of individual patient data from EORTC clinical trials. *Lancet Oncol.* 2009;10(9):865–871.
28. Mao M, Li Y, Wang L, et al. Aitongxiao improves pain symptoms of rats with cancer pain by reducing IL-1, TNF-α, and PGE2. *Int J Clin Exp Pathol.* 2021;14(1):133–139.
29. Heussner MJ, Folger JK, Dias C, et al. A Novel Syngeneic Immunocompetent Mouse Model of Head and Neck Cancer Pain Independent of Interleukin-1 Signaling. *Anesth Analg.* 2021;132(4):1156–1163.
30. Grayson M, Arris D, Wu P, et al. Oral squamous cell carcinoma-released brain-derived neurotrophic factor contributes to oral cancer pain by peripheral tropomyosin receptor kinase B activation. *Pain.* 2022;163(3):496–507.
31. Salvo E, Tu NH, Scheff NN, et al. TNFα promotes oral cancer growth, pain, and Schwann cell activation. *Sci Rep.* 2021;11(1):1840. Published 2021 Jan 19.
32. Tang Y, Chen Y, Yang M, Zheng Q, Li Y, Bao Y. Knockdown of PAR2 alleviates cancer-induced bone pain by inhibiting the activation of astrocytes and the ERK pathway. *BMC Musculoskelet Disord.* 2022;23(1):514.
33. Schliebs, R., & Arendt, T. (2011). The cholinergic system in aging and neuronal degeneration. *Behavioural brain research, 221*(2), 555–563.
34. Renz BW, Tanaka T, Sunagawa M, et al. Cholinergic Signaling via Muscarinic Receptors Directly and Indirectly Suppresses Pancreatic Tumorigenesis and Cancer Stemness. *Cancer Discov.* 2018;8(11):1458–1473.
35. Hanoun M, Zhang D, Mizoguchi T, et al. Acute myelogenous leukemia-induced sympathetic neuropathy promotes malignancy in an altered hematopoietic stem cell niche. *Cell Stem Cell.* 2014;15(3):365–375.
36. Arranz L, Sanchez-Aguilera A, Martin-Perez D, et al. Neuropathy of haematopoietic stem cell niche is essential for myeloproliferative neoplasms. *Nature.* 2014;512(7512):78–81.
37. Katayama Y, Battista M, Kao WM, et al. Signals from the sympathetic nervous system regulate hematopoietic stem cell egress from bone marrow. *Cell.* 2006;124(2):407–421.
38. Gao, X., Zhang, D., Xu, C., Li, H., Caron, K. M., & Frenette, P. S. (2021). Nociceptive nerves regulate haematopoietic stem cell mobilization. *Nature, 589*(7843), 591–596.
39. Maryanovich M, Zahalka AH, Pierce H, et al. Adrenergic nerve degeneration in bone marrow drives aging of the hematopoietic stem cell niche. *Nat Med.* 2018;24(6):782–791.
40. Kumar A, Brockes JP. Nerve dependence in tissue, organ, and appendage regeneration. *Trends Neurosci.* 2012;35(11):691–699.

41. Rinkevich Y, Montoro DT, Muhonen E, et al. Clonal analysis reveals nerve-dependent and independent roles on mammalian hind limb tissue maintenance and regeneration. *Proc Natl Acad Sci U S A*. 2014;111(27):9846–9851.
42. Satoh A, Bryant SV, Gardiner DM. Nerve signaling regulates basal keratinocyte proliferation in the blastema apical epithelial cap in the axolotl (Ambystoma mexicanum). *Dev Biol*. 2012;366(2):374–381.
43. Albo D, Akay CL, Marshall CL, et al. Neurogenesis in colorectal cancer is a marker of aggressive tumor behavior and poor outcomes. *Cancer*. 2011;117(21):4834–4845.
44. Peterson SC, Eberl M, Vagnozzi AN, et al. Basal cell carcinoma preferentially arises from stem cells within hair follicle and mechanosensory niches. *Cell Stem Cell*. 2015;16(4):400–412.
45. Horvathova L, Mravec B. Effect of the autonomic nervous system on cancer progression depends on the type of tumor: solid are more affected then ascitic tumors. *Endocr Regul*. 2016;50(4):215–224.
46. Yu K, Lin CJ, Hatcher A, et al. PIK3CA variants selectively initiate brain hyperactivity during gliomagenesis. *Nature*. 2020;578(7793):166–171.
47. Amit M, Takahashi H, Dragomir MP, et al. Loss of p53 drives neuron reprogramming in head and neck cancer. *Nature*. 2020;578(7795):449–454.
48. Lykissas MG, Batistatou AK, Charalabopoulos KA, Beris AE. The role of neurotrophins in axonal growth, guidance, and regeneration. *Curr Neurovasc Res*. 2007;4(2):143–151.
49. Pundavela J, Roselli S, Faulkner S, et al. Nerve fibers infiltrate the tumor microenvironment and are associated with nerve growth factor production and lymph node invasion in breast cancer. *Mol Oncol*. 2015;9(8):1626–1635.
50. Pundavela J, Demont Y, Jobling P, et al. ProNGF correlates with Gleason score and is a potential driver of nerve infiltration in prostate cancer. *Am J Pathol*. 2014;184(12):3156–3162.
51. Renz BW, Takahashi R, Tanaka T, et al. β2 Adrenergic-Neurotrophin Feedforward Loop Promotes Pancreatic Cancer. *Cancer Cell*. 2018;34(5):863–867.
52. Faulkner S, Roselli S, Demont Y, et al. ProNGF is a potential diagnostic biomarker for thyroid cancer. *Oncotarget*. 2016;7(19):28488–28497.
53. Jimenez-Andrade JM, Ghilardi JR, Castañeda-Corral G, Kuskowski MA, Mantyh PW. Preventive or late administration of anti-NGF therapy attenuates tumor-induced nerve sprouting, neuroma formation, and cancer pain. *Pain*. 2011;152(11):2564–2574.
54. Zhang S, Qi L, Li M, et al. Chemokine CXCL12 and its receptor CXCR4 expression are associated with perineural invasion of prostate cancer. *J Exp Clin Cancer Res*. 2008;27(1):62.
55. Hirth, M., Gandla, J., Höper, C., Gaida, M. M., Agarwal, N., Simonetti, M., Demir, A., Xie, Y., Weiss, C., Michalski, C. W., Hackert, T., Ebert, M. P., & Kuner, R. (2020). CXCL10 and CCL21 Promote Migration of Pancreatic Cancer Cells Toward Sensory Neurons and Neural Remodeling in Tumors in Mice, Associated With Pain in Patients. *Gastroenterology, 159*(2), 665–681.e13.
56. Sawai H, Okada Y, Kazanjian K, et al. The G691S RET polymorphism increases glial cell line-derived neurotrophic factor-induced pancreatic cancer cell invasion by amplifying mitogen-activated protein kinase signaling. *Cancer Res*. 2005;65(24):11536–11544.
57. He S, Chen CH, Chernichenko N, et al. GFRα1 released by nerves enhances cancer cell perineural invasion through GDNF-RET signaling. *Proc Natl Acad Sci U S A*. 2014;111(19):E2008–E2017.
58. Cavel O, Shomron O, Shabtay A, et al. Endoneurial macrophages induce perineural invasion of pancreatic cancer cells by secretion of GDNF and activation of RET tyrosine kinase receptor. *Cancer Res*. 2012;72(22):5733–5743.
59. Amit M, Na'ara S, Leider-Trejo L, et al. Upregulation of RET induces perineurial invasion of pancreatic adenocarcinoma. *Oncogene*. 2017;36(23):3232–3239.
60. Hengst U, Cox LJ, Macosko EZ, Jaffrey SR. Functional and selective RNA interference in developing axons and growth cones. *J Neurosci*. 2006;26(21):5727–5732.
61. Schratt GM, Tuebing F, Nigh EA, et al. A brain-specific microRNA regulates dendritic spine development [published correction appears in Nature. 2006 Jun 15;441(7095):902]. *Nature*. 2006;439(7074):283–289.

62. Wu D, Raafat A, Pak E, Clemens S, Murashov AK. Dicer-microRNA pathway is critical for peripheral nerve regeneration and functional recovery in vivo and regenerative axonogenesis in vitro. *Exp Neurol.* 2012;233(1):555–565.
63. Yu YM, Gibbs KM, Davila J, et al. MicroRNA miR-133b is essential for functional recovery after spinal cord injury in adult zebrafish. *Eur J Neurosci.* 2011;33(9):1587–1597.
64. Zhang HY, Zheng SJ, Zhao JH, et al. MicroRNAs 144, 145, and 214 are down-regulated in primary neurons responding to sciatic nerve transection. *Brain Res.* 2011;1383:62–70.
65. Zhu M, Zhang N, He S, Lu X. Exosomal miR-106a derived from gastric cancer promotes peritoneal metastasis via direct regulation of Smad7. *Cell Cycle.* 2020;19(10):1200–1221.
66. Zhou Y, Chen F, Xie X, et al. Tumor-derived Exosome Promotes Metastasis via Altering its Phenotype and Inclusions. *J Cancer.* 2021;12(14):4240–4246
67. Zhou Y, Ren H, Dai B, et al. Hepatocellular carcinoma-derived exosomal miRNA-21 contributes to tumor progression by converting hepatocyte stellate cells to cancer-associated fibroblasts. *J Exp Clin Cancer Res.* 2018;37(1):324. Published 2018
68. Li X, Zheng J, Chen L, Diao H, Liu Y. Predictive and Prognostic Roles of Abnormal Expression of Tissue miR-125b, miR-221, and miR-222 in Glioma. *Mol Neurobiol.* 2016;53(1):577–583.
69. Igarashi H, Kurihara H, Mitsuhashi K, et al. Association of MicroRNA-31-5p with Clinical Efficacy of Anti-EGFR Therapy in Patients with Metastatic Colorectal Cancer. *Ann Surg Oncol.* 2015;22(8):2640–2648.
70. Nagy ZB, Barták BK, Kalmár A, et al. Comparison of Circulating miRNAs Expression Alterations in Matched Tissue and Plasma Samples During Colorectal Cancer Progression. *Pathol Oncol Res.* 2019;25(1):97–105.
71. Kye MJ, Liu T, Levy SF, et al. Somatodendritic microRNAs identified by laser capture and multiplex RT-PCR. *RNA.* 2007;13(8):1224–1234.
72. Natera-Naranjo O, Aschrafi A, Gioio AE, Kaplan BB. Identification and quantitative analyses of microRNAs located in the distal axons of sympathetic neurons. *RNA.* 2010;16(8):1516–1529.
73. Martin KC, Kosik KS. Synaptic tagging – who's it?. *Nat Rev Neurosci.* 2002;3(10):813–820.
74. Jin P, Zarnescu DC, Ceman S, et al. Biochemical and genetic interaction between the fragile X mental retardation protein and the microRNA pathway. *Nat Neurosci.* 2004;7(2):113–117.
75. Hunt PJ, Kabotyanski KE, Calin GA, Xie T, Myers JN, Amit M. Interrupting Neuron-Tumor Interactions to Overcome Treatment Resistance. *Cancers (Basel).* 2020;12(12):3741.
76. Liu S, Jia J, Zhou H, et al. PTEN modulates neurites outgrowth and neuron apoptosis involving the PI3K/Akt/mTOR signaling pathway. *Mol Med Rep.* 2019;20(5):4059–4066.
77. Choi YJ, Lin CP, Ho JJ, et al. miR-34 miRNAs provide a barrier for somatic cell reprogramming. *Nat Cell Biol.* 2011;13(11):1353–1360.
78. Kim NH, Kim HS, Li XY, et al. A p53/miRNA-34 axis regulates Snail1-dependent cancer cell epithelial-mesenchymal transition. *J Cell Biol.* 2011;195(3):417–433.
79. Jiang CC, Marsland M, Wang Y, et al. Tumor innervation is triggered by endoplasmic reticulum stress. *Oncogene.* 2022;41(4):586–599.
80. Yadav RK, Chae SW, Kim HR, Chae HJ. Endoplasmic reticulum stress and cancer. *J Cancer Prev.* 2014;19(2):75–88.
81. Li J, Ma Q. Hyperglycemia promotes the perineural invasion in pancreatic cancer. *Med Hypotheses.* 2008;71(3):386–389.
82. Walker WH 2nd, Borniger JC. Molecular Mechanisms of Cancer-Induced Sleep Disruption. *Int J Mol Sci.* 2019;20(11):2780.
83. Borniger JC, Walker Ii WH, Surbhi, et al. A Role for Hypocretin/Orexin in Metabolic and Sleep Abnormalities in a Mouse Model of Non-metastatic Breast Cancer. *Cell Metab.* 2018;28(1):118–129.e5.
84. Barr JL, Kruse A, Restaino AC, et al. Intra-Tumoral Nerve-Tracing in a Novel Syngeneic Model of High-Grade Serous Ovarian Carcinoma. *Cells.* 2021;10(12):3491.

Chapter 6
Neuroimmune Interactions and Their Role in Carcinogenesis

Shahrukh Ali, Dan Yaniv, and Moran Amit

Abbreviations

CD	Cluster of differentiation
CGRP	Calcitonin gene-related peptide
GABA	Gamma-aminobutyric acid
HNSCC	Head and neck squamous cell cancer
IL	Interleukin
MDSC	Myeloid-derived suppressor cells
NK-1R	Neurokinin-1 receptor
PD1	Programmed cell death protein 1
PD-L1	Programmed death-ligand 1
RAMP	Receptor activity-modifying protein
TFF2	Trefoil factor 2
TME	Tumor microenvironment
Treg	Regulatory T
VIP	Vasoactive intestinal peptide

S. Ali
University of Texas Medical Branch John Sealy School of Medicine, Galveston, TX, USA

D. Yaniv · M. Amit (✉)
Department of Head and Neck Surgery, The University of Texas MD Anderson Cancer Center, Houston, TX, USA
e-mail: MAmit@mdanderson.org

M. Amit, N. N. Scheff (eds.), *Cancer Neuroscience*,
https://doi.org/10.1007/978-3-031-32429-1_6

Introduction

The nervous system and the immune system are complex and interdependent when reacting to internal and external stimuli to maintain homeostasis in the body. For example, when pathogens infiltrate into the bloodstream, the neurological system activates the sympathetic response, which causes vasoconstriction and releases endogenous hormones that signal immune cells to migrate toward the site of invasion. This type of interaction, in which the neurological and immune systems collaborate with each other to prevent harm to the host, has been described as neuroimmune cross talk. Moreover, this interaction plays an important role in preventing malignancies, and disruption of this system can promote tumorigenesis in certain types of cancer [1–7]. Two specific types of neuroimmune interactions that have been documented extensively in the literature include regulation via neuron stimuli (i.e., neural transmission) and checkpoint-mediated neuronal immune regulation. The latter relies upon the expression of immune checkpoints on the membranes of intratumoral nerves and glial cells.

Neuronal-Dependent Immune Regulation

Sensory Nerve-Mediated Immune Reactions

Sensory afferent fibers respond to external stimuli (i.e., chemical, thermal, and mechanical stimuli) and are involved in regulating both immune and inflammatory reactions [8–10]. Stimulation of the peripheral sensory nerve fibers can cause the release of neuropeptides such as calcitonin gene-related peptide (CGRP) and substance P. Substance P supports the survival of active T lymphocytes, upregulates the production of pro-inflammatory cytokines from macrophages, and prepares neutrophils for chemokine ligand 5 (CCL5)-induced chemotaxis and migration [11]. Neuropeptides lead to leukocyte influx and chemotaxis of immune cells at the inflammation site by inducing vasodilatation and subsequent edema and plasma extravasation. CGRP evokes neurogenic vasodilatation and inflammation. The receptors for CGRP are expressed by myeloid immune subsets and inhibit immunity [12, 13]. In inflammatory states, CGRP restricts tissue damage by limiting the release of pro-inflammatory molecules produced by macrophages and dendritic cells. On the molecular level, CGRP limits the inflammatory reaction by decreasing the secretion of interleukin (IL)-10 and inhibiting nuclear factor kappa B activity [14, 15]. CGRP also dampens tumor necrosis factor-alpha (TNF-α) transcription in dendritic cells through a cyclic adenosine monophosphate (cAMP)-dependent repressor mechanism [16].

Cancer cells spread through the use of efferent lymphatic vessels, primarily to regional lymph nodes. Research has shown that lymph nodes are richly supplied with nerves, specifically from sympathetic neurons that produce tyrosine hydroxylase [16–19]. However, a recent study investigating lymph node innervation discovered that sensory nerve fibers, mostly from peptidergic nociceptive sensory neurons, also reach the popliteal lymph nodes and surrounding tissue [20]. In addition, the study revealed an increase in the density of sensory nerve fibers in the lymph nodes during inflammatory conditions, during which sensory neurons innervating the lymph nodes expressed toll-like receptors 1, 2, and 4, resulting in the expansion of nerve fibers in the lymphoid tissue. Using optogenetics and single-cell RNA sequencing, the researchers found that the genes for the CGRP receptors (i.e., *Calr*, *Calcrl*, *Ramp1*) were primarily expressed in innate immune cells in lymph nodes, such as dendritic cells and mast cells, which were influenced by sensory neurons that regulate the myeloid lineage via CGRP. The activation of these sensory neurons through optogenetics also led to changes in neutrophils and natural killer cells, but no significant changes were observed in the T and B lymphocytes in the lymph nodes. Further investigation is required to understand the relationship between the nerve fibers and the structure of the cells in the lymph nodes, as well as their potential role in the cancer microenvironment.

Two recent studies have begun to uncover a role for CGRP neurotransmission in cancer immunosurveillance and have suggested that sensory neuron signaling might be a viable target for improving antitumor immunity. Head and neck squamous cell cancer (HNSCC) in the oral cavity has dense sensory neural innervation, and a main source of these sensory nerve fibers is the trigeminal ganglia [21, 22]. CGRP is a major neurotransmitter in the trigeminal ganglia, with two isoforms, αCGRP, and βCGRP. The αCGRP-containing sensory nerve fibers and increased CGRP receptor expression on the lymphocytes infiltrating the tumor microenvironment (TME) have been found in human HNSCC tissue samples as well as in rodent models, and αCGRP is thought to impact the immune response in tumor sites via the receptor activity-modifying protein (RAMP)1 signaling pathway [21, 22]. The RAMP1 pathway plays a substantial role in the innate and adaptive immune responses. In terms of its impact in HNSCC specifically, it has been shown that, compared to controls, syngeneic CGRP-knockout mouse models for oral cancer had reduced tumor sizes and more infiltration by cluster of differentiation (CD)4^+ and $CD8^+$ T lymphocytes and natural killer cells [23, 24], suggesting that CGRP in intact animals modulated antitumor immunity. These findings were recently corroborated by an additional study using a melanoma mouse model; the lymphocytes infiltrating the tumor sites were shown to have decreased efficacy because of CGRP activation and subsequent RAMP1 signaling. Using pharmacologic agents that block nociceptors, researchers suppressed CGRP neurotransmission and signaling and saw decreased $CD8^+$ T-cell exhaustion. In turn, this increased overall survival and decreased tumor growth. Single-cell RNA sequencing of melanoma patients' samples also showed that RAMP1-expressing $CD8^+$ T cells were more prone to exhaustion than $RAMP1^-$ $CD8^+$ T cells and that patients with $RAMP1^-$ $CD8^+$ T cells had better clinical prognoses [25–29].

Substance P

Substance P is a tachykinin neuropeptide released either from nerve terminals within the neural tumor microenvironment (NTME) or directly by cancer cells into blood vessels. Primarily, it mediates inflammation in response to noxious stimuli

[30]. Its primary receptor, neurokinin-1 receptor (NK-1R), is expressed in various human cancer cell lines. When activated, the NK-1R downstream signaling pathways promote cell proliferation and migration [31]. Clinically, high substance P expression levels correlate with increased Ki-67 expression in oral cavity squamous cell carcinoma samples, strongly suggesting a role for substance P and NK-1R in tumor development and progression, as well as in advanced tumor grades for head and neck and gastric cancers [32–34]. Studies have demonstrated that substance P is a universal mitogen in NK-1R-expressing tumor cells [34]. Tumor cells may strongly depend on the potent mitotic signals mediated by substance P for survival, as the overexpression of NK-1R in tumor cells neutralizes its normally activated pathways, resulting in cell death and the knockdown of NK-1R-induced apoptosis evasion. Of note, the pharmacological blockade of NK-1R via treatment with aprepitant (an antiemetic medicine) in the presence of NK-1R activation increased apoptosis in lung cancer cells in vitro to a greater degree than did aprepitant-based treatment alone [35]. Additionally, the effects of NK-1R inhibition were different in various metastatic cell populations. Specifically, the inhibition of NK-1R markedly increased liver metastasis of tumors formed by breast cancer cells but not breast cancer cells which metastasized to the brain. The activation of NK-1R can increase DNA synthesis in tumor cells and activate the mitogen-activated protein kinase (MAPK) pathways, including the extracellular signal-regulated kinase (ERK)1/2 and p38/MAPK pathways [36, 37]. Once activated, ERK1/2 translocates from the cytoplasm to the nucleus to induce cell proliferation and antiapoptotic signaling pathways [37]. Conversely, lymphokine-associated killer cells have a stronger cytotoxic effect against fresh colonic cancer cells when incubated in the presence of substance P. A low dose of substance P also improved the effect of radiotherapy by 50% in a mouse model of poorly differentiated breast carcinoma by decreasing the number of tumor-infiltrating, myeloid-derived suppressor cells while enhancing interferon-γ secretion from leukocytes [38–40].

Adrenergic Signaling-Mediated Immune Signaling

The activation of the sympathetic nervous system from endogenous and exogenous stimuli maintains homeostasis in the body. Sympathetic nerve fibers innervate aspects of the immune system, including the colorectal system and primary and secondary lymphoid structures such as the bone marrow, thymus, spleen, and lymph nodes [41–43]. Sympathetic nerves signal information through the lymph nodes, including the lymph node cortex, medulla, hilum, and capsule, with varying distribution. Most of this signaling occurs close to the outer surface of the capsule, where the T cells are packed densely [33]. T cells display β-adrenergic receptors on their surfaces and are regulated by the sympathetic innervation of lymph nodes, but the particular mechanism that modulates the function of these cells within the lymph nodes has yet to be elucidated. In the spleen, in which sympathetic fibers innervate both the red pulp and the white pulp [44–48], sympathetic system activation has a significant impact on cytokine gene expression [49, 50]. In one study that observed this affect, rats whose sympathetic systems were activated by exposure to higher temperatures had higher levels of *Il1b*, *Il6*, and *Gro1* (growth-regulated oncogene 1 mRNA) compared with normothermic controls [51]. This effect was attenuated following splenic nerve denervation, suggesting that the activation of the sympathetic system increased the expression of genes encoding for pro-inflammatory cytokines [52].

Chronic stress promotes tumorigenesis, but the precise effect of neurotransmitters like norepinephrine on the tumor immune microenvironment is not fully understood [53, 54]. The activation of adrenergic receptors in tumor cells has been shown to prevent cellular apoptosis and promote apoptosis in $CD4^+$ T cells and B cells [55–58]. β-Adrenergic stimulation also increases the release of cytokines like IL-6 from myeloid lineage-derived immune cells, attenuating the immune response [59–61]. In addition, β-adrenergic signaling activates adrenergic receptors on macrophages, which release immunosuppressive molecules like IL-6, transforming growth factor-β (TGF-β), vascular endothelial growth factor (VEGF), matrix metalloproteinase (MMP), and prostaglandin-endoperoxide synthase 2 (PGE-2) [62]. Furthermore, β-adrenergic signaling causes the release of granulocyte colony-stimulating factor (GCSF) from tumor cells, stromal cells, and other immune cells in the TME, which leads to increased expression of β2 adrenergic receptors on myeloid-derived suppressor cells (MDSCs). Following the activation of these β2 adrenergic receptors by sympathetic stimulation, MDSCs undergo downregulation of glycolysis and enhanced fatty acid oxidation, with higher expression of carnitine palmitoyltransferase 1A (CPT1A, a mitochondrial enzyme) and oxidative phosphorylation. CPT1A is necessary for the fatty acid oxidation-mediated suppression of the immune response mediated by MDSCs [63–67]. Stress also affects dendritic cell function, which regulates T lymphocytes and antitumor immunity in the TME. A study that analyzed temperature as a form of stress in mouse models found that chronic stress may promote tumor survival by increasing the percentage of $CD11b^+$ myeloid cells and interferon-producing plasmacytoid dendritic cells [68]. Tumor growth is slower in mice with low percentages of $CD11b^+$ myeloid cells than the growth seen in mice with higher percentages of CD11B+, indicating that chronic stress may promote tumor survival [68–73]. In breast cancer patients, high stress levels correlate with high levels of immunosuppressive MDSCs, suggesting that stress promotes tumor survival [73–76]. Stress also increases the percentage of tumor-associated macrophages recruited to the TME, where these macrophages release catecholamines that promote tumor development [77, 78].

Cholinergic Signaling-Mediated Immune Signaling

Parasympathetic nervous system signaling also plays an important role in the neuroimmune network. Acetylcholine is the main neurotransmitter of the parasympathetic nervous system and exerts its autocrine and paracrine effects via the classical cholinergic system. The cholinergic system consists of two major categories of receptors: ionotropic ligand-gated channels, called nicotinic acetylcholine receptors, and metabotropic G protein-coupled receptors, called muscarinic acetylcholine receptors [77]. Acetylcholine is released by various nerve fibers, including the vagus nerve. The vagus nerve regulates splanchnic immunity by innervating the spleen, and vagal stimulation prevents organ injury and systemic inflammation by inhibiting the production of inflammatory cytokines in that organ. Surprisingly, splanchnic vagal nerve endings do not produce acetylcholine; instead, the vagal nerve endings form synapse-like structures on T lymphocytes in the spleen and stimulate the release of acetylcholine from splenic memory $CD4^+$ $ChAT^+$ T cells via adrenergic receptor signaling [78, 79]. The acetylcholine released from these cells binds to nicotinic receptors expressed on macrophages and inhibits the release of pro-inflammatory cytokines such as tumor necrosis factor-alpha (TNF-α) to downregulate the inflammatory reaction [79].

Vagal stimulation not only downregulates the inflammatory reaction but also drives an antitumor effect. Specifically, it has been shown that the stimulation of muscarinic receptors on T cells in mice and human colony cells causes aldehyde gene expression in T cells [80, 81]. This expression puts the T cells into an active regulatory state and allows them to invade into peripheral tissue to destroy tumor cells. Researchers have further shown that, in a pancreatic ductal adenocarcinoma mouse model, bilateral subdiaphragmatic vagotomy allowed for accelerated tumor cell proliferation and invasion, whereas the systemic muscarinic agonist bethanechol restored the normal phenotype [82, 83]. This research shows that cholinergic signaling may prevent tumorigenesis in some scenarios and may also prevent excess damage from overactivation of the inflammatory network.

Cholinergic vagal signaling also increases the release of trefoil factor 2 (TFF2) from splenic memory T cells, which further downregulate inflammation and colon cancer development [84]. In mice, researchers have observed a decrease in the splenic expression of TFF2 after subdiaphragmatic bilateral vagotomy. After 5 months of the application of the carcinogens azoxymethane and dextran sodium sulfate, compared to TFF2-intact mice, TFF2-null mice developed more colonic tumors, higher-grade dysplasia, and greater MDSC infiltration to the tumor tissue with no detectable colonic $CD8^+$ T cells. There was also an increase in programmed death-ligand 1 (PD-L1) expression in the MDSCs from the TFF2-null mice, and both interferon-γ production and $CD4^+$ T-cell function suppressed. Moreover, the TFF2-null mice had higher numbers of C-X-C chemokine receptor type 4 (CXCR4)-expressing MDSCs expressing protumorigenic factors such as IL-17A and IL-1β. The researchers hypothesized that the activation of the vagal network and of T cells expressing TFF2 molecules may decrease the infiltration of MDSCs into the TME, thus decreasing tumorigenesis [85].

Vasoactive intestinal peptide (VIP) is another cholinergic molecule that, upon binding to cholinergic receptors, leads to decreased tumorigenesis. VIP receptor activation on T cells triggers T cells' differentiation into T helper 2 (Th2) and regulatory T (Treg) cells [86–90]. This phenomenon has been described extensively in pancreatic ductal adenocarcinoma mouse models, in which the activation of VIP receptors was associated with an increase of Treg and Th2 cells in the TME [91]. When cells isolated from a patient with pancreatic ductal adenocarcinoma were treated in vitro with a VIP receptor antagonist, the proportion of Tregs and of T cells with PD-1 expression decreased. Thus, VIP signaling activation may promote tumorigenesis, and VIP receptor blockers may produce an antitumor immune response. [91, 92]

The Nonneuronal Release of Neurotransmitters

The neurotransmitters most commonly associated with the central nervous system, including gamma-aminobutyric acid (GABA) and serotonin, also mediate the immune environment in the TME. Researchers have identified three receptors for GABA: the ionotropic $GABA_A$ and $GABA_C$ receptors and the metabotropic $GABA_B$ receptor. Investigators have observed increased GABA levels in certain solid tumors, such as colon cancer, sarcoma, and ovarian cancer, and have also determined that GABA-positive tumors are more aggressive [93–97]. The production and release of GABA from B cells, and the subsequent activation of $GABA_A$ receptors on $CD8^+$ T cells and the enhanced recruitment of IL-10-secreting macrophages, downregulate the antitumor response [98]. Most of the body's serotonin is generated by enterochromaffin cells, which mediate between the gut and the brain and communicate with the immune and enteric nervous systems. Macrophages, dendritic cells, and lymphocytes all express serotonin receptor subtypes. Moreover, in pancreatic and gastric cancers, patients with higher expression levels of the intratumoral 5-hydroxytryptamine (5-HT) receptors 1A and 1B exhibit more CD4+CD25+FoxP3+ Treg cells, higher levels of expression of PD-L1, and more phosphorylated signal transducer and activator of transcription-3 (STAT-3) molecules in the TME [99].

Checkpoint-Mediated Neuronal Immune Regulation

Checkpoint molecules in the immune system play a crucial role in the fight against cancer [100]. In some cases, these molecules are expressed in the nerves surrounding the tumor and result in the inhibition of T cells, promoting the growth of cancer. To counter this, drugs that target checkpoint molecules aim to boost antitumor immunity by preventing the inhibition or exhaustion of immune cells and increasing their ability to identify and destroy cancer cells [87].

It has been previously thought that programmed cell death protein 1 (PD1) only exists on immune and tumor cells, whereas the expression of its binding partner, PD-L1 (encoded by CD274), has been detected in several cell types, including cortical and trigeminal ganglion neurons [101–103]. Several single-cell RNA sequencing studies in mice have detected mRNA from immune checkpoint genes (e.g., *Cd274*) in sensory and sympathetic neurons; checkpoint protein expression in postmitotic cells whose axons are exposed continuously to the immune system is likely needed to induce immune tolerance to avoid damage to peripheral neuron processes. However, the functional impact of checkpoint-mediated neuroimmune communication is currently unknown.

Studies have shown that the density of the checkpoint molecule PD-L1 expressed on nerve fibers within a tumor predicts a poor prognosis in prostate cancer [104]. Reducing the expression of checkpoint molecules improves the immune response and slows down tumor progression, as demonstrated in breast cancer patients whose sympathetic nerves were genetically inhibited [105].

As our understanding of the cross talk between the nervous system, the immune system, and cancer cells improves, the nervous system will offer new possibilities for therapeutic targets to improve patient outcomes. For instance, reducing adrenergic stress has been shown to increase the efficacy of anti-PD1 checkpoint inhibition by increasing CD8+ T-cell effector levels and decreasing PD1 receptor levels in the TME [106, 107].

In conclusion, checkpoint molecules play a major role in the immune escape of cancer. Drugs targeting these molecules aim to enhance antitumor immunity in the TME to inhibit cancer progression and shrink tumors. Further research into the nervous system and its relationship with cancer immunology may offer new therapeutic targets to improve patient outcomes.

Glial Cells in Cancer

The neuroimmunology axis in cancer involves the participation of glial cells, including nerve-ensheathing Schwann cells. Schwann cells, particularly the subtypes involved in nerve repair, boost the chemotaxis of immune cells by releasing chemokines [108]. When Schwann cells are activated and exposed to a melanoma-conditioned environment, they increase the mRNA levels of genes associated with immune surveillance and chemotaxis, including *Il6*, *Tgfb*, and *Vgef* [109]. Experiments with cervical cancer cell lines have shown that chemokine (C-C motif) ligand 2 (CCL2) released from Schwann cells can affect the growth of cancer cells and modify the immune microenvironment, leading to larger tumors and worse overall survival [110, 111]. Schwann cells also regulate the behavior of M2 macrophages and MDSCs, promoting their inhibitory effects and contributing to cancer progression [112]. Schwann cells have toll-like receptors and are capable of recognizing antigens and activating the innate immune response and T cells [113]. Danger-associated molecular patterns (DAMPs) produced by various cancers, including pancreatic, breast, and colon cancers and glioblastoma, are sensed by the toll-like receptors on Schwann cells. Schwann cells also have cell-surface receptors CD74, CD1a, CD1b, CD1d, B7-1, BB-1, and CD58, suggesting that they play a role as antigen-presenting cells in the TME [114, 115].

Emerging Therapeutics Targeting the Neuroimmune Network

As more data emerges about the interconnectedness between the nervous system and the immune system within the TME, exploring the nervous system as a potential therapeutic target appears to be a promising approach for enhancing existing treatments and improving patient outcomes. Studies in mice have shown that reducing adrenergic stress through physiologic means (i.e., temperature), pharmacologic means (i.e., adrenergic pharmacology), and genetic modulation (i.e., *Adrb2* knockout mice) results in increased $CD8^+$ effector T-cell activity and decreased numbers of PD1 receptors in the TME [116, 117]. A preclinical trial combining propranolol with an immunotherapy agent (an anti-CTLA-4 antibody) showed improved T lymphocyte activity and decreased MDSC invasion in the TME in mice with fibrosarcoma and colon cancer and led to improvements in the efficacy of anti-CTLA-4 treatment [117].

The first phase I clinical trial of propranolol and the anti-PD1 checkpoint inhibitor pembrolizumab was conducted to evaluate the effects of immunotherapy in melanoma patients ($n = 9$) and determine the estimated phase II dose. The trial showed no dose-limiting toxicities and a response rate of 78% [118]. Patients who responded to the treatment regimen showed higher levels of IFNγ and lower levels of IL-6. However, a subsequent phase III clinical trial in patients with high-risk stage III melanoma treated with pembrolizumab as adjuvant therapy found that β-blockers had no independent effect on recurrence-free survival, highlighting the need for further investigation of the exact mechanisms involved [119]. The impact of lidocaine, a sodium channel blocker that also has anticholinergic properties, was also analyzed in clinical trials for breast cancer. In the trial, 120 patients underwent surgical resection and received intravenous lidocaine as part of their cancer treatment [120]. Lidocaine decreased the production of MMP3, neutrophil extracellular molecular trapping biomarkers, and myeloperoxidase, but further investigation is needed to prove the drug's clinical efficacy and possible beneficial effects on patient survival [121].

Conclusion

The neuroimmune system is a complex network that fights against both external and internal dangers. As research has progressed, it has become clear that neuroimmune regulation plays a significant role in preventing tumor formation and that a breakdown in this system can contribute to tumor growth. In this chapter, we discussed how sensory, adrenergic, and cholinergic stimulation impact the neuroimmune network, the role of peripheral nerves and neurotransmitters in tumorigenesis, and new therapies being developed to tackle malignancies.

References

1. Felten DL, Felten SY. Immune interactions with specific neural structures. Brain Behav Immun. 1987;1(4):279–283. https://doi.org/10.1016/0889-1591(87)90030-4
2. Klein RS, Garber C, Howard N. Infectious immunity in the central nervous system and brain function. Nat Immunol. 2017;18(2):132–141. https://doi.org/10.1038/ni.3656
3. Norris GT, Kipnis J. Immune cells and CNS physiology: Microglia and beyond. J Exp Med. 2019;216(1):60–70. https://doi.org/10.1084/jem.20180199
4. Prinz M, Priller J. The role of peripheral immune cells in the CNS in steady state and disease. Nat Neurosci. 2017;20(2):136–144. https://doi.org/10.1038/nn.4475
5. Baral P, Udit S, Chiu IM. Pain and immunity: implications for host defence. Nat Rev Immunol. 2019;19(7):433–447. https://doi.org/10.1038/s41577-019-0147-2
6. Godinho-Silva C, Cardoso F, Veiga-Fernandes H. Neuro-Immune Cell Units: A New Paradigm in Physiology. Annu Rev Immunol. 2019;37:19–46. https://doi.org/10.1146/annurev-immunol-042718-041812
7. Emery, E.C., and Ernfors, P. Dorsal root ganglion neuron types and their functional specialization. In The Oxford Handbook of the Neurobiology of Pain, 2018 pp. 129–155. https://doi.org/10.1093/oxfordhb/9780190860509.013.4.
8. Kupari J, Häring M, Agirre E, Castelo-Branco G, Ernfors P. An Atlas of Vagal Sensory Neurons and Their Molecular Specialization. Cell Rep. 2019;27(8):2508-2523.e4. https://doi.org/10.1016/j.celrep.2019.04.096
9. Sharma N, Flaherty K, Lezgiyeva K, Wagner DE, Klein AM, Ginty DD. The emergence of transcriptional identity in somatosensory neurons. Nature. 2020;577(7790):392–398. https://doi.org/10.1038/s41586-019-1900-1
10. Nowicki M, Ostalska-Nowicka D, Kondraciuk B, Miskowiak B. The significance of substance P in physiological and malignant haematopoiesis. J Clin Pathol. 2007;60(7):749–755. https://doi.org/10.1136/jcp.2006.041475
11. Holzmann B. Modulation of immune responses by the neuropeptide CGRP. Amino Acids. 2013;45(1):1–7. https://doi.org/10.1007/s00726-011-1161-2
12. Chiu IM, Heesters BA, Ghasemlou N, et al. Bacteria activate sensory neurons that modulate pain and inflammation. Nature. 2013;501(7465):52–57. https://doi.org/10.1038/nature12479
13. Chernova I, Lai JP, Li H, et al. Substance P (SP) enhances CCL5-induced chemotaxis and intracellular signaling in human monocytes, which express the truncated neurokinin-1 receptor (NK1R). J Leukoc Biol. 2009;85(1):154–164. https://doi.org/10.1189/jlb.0408260
14. Mashaghi A, Marmalidou A, Tehrani M, Grace PM, Pothoulakis C, Dana R. Neuropeptide substance P and the immune response. Cell Mol Life Sci. 2016;73(22):4249–4264. https://doi.org/10.1007/s00018-016-2293-z
15. Holzmann B. Antiinflammatory activities of CGRP modulating innate immune responses in health and disease. Curr Protein Pept Sci. 2013;14(4):268–274. https://doi.org/10.2174/13892037113149990046
16. Harzenetter MD, Novotny AR, Gais P, Molina CA, Altmayr F, Holzmann B. Negative regulation of TLR responses by the neuropeptide CGRP is mediated by the transcriptional repressor ICER. J Immunol. 2007;179(1):607–615. https://doi.org/10.4049/jimmunol.179.1.607
17. Duan JX, Zhou Y, Zhou AY, et al. Calcitonin gene-related peptide exerts anti-inflammatory property through regulating murine macrophages polarization in vitro. Mol Immunol. 2017;91:105–113. https://doi.org/10.1016/j.molimm.2017.08.020
18. Bellinger DL, Lorton D, Felten SY, Felten DL. Innervation of lymphoid organs and implications in development, aging, and autoimmunity. Int J Immunopharmacol. 1992;14(3):329–344. https://doi.org/10.1016/0192-0561(92)90162-e
19. Felten DL, Felten SY, Carlson SL, Olschowka JA, Livnat S. Noradrenergic and peptidergic innervation of lymphoid tissue. J Immunol. 1985;135(2 Suppl):755s–765s.

20. Huang S, Ziegler CGK, Austin J, et al. Lymph nodes are innervated by a unique population of sensory neurons with immunomodulatory potential. Cell. 2021;184(2):441–459.e25. https://doi.org/10.1016/j.cell.2020.11.028
21. Amit M, Takahashi H, Dragomir MP, et al. Loss of p53 drives neuron reprogramming in head and neck cancer. Nature. 2020;578(7795):449–454. https://doi.org/10.1038/s41586-020-1996-3
22. Scheff NN, Ye Y, Bhattacharya A, et al. Tumor necrosis factor alpha secreted from oral squamous cell carcinoma contributes to cancer pain and associated inflammation. Pain. 2017;158(12):2396–2409. https://doi.org/10.1097/j.pain.0000000000001044
23. McIlvried LA, Atherton MA, Horan NL, Goch TN, Scheff NN. Sensory Neurotransmitter Calcitonin Gene-Related Peptide Modulates Tumor Growth and Lymphocyte Infiltration in Oral Squamous Cell Carcinoma. Adv Biol (Weinh). 2022;6(9):e2200019. https://doi.org/10.1002/adbi.202200019
24. Honda M, Ito Y, Hattori K, et al. Inhibition of receptor activity-modifying protein 1 suppresses the development of endometriosis and the formation of blood and lymphatic vessels. J Cell Mol Med. 2020;24(20):11984–11997. https://doi.org/10.1111/jcmm.15823
25. Tsuru S, Ito Y, Matsuda H, et al. RAMP1 signaling in immune cells regulates inflammation-associated lymphangiogenesis. Lab Invest. 2020;100(5):738–750. https://doi.org/10.1038/s41374-019-0364-0
26. McIlvried, L.A., Atherton, M.A., Horan, N.L., Goch, T.N., and Scheff, N.N. (2022). Sensory Neurotransmitter Calcitonin Gene-Related Peptide Modulates Tumor Growth and Lymphocyte Infiltration in Oral Squamous Cell Carcinoma. Adv. Biol. 6. https://doi.org/10.1002/adbi.202200019.
27. Toda M, Suzuki T, Hosono K, et al. Neuronal system-dependent facilitation of tumor angiogenesis and tumor growth by calcitonin gene-related peptide. *Proc Natl Acad Sci U S A*. 2008;105(36):13550–13555. https://doi.org/10.1073/pnas.0800767105
28. Balood M, Ahmadi M, Eichwald T, Ahmadi A, Majdoubi A, Roversi K, Roversi K, Lucido CT, Restaino AC, Huang S, Ji L, Huang KC, Semerena E, Thomas SC, Trevino AE, Merrison H, Parrin A, Doyle B, Vermeer DW, Spanos WC, Williamson CS, Seehus CR, Foster SL, Dai H, Shu CJ, Rangachari M, Thibodeau JV, Del Rincon S, Drapkin R, Rafei M, Ghasemlou N, Vermeer PD, Woolf CJ, Talbot S. Nociceptor neurons affect cancer immunosurveillance. Nature. 2022 Nov;611(7935):405–412. https://doi.org/10.1038/s41586-022-05374-w. Epub 2022 Nov 2. PMID: 36323780; PMCID: PMC9646485.
29. McIlvried LA, Atherton MA, Horan NL, Goch TN, Scheff NN. Sensory Neurotransmitter Calcitonin Gene-Related Peptide Modulates Tumor Growth and Lymphocyte Infiltration in Oral Squamous Cell Carcinoma. Adv Biol (Weinh). 2022 Sep;6(9):e2200019. https://doi.org/10.1002/adbi.202200019. Epub 2022 Apr 7. PMID: 35388989; PMCID: PMC9474661.
30. Muñoz M, Coveñas R. Involvement of substance P and the NK-1 receptor in cancer progression. Peptides. 2013 Oct;48:1–9. https://doi.org/10.1016/j.peptides.2013.07.024. Epub 2013 Aug 7. PMID: 23933301.
31. Restaino AC, Vermeer PD. Neural regulations of the tumor microenvironment. FASEB Bioadv. 2021 Sep 12;4(1):29–42. https://doi.org/10.1096/fba.2021-00066. PMID: 35024571; PMCID: PMC8728107.
32. Brener S, González-Moles MA, Tostes D, Esteban F, Gil-Montoya JA, Ruiz-Avila I, Bravo M, Muñoz M. A role for the substance P/NK-1 receptor complex in cell proliferation in oral squamous cell carcinoma. Anticancer Res. 2009 Jun;29(6):2323–9. PMID: 19528498.
33. Mehboob R, Tanvir I, Warraich RA, Perveen S, Yasmeen S, Ahmad FJ. Role of neurotransmitter Substance P in progression of oral squamous cell carcinoma. Pathol Res Pract. 2015 Mar;211(3):203–7. https://doi.org/10.1016/j.prp.2014.09.016. Epub 2014 Oct 13. PMID: 25433994.
34. Feng F, Yang J, Tong L, Yuan S, Tian Y, Hong L, Wang W, Zhang H. Substance P immunoreactive nerve fibres are related to gastric cancer differentiation status and could promote

proliferation and migration of gastric cancer cells. Cell Biol Int. 2011 Jun;35(6):623–9. https://doi.org/10.1042/CBI20100229. PMID: 21091434.
35. Muñoz M, Rosso M. The NK-1 receptor antagonist aprepitant as a broad spectrum antitumor drug. Invest New Drugs. 2010 Apr;28(2):187–93. https://doi.org/10.1007/s10637-009-9218-8. Epub 2009 Jan 17. PMID: 19148578.
36. Muñoz M, Rosso M, Robles-Frias MJ, Salinas-Martín MV, Rosso R, González-Ortega A, Coveñas R. The NK-1 receptor is expressed in human melanoma and is involved in the antitumor action of the NK-1 receptor antagonist aprepitant on melanoma cell lines. Lab Invest. 2010 Aug;90(8):1259–69. https://doi.org/10.1038/labinvest.2010.92. Epub 2010 May 10. PMID: 20458280.
37. Muñoz M, González-Ortega A, Rosso M, Robles-Frias MJ, Carranza A, Salinas-Martín MV, Coveñas R. The substance P/neurokinin-1 receptor system in lung cancer: focus on the antitumor action of neurokinin-1 receptor antagonists. Peptides. 2012 Dec;38(2):318–25. https://doi.org/10.1016/j.peptides.2012.09.024. Epub 2012 Sep 28. PMID: 23026680.
38. Nizam E, Köksoy S, Erin N. NK1R antagonist decreases inflammation and metastasis of breast carcinoma cells metastasized to liver but not to brain; phenotype-dependent therapeutic and toxic consequences. Cancer Immunol Immunother. 2020 Aug;69(8):1639–1650. https://doi.org/10.1007/s00262-020-02574-z. Epub 2020 Apr 22. PMID: 32322911.
39. Luo W, Sharif TR, Sharif M. Substance P-induced mitogenesis in human astrocytoma cells correlates with activation of the mitogen-activated protein kinase signaling pathway. Cancer Res. 1996 Nov 1;56(21):4983–91. PMID: 8895754.
40. DeFea KA, Vaughn ZD, O'Bryan EM, Nishijima D, Déry O, Bunnett NW. The proliferative and antiapoptotic effects of substance P are facilitated by formation of a beta -arrestin-dependent scaffolding complex. Proc Natl Acad Sci U S A. 2000 Sep 26;97(20):11086–91. https://doi.org/10.1073/pnas.190276697. PMID: 10995467; PMCID: PMC27152.
41. Hosking KG, Fels RJ, Kenney MJ. Inhibition of RVLM synaptic activation at peak hyperthermia reduces visceral sympathetic nerve discharge. Auton Neurosci. 2009 Oct 5;150(1–2):104–10. https://doi.org/10.1016/j.autneu.2009.06.004. Epub 2009 Jul 8. PMID: 19589733; PMCID: PMC2739272.
42. Kenney MJ, Barney CC, Hirai T, Gisolfi CV. Sympathetic nerve responses to hyperthermia in the anesthetized rat. J Appl Physiol (1985). 1995 Mar;78(3):881–9. https://doi.org/10.1152/jappl.1995.78.3.881. PMID: 7775333.
43. Kenney MJ, Claassen DE, Bishop MR, Fels RJ. Regulation of the sympathetic nerve discharge bursting pattern during heat stress. Am J Physiol. 1998 Dec;275(6):R1992–R2001. https://doi.org/10.1152/ajpregu.1998.275.6.R1992. PMID: 9843889.
44. Fink T, Weihe E. Multiple neuropeptides in nerves supplying mammalian lymph nodes: messenger candidates for sensory and autonomic neuroimmunomodulation?. Neurosci Lett. 1988;90(1–2):39–44. https://doi.org/10.1016/0304-3940(88)90783-5
45. Cleypool CGJ, Mackaaij C, Lotgerink Bruinenberg D, Schurink B, Bleys RLAW. Sympathetic nerve distribution in human lymph nodes. J Anat. 2021 Aug;239(2):282–289. https://doi.org/10.1111/joa.13422. Epub 2021 Mar 6. PMID: 33677834; PMCID: PMC8273593.
46. Ackerman KD, Felten SY, Bellinger DL, Felten DL. Noradrenergic sympathetic innervation of the spleen: III. Development of innervation in the rat spleen. J Neurosci Res. 1987;18(1):49–54, 123–5. https://doi.org/10.1002/jnr.490180109. PMID: 3682027.
47. Felten DL, Ackerman KD, Wiegand SJ, Felten SY. Noradrenergic sympathetic innervation of the spleen: I. Nerve fibers associate with lymphocytes and macrophages in specific compartments of the splenic white pulp. J Neurosci Res. 1987;18(1):28–36, 118–21. https://doi.org/10.1002/jnr.490180107. PMID: 3316680.
48. Livnat S, Felten SY, Carlson SL, Bellinger DL, Felten DL. Involvement of peripheral and central catecholamine systems in neural-immune interactions. J Neuroimmunol. 1985 Nov;10(1):5–30. https://doi.org/10.1016/0165-5728(85)90031-1. PMID: 3902888.

49. Straub RH. Complexity of the bi-directional neuroimmune junction in the spleen. Trends Pharmacol Sci. 2004 Dec;25(12):640–6. https://doi.org/10.1016/j.tips.2004.10.007. PMID: 15530642.
50. Felten SY, Olschowka J. Noradrenergic sympathetic innervation of the spleen: II. Tyrosine hydroxylase (TH)-positive nerve terminals form synapticlike contacts on lymphocytes in the splenic white pulp. J Neurosci Res. 1987;18(1):37–48. https://doi.org/10.1002/jnr.490180108. PMID: 2890771.
51. Qiao G, Bucsek MJ, Winder NM, Chen M, Giridharan T, Olejniczak SH, Hylander BL, Repasky EA. β-Adrenergic signaling blocks murine CD8+ T-cell metabolic reprogramming during activation: a mechanism for immunosuppression by adrenergic stress. Cancer Immunol Immunother. 2019 Jan;68(1):11–22. https://doi.org/10.1007/s00262-018-2243-8. Epub 2018 Sep 18. PMID: 30229289; PMCID: PMC6326964.
52. Ganta CK, Blecha F, Ganta RR, Helwig BG, Parimi S, Lu N, Fels RJ, Musch TI, Kenney MJ. Hyperthermia-enhanced splenic cytokine gene expression is mediated by the sympathetic nervous system. Physiol Genomics. 2004 Oct 4;19(2):175–83. https://doi.org/10.1152/physiolgenomics.00109.2004. Epub 2004 Aug 3. PMID: 15292487.
53. Erin N, Korcum AF, Tanrıöver G, Kale Ş, Demir N, Köksoy S. Activation of neuroimmune pathways increases therapeutic effects of radiotherapy on poorly differentiated breast carcinoma. Brain Behav Immun. 2015 Aug;48:174–85. https://doi.org/10.1016/j.bbi.2015.02.024. Epub 2015 Feb 28. PMID: 25736062.
54. Ganta CK, Lu N, Helwig BG, Blecha F, Ganta RR, Zheng L, Ross CR, Musch TI, Fels RJ, Kenney MJ. Central angiotensin II-enhanced splenic cytokine gene expression is mediated by the sympathetic nervous system. Am J Physiol Heart Circ Physiol. 2005 Oct;289(4):H1683–91. https://doi.org/10.1152/ajpheart.00125.2005. Epub 2005 May 20. PMID: 15908469.
55. Podojil JR, Sanders VM. Selective regulation of mature IgG1 transcription by CD86 and beta 2-adrenergic receptor stimulation. J Immunol. 2003 May 15;170(10):5143–51. https://doi.org/10.4049/jimmunol.170.10.5143. PMID: 12734361.
56. Kohm AP, Tang Y, Sanders VM, Jones SB. Activation of antigen-specific CD4+ Th2 cells and B cells in vivo increases norepinephrine release in the spleen and bone marrow. J Immunol. 2000 Jul 15;165(2):725–33. https://doi.org/10.4049/jimmunol.165.2.725. PMID: 10878345.
57. Swanson MA, Lee WT, Sanders VM. IFN-gamma production by Th1 cells generated from naive CD4+ T cells exposed to norepinephrine. J Immunol. 2001 Jan 1;166(1):232–40. https://doi.org/10.4049/jimmunol.166.1.232. Erratum in: J Immunol 2001 Jun 1;166(11):6992. PMID: 11123297.
58. Kohm AP, Sanders VM. Norepinephrine and beta 2-adrenergic receptor stimulation regulate CD4+ T and B lymphocyte function in vitro and in vivo. Pharmacol Rev. 2001 Dec;53(4):487–525. PMID: 11734616.
59. Zhang Y, Guan Z, Reader B, et al. Autonomic dysreflexia causes chronic immune suppression after spinal cord injury. J Neurosci. 2013;33(32):12970–12981. https://doi.org/10.1523/JNEUROSCI.1974-13.2013
60. Frohman EM, Vayuvegula B, Gupta S, van den Noort S. Norepinephrine inhibits gamma-interferon-induced major histocompatibility class II (Ia) antigen expression on cultured astrocytes via beta-2-adrenergic signal transduction mechanisms. Proc Natl Acad Sci U S A. 1988 Feb;85(4):1292–6. https://doi.org/10.1073/pnas.85.4.1292. PMID: 2829222; PMCID: PMC279753.
61. Nilsson MB, Armaiz-Pena G, Takahashi R, Lin YG, Trevino J, Li Y, Jennings N, Arevalo J, Lutgendorf SK, Gallick GE, Sanguino AM, Lopez-Berestein G, Cole SW, Sood AK. Stress hormones regulate interleukin-6 expression by human ovarian carcinoma cells through a Src-dependent mechanism. J Biol Chem. 2007 Oct 12;282(41):29919–26. https://doi.org/10.1074/jbc.M611539200. Epub 2007 Aug 23. PMID: 17716980.
62. Shahzad MM, Arevalo JM, Armaiz-Pena GN, Lu C, Stone RL, Moreno-Smith M, Nishimura M, Lee JW, Jennings NB, Bottsford-Miller J, Vivas-Mejia P, Lutgendorf SK, Lopez-Berestein

G, Bar-Eli M, Cole SW, Sood AK. Stress effects on FosB- and interleukin-8 (IL8)-driven ovarian cancer growth and metastasis. J Biol Chem. 2010 Nov 12;285(46):35462–70. https://doi.org/10.1074/jbc.M110.109579. Epub 2010 Sep 8. Erratum in: J Biol Chem. 2018 Jun 29;293(26):10041. PMID: 20826776; PMCID: PMC2975170.

63. Yang R, Lin Q, Gao HB, Zhang P. Stress-related hormone norepinephrine induces interleukin-6 expression in GES-1 cells. Braz J Med Biol Res. 2014 Feb;47(2):101–9. https://doi.org/10.1590/1414-431X20133346. Epub 2014 Jan 17. PMID: 24519125; PMCID: PMC4051180.
64. Armaiz-Pena GN, Gonzalez-Villasana V, Nagaraja AS, Rodriguez-Aguayo C, Sadaoui NC, Stone RL, Matsuo K, Dalton HJ, Previs RA, Jennings NB, Dorniak P, Hansen JM, Arevalo JM, Cole SW, Lutgendorf SK, Sood AK, Lopez-Berestein G. Adrenergic regulation of monocyte chemotactic protein 1 leads to enhanced macrophage recruitment and ovarian carcinoma growth. Oncotarget. 2015 Feb 28;6(6):4266–73. https://doi.org/10.18632/oncotarget.2887. PMID: 25738355; PMCID: PMC4414188.
65. Kuol N, Stojanovska L, Apostolopoulos V, Nurgali K. Crosstalk between cancer and the neuro-immune system. J Neuroimmunol. 2018 Feb 15;315:15–23. https://doi.org/10.1016/j.jneuroim.2017.12.016. Dantzer, R. (2018). Neuroimmune interactions: From the brain to the immune system and vice versa. Physiol. Rev. 98, 477–504. https://doi.org/10.1152/physrev.00039.2016.
66. Dantzer R. Neuroimmune Interactions: From the Brain to the Immune System and Vice Versa. *Physiol Rev*. 2018;98(1):477–504. https://doi.org/10.1152/physrev.00039.2016
67. Onaga T. Tachykinin: recent developments and novel roles in health and disease. Biomol Concepts. 2014 Jun;5(3):225–43. https://doi.org/10.1515/bmc-2014-0008. PMID: 25372755.
68. Sarkar C, Chakroborty D, Basu S. Neurotransmitters as regulators of tumor angiogenesis and immunity: the role of catecholamines. J Neuroimmune Pharmacol. 2013 Mar;8(1):7–14. https://doi.org/10.1007/s11481-012-9395-7. Epub 2012 Aug 11. PMID: 22886869; PMCID: PMC3869381.
69. St-Pierre S, Jiang W, Roy P, et al. Nicotinic Acetylcholine Receptors Modulate Bone Marrow-Derived Pro-Inflammatory Monocyte Production and Survival. PLoS One. 2016;11(2):e0150230. 1–8 Published 2016 Feb 29. https://doi.org/10.1371/journal.pone.0150230
70. Mohammadpour H, MacDonald CR, Qiao G, Chen M, Dong B, Hylander BL, McCarthy PL, Abrams SI, Repasky EA. β2 adrenergic receptor-mediated signaling regulates the immunosuppressive potential of myeloid-derived suppressor cells. J Clin Invest. 2019 Dec 2;129(12):5537–5552. https://doi.org/10.1172/JCI129502. PMID: 31566578; PMCID: PMC6877316.
71. Hylander BL, Eng JW, Repasky EA. The Impact of Housing Temperature-Induced Chronic Stress on Preclinical Mouse Tumor Models and Therapeutic Responses: An Important Role for the Nervous System. Adv Exp Med Biol. 2017;1036:173–189. https://doi.org/10.1007/978-3-319-67577-0_12. PMID: 29275472; PMCID: PMC9423006.
72. An J, Feng L, Ren J, Li Y, Li G, Liu C, Yao Y, Yao Y, Jiang Z, Gao Y, Xu Y, Wang Y, Li J, Liu J, Cao L, Qi Z, Yang L. Chronic stress promotes breast carcinoma metastasis by accumulating myeloid-derived suppressor cells through activating β-adrenergic signaling. Oncoimmunology. 2021 Nov 23;10(1):2004659. https://doi.org/10.1080/2162402X.2021.2004659. PMID: 34858728; PMCID: PMC8632282.
73. Dai S, Mo Y, Wang Y, Xiang B, Liao Q, Zhou M, Li X, Li Y, Xiong W, Li G, Guo C, Zeng Z. Chronic Stress Promotes Cancer Development. Front Oncol. 2020 Aug 19;10:1492. https://doi.org/10.3389/fonc.2020.01492. PMID: 32974180; PMCID: PMC7466429.
74. Cole SW, Korin YD, Fahey JL, Zack JA. Norepinephrine accelerates HIV replication via protein kinase A-dependent effects on cytokine production. J Immunol. 1998 Jul 15;161(2):610–6. PMID: 9670934.
75. Kokolus KM, Spangler HM, Povinelli BJ, Farren MR, Lee KP, Repasky EA. Stressful presentations: mild cold stress in laboratory mice influences phenotype of dendritic cells in naïve and tumor-bearing mice. Front Immunol. 2014 Feb 10;5:23. https://doi.org/10.3389/fimmu.2014.00023. PMID: 24575090; PMCID: PMC3918933.

76. Sloan EK, Priceman SJ, Cox BF, Yu S, Pimentel MA, Tangkanangnukul V, Arevalo JM, Morizono K, Karanikolas BD, Wu L, Sood AK, Cole SW. The sympathetic nervous system induces a metastatic switch in primary breast cancer. Cancer Res. 2010 Sep 15;70(18):7042–52. https://doi.org/10.1158/0008-5472.CAN-10-0522. Epub 2010 Sep 7. PMID: 20823155; PMCID: PMC2940980.
77. Mundy-Bosse BL, Thornton LM, Yang HC, Andersen BL, Carson WE. Psychological stress is associated with altered levels of myeloid-derived suppressor cells in breast cancer patients. Cell Immunol. 2011;270(1):80–7. https://doi.org/10.1016/j.cellimm.2011.04.003. Epub 2011 Apr 23. PMID: 21600570; PMCID: PMC3129455.
78. Carlson AB, Kraus GP. Physiology, Cholinergic Receptors. 2022 Aug 22. In: StatPearls [Internet]. Treasure Island (FL): StatPearls Publishing; 2022 Jan–. PMID: 30252390.
79. Rosas-Ballina M, Olofsson PS, Ochani M, Valdés-Ferrer SI, Levine YA, Reardon C, Tusche MW, Pavlov VA, Andersson U, Chavan S, Mak TW, Tracey KJ. Acetylcholine-synthesizing T cells relay neural signals in a vagus nerve circuit. Science. 2011 Oct 7;334(6052):98–101. https://doi.org/10.1126/science.1209985. Epub 2011 Sep 15. PMID: 21921156; PMCID: PMC4548937.
80. Elkhatib SK, Case AJ. Autonomic regulation of T-lymphocytes: Implications in cardiovascular disease. Pharmacol Res. 2019 Aug;146:104293. https://doi.org/10.1016/j.phrs.2019.104293. Epub 2019 Jun 6. PMID: 31176794; PMCID: PMC6679768.
81. Teratani T, Mikami Y, Nakamoto N, Suzuki T, Harada Y, Okabayashi K, Hagihara Y, Taniki N, Kohno K, Shibata S, Miyamoto K, Ishigame H, Chu PS, Sujino T, Suda W, Hattori M, Matsui M, Okada T, Okano H, Inoue M, Yada T, Kitagawa Y, Yoshimura A, Tanida M, Tsuda M, Iwasaki Y, Kanai T. The liver-brain-gut neural arc maintains the Treg cell niche in the gut. Nature. 2020 Sep;585(7826):591–596. https://doi.org/10.1038/s41586-020-2425-3. Epub 2020 Jun 11. PMID: 32526765.
82. Renz BW, Tanaka T, Sunagawa M, Takahashi R, Jiang Z, Macchini M, Dantes Z, Valenti G, White RA, Middelhoff MA, Ilmer M, Oberstein PE, Angele MK, Deng H, Hayakawa Y, Westphalen CB, Werner J, Remotti H, Reichert M, Tailor YH, Nagar K, Friedman RA, Iuga AC, Olive KP, Wang TC. Cholinergic Signaling via Muscarinic Receptors Directly and Indirectly Suppresses Pancreatic Tumorigenesis and Cancer Stemness. Cancer Discov. 2018 Nov;8(11):1458–1473. https://doi.org/10.1158/2159-8290.CD-18-0046. Epub 2018 Sep 5. PMID: 30185628; PMCID: PMC6214763.
83. Partecke LI, Käding A, Trung DN, Diedrich S, Sendler M, Weiss F, Kühn JP, Mayerle J, Beyer K, von Bernstorff W, Heidecke CD, Keßler W. Subdiaphragmatic vagotomy promotes tumor growth and reduces survival via TNFα in a murine pancreatic cancer model. Oncotarget. 2017 Apr 4;8(14):22501–22512. https://doi.org/10.18632/oncotarget.15019. PMID: 28160574; PMCID: PMC5410240.
84. Dubeykovskaya Z, Si Y, Chen X, Worthley DL, Renz BW, Urbanska AM, Hayakawa Y, Xu T, Westphalen CB, Dubeykovskiy A, et al. Neural innervation stimulates splenic TFF2 to arrest myeloid cell expansion and cancer. Nat Commun. 2016;7:10517. https://doi.org/10.1038/ncomms10517.
85. Anderson P, Gonzalez-Rey E. Vasoactive Intestinal Peptide Induces Cell Cycle Arrest and Regulatory Functions in Human T Cells at Multiple Levels. Mol Cell Biol. 2010;30:2537–2551. https://doi.org/10.1128/MCB.01282-09.
86. Gonzalez-Rey E, Delgado M. Vasoactive intestinal peptide and regulatory T-cell induction: a new mechanism and therapeutic potential for immune homeostasis. Trends Mol Med. 2007;13:241–251. https://doi.org/10.1016/j.molmed.2007.04.003.
87. Zhang B, Vogelzang A, Miyajima M, Sugiura Y, Wu Y, Chamoto K, Nakano R, Hatae R, Menzies RJ, Sonomura K, et al. B cell-derived GABA elicits IL-10+ macrophages to limit anti-tumour immunity. Nature. 2021;599:471–476. https://doi.org/10.1038/s41586-021-04082-1.
88. Ravindranathan S, Passang T, Li JM, Wang S, Dhamsania R, Ware MB, et al. Targeting vasoactive intestinal peptide-mediated signaling enhances response to immune checkpoint therapy in pancreatic ductal adenocarcinoma. Nat Commun. 2022;13:6418. https://doi.org/10.1038/s41467-022-34242-4.

89. Beatty GL, Gladney WL. Immune escape mechanisms as a guide for cancer immunotherapy. Clin Cancer Res. 2015;21:687–692. https://doi.org/10.1158/1078-0432.CCR-14-1860.
90. Ravindranathan S, Passang T, Li JM, Wang S, Dhamsania R, Ware MB, Zaidi MY, Zhu J, Cardenas M, Liu Y, et al. Targeting vasoactive intestinal peptide-mediated signaling enhances response to immune checkpoint therapy in pancreatic ductal adenocarcinoma. Nat Commun. 2022;13. https://doi.org/10.1038/s41467-022-34242-4.
91. Jiang SH, Hu LP, Wang X, Li J, Zhang ZG. Neurotransmitters: emerging targets in cancer. Oncogene. 2020;39:503–515. https://doi.org/10.1038/s41388-019-1052-3.
92. Jung HY, Yang SD, Ju W, Ahn JH. Aberrant epigenetic regulation of GABRP associates with aggressive phenotype of ovarian cancer. Exp Mol Med. 2017;49:e335. https://doi.org/10.1038/emm.2017.16.
93. Kanbara K, Hirasawa N, Ueno K, et al. GABAB receptor regulates proliferation in the high-grade chondrosarcoma cell line OUMS-27 via apoptotic pathways. BMC Cancer. 2018;18:263. https://doi.org/10.1186/s12885-018-4173-3.
94. Maemura K, Nishimura J, Kubota E, et al. Gamma-aminobutyric acid immunoreactivity in intramucosal colonic tumors. J Gastroenterol Hepatol. 2003;18(9):1089–1094. https://doi.org/10.1046/j.1440-1746.2003.03160.x
95. Liu Y, Li Y, Liang Y, et al. Gamma-aminobutyric acid promotes human hepatocellular carcinoma growth through overexpressed gamma-aminobutyric acid A receptor α3 subunit. World J Gastroenterol. 2008;14(46):7175–7182. https://doi.org/10.3748/wjg.14.7175
96. Liu Y, Zhang H, Wang Z, Wu P, Gong W. 5-Hydroxytryptamine1a receptors on tumour cells induce immune evasion in lung adenocarcinoma patients with depression via autophagy/pSTAT3. Eur J Cancer. 2019;114:8–24. https://doi.org/10.1016/j.ejca.2019.02.019
97. Karmakar S, Lal G. Role of serotonin receptor signaling in cancer cells and anti-tumor immunity. Theranostics. 2021;11(11):5296–5312. https://doi.org/10.7150/thno.55986
98. He X, Xu C. Immune checkpoint signaling and cancer immunotherapy. Cell Res. 2020;30(8):660–669. https://doi.org/10.1038/s41422-020-0343-4
99. Sharpe AH, Pauken KE. The diverse functions of the PD1 inhibitory pathway. Nat Rev Immunol. 2018;18(3):153–167. https://doi.org/10.1038/nri.2017.108
100. Shi S, Han Y, Wang D, Guo P, Wang J, Ren T, Wang W. PD-L1 and PD-1 expressed in trigeminal ganglia may inhibit pain in an acute migraine model. Cephalalgia. 2020;40(3):288–298. https://doi.org/10.1177/0333102420974199
101. Meerschaert KA, Edwards BS, Epouhe AY, et al. Neuronally expressed PDL1, not PD1, suppresses acute nociception. Brain Behav Immun. 2022;106:233–246. https://doi.org/10.1016/j.bbi.2022.09.001
102. Mo RJ, Han ZD, Liang YK, et al. Expression of PD-L1 in tumor-associated nerves correlates with reduced CD8+ tumor-associated lymphocytes and poor prognosis in prostate cancer. Int J Cancer. 2019;144(13):3099–3110. https://doi.org/10.1002/ijc.32061
103. Kamiya A, Hayama Y, Kato S, et al. Genetic manipulation of autonomic nerve fiber innervation and activity and its effect on breast cancer progression. Nat Neurosci. 2019;22(8):1289–1305. https://doi.org/10.1038/s41593-019-0430-3
104. Chen M, Qiao G, Hylander BL, Mohammadpour H, Wang XY, Subjeck JR, Singh AK, Repasky EA. Adrenergic stress constrains the development of anti-tumor immunity and abscopal responses following local radiation. Nat Commun. 2020;11(1):1821. https://doi.org/10.1038/s41467-020-15676-0. PMID: 32286326; PMCID: PMC7156731.
105. Qiao G, Bucsek MJ, Winder NM, Chen M, Giridharan T, Olejniczak SH, Hylander BL, Repasky EA. β-Adrenergic signaling blocks murine CD8+ T-cell metabolic reprogramming during activation: a mechanism for immunosuppression by adrenergic stress. Cancer Immunol Immunother. 2019;68(1):11–22. https://doi.org/10.1007/s00262-018-2243-8. PMID: 30229289; PMCID: PMC6326964.
106. Jessen KR, Mirsky R. The success and failure of the Schwann cell response to nerve injury. Front Cell Neurosci. 2019;13:33. https://doi.org/10.3389/fncel.2019.00033.

107. Deborde S, Wong RJ. The Role of Schwann Cells in Cancer. Adv Biol. 2022;6:2200089. https://doi.org/10.1002/adbi.202200089.
108. Huang T, Fan Q, Wang Y, Cui Y, Wang Z, Yang L, Sun X, Wang Y. Schwann Cell-Derived CCL2 Promotes the Perineural Invasion of Cervical Cancer. Front Oncol. 2020;10:19. https://doi.org/10.3389/fonc.2020.00019.
109. Shurin GV, Kruglov O, Ding F, Lin Y, Hao X, Keskinov AA, You Z, Lokshin AE, LaFramboise WA, Falo LD Jr, et al. Melanoma-induced reprogramming of Schwann cell signaling aids tumor growth. Cancer Res. 2019;79(11):2736–2747. https://doi.org/10.1158/0008-5472.CAN-18-3872.
110. Oltz EM. Neuroimmunology: To Sense and Protect. J Immunol. 2020;204(2):239–240. https://doi.org/10.4049/jimmunol.1990024.
111. Martyn GV, Shurin GV, Keskinov AA, Bunimovich YL, Shurin MR. Schwann cells shape the neuro-immune environs and control cancer progression. Cancer Immunol Immunother. 2019;68(11):1819–1829. https://doi.org/10.1007/s00262-018-02296-3.
112. Urban-Wojciuk Z, Khan MM, Oyler BL, Fåhraeus R, Marek-Trzonkowska N, Nita-Lazar A, Hupp TR, Goodlett DR. The role of TLRs in anti-cancer immunity and tumor rejection. Front Immunol. 2019;10:2388. https://doi.org/10.3389/fimmu.2019.02388.
113. Yang D, Han Z, Oppenheim JJ. Alarmins and immunity. Immunol Rev. 2017;280:41–56. https://doi.org/10.1111/imr.12577.
114. Im JS, Tapinos N, Chae GT, Illarionov PA, Besra GS, DeVries GH, Modlin RL, Sieling PA, Rambukkana A, Porcelli SA. Expression of CD1d molecules by human Schwann cells and potential interactions with immunoregulatory invariant NK T cells. J Immunol. 2006;177:5226–5235. https://doi.org/10.4049/jimmunol.177.8.5226.
115. Van Rhijn I, Van Den Berg LH, Bosboom WMJ, Otten HG, Logtenberg T. Expression of accessory molecules for T-cell activation in peripheral nerve of patients with CIDP and vasculitic neuropathy. Brain. 2000;123:2020–2029. https://doi.org/10.1093/brain/123.10.2020.
116. Murata KY, Dalakas MC. Expression of the co-stimulatory molecule BB-1, the ligands CTLA-4 and CD28 and their mRNAs in chronic inflammatory demyelinating polyneuropathy. Brain. 2000;123:1660–1666. https://doi.org/10.1093/brain/123.8.1660.
117. Bucsek MJ, Qiao G, MacDonald CR, Giridharan T, Evans L, Niedzwecki B, Liu H, Kokolus KM, Eng JWL, Messmer MN, et al. β-Adrenergic signaling in mice housed at standard temperatures suppresses an effector phenotype in CD8+ T cells and undermines checkpoint inhibitor therapy. Cancer Res. 2017;77:5639–5651. https://doi.org/10.1158/0008-5472.CAN-17-0546.
118. Gandhi S, Pandey MR, Attwood K, Ji W, Witkiewicz AK, Knudsen ES, Allen C, Tario JD, Wallace PK, Cedeno CD, et al. Phase I clinical trial of combination propranolol and pembrolizumab in locally advanced and metastatic melanoma: safety, tolerability, and preliminary evidence of antitumor activity. Clin Cancer Res. 2021;27:87–95. https://doi.org/10.1158/1078-0432.CCR-20-2381.
119. Kennedy OJ, Kicinski M, Valpione S, Gandini S, Suciu S, Blank CU, Long GV, Atkinson VG, Dalle S, Haydon AM, et al. Prognostic and predictive value of β-blockers in the EORTC 1325/KEYNOTE-054 phase III trial of pembrolizumab versus placebo in resected high-risk stage III melanoma. Eur J Cancer. 2022;165:97–112. https://doi.org/10.1016/j.ejca.2022.01.017.
120. Galoş EV, Tat TF, Popa R, Efrimescu CI, Finnerty D, Buggy DJ, Ionescu DC, Mihu CM. Neutrophil extracellular trapping and angiogenesis biomarkers after intravenous or inhalation anaesthesia with or without intravenous lidocaine for breast cancer surgery: a prospective, randomised trial. Br J Anaesth. 2020;125(5):712–721. https://doi.org/10.1016/j.bja.2020.05.003.
121. Tsai W, Morielli AD, Peralta EG. The m1 muscarinic acetylcholine receptor transactivates the EGF receptor to modulate ion channel activity. EMBO J. 1997;16(15):4597–4605. https://doi.org/10.1093/emboj/16.15.4597.

Chapter 7
Neuroimmunoregulation of Cancer: The Case for Multiple Myeloma

Sheeba Ba Aqeel, Caitlin James, Jens Hillengass, and Elizabeth Repasky

Introduction

Cancer cells usually acquire various forms of antigens that distinguish them from the normal cell precursors. We know that the surveilling immune system is capable of detecting and eliminating cancerous cells based upon the expression of these antigens; however, when cancer cells are able to evade immune detection or killing, they are able to continue to divide and progress into a malignancy. This is the concept underlying the basis of immuno-oncology and its goal of stimulating antitumor immunity to eliminate tumors and prevent their future recurrence [1, 2]. Whether stress can influence this balance between effective surveillance by the immune system and the ability of tumors to evade immunity has been of long-standing interest to researchers [3–5].

The "fight-or-flight" mechanism in which the brain perceives stress, warns of danger, and enables an organism to deal with the consequences is accomplished through the release of stress-responsive neurotransmitters and hormones [6]. This mechanism gets the body ready for exogenous harm, whether injury or infection as well as primes the immune system to fight against cancer. While the field of "psychoneuroimmunology" has revealed many negative impacts of stress on the immune system and on cancer progression, many mechanistic details remain undefined [5, 7, 8].

It is generally agreed that any physical or psychological stimuli that disrupt cellular or systemic homeostasis result in various degrees of a "stress response" that can be seen in biological, psychological, and physiological changes [9]. A stress

S. B. Aqeel · J. Hillengass
Department of Medicine, Roswell Park Comprehensive Cancer Center, Buffalo, NY, USA

C. James · E. Repasky (✉)
Department of Immunology, Roswell Park Comprehensive Cancer Center, Buffalo, NY, USA
e-mail: Elizabeth.repasky@roswellpark.org

M. Amit, N. N. Scheff (eds.), *Cancer Neuroscience*,
https://doi.org/10.1007/978-3-031-32429-1_7

response can occur in cells and even subcellular organelles in response to molecular changes in the microenvironment, or it can involve the entire organism, including a dynamic and interdependent activation of the brain and peripheral nervous systems, endocrine and immune responses, vascular and muscular activity, and more. The precise mechanistic integrations that occur between immune surveillance and cancer and how they are each affected by these stress responses remain unclear, although considerable progress has been made in the past 10–15 years. As these interactions begin to emerge, there is considerable excitement that the study of stress responses could provide new avenues for therapeutic targets in a variety of different cancers.

From an oncological vantage point, the dynamic neurobiochemical pathways associated with chronic stress via increased autonomic nerve conduction in β2-adrenergic receptors (AR) and β3-AR signaling have been repeatedly observed to directly and indirectly contribute to the hallmarks of cancer [10–13]. This includes significant alterations within the tumor microenvironment (TME) to promote tumor growth and metastasis while driving the immune system toward an exhausted immunophenotype which ultimately compromises antitumoral immunity and the effectiveness of immune-modulating therapies [12, 14–21]. We have previously shown that CD8+ T cells, which are known to express β2-AR on their surface, exhibit decreased glycolytic metabolism and mitochondrial respiration when treated with the pan-β-AR agonist, isoproterenol. Additionally, CD8+ T cells isolated from the tumors of mice treated with the nonselective beta-adrenergic antagonist, propranolol, exhibited increased mitochondrial respiration and decreased surface expression of key surface proteins indicative of T cell exhaustion. Thus, β2-AR signaling in CD8+ T cells results in impaired antitumor activity [17, 22]. Conversely, our group has also shown that β-AR signaling in myeloid derived suppressor cells (MDSCs) increases their immunosuppressive activity. In vitro treatment with isoproterenol resulted in greater suppression of CD4+ and CD8+ T cell proliferation. In vivo, global β2-AR knockout mice exhibited decreased accumulation of MDSCs within tumors compared to WT mice. Additionally, adoptive transfer of β2-AR$^{-/-}$ MDSCs in WT tumor-bearing mice resulted in better tumor control compared to WT MDSCs, together demonstrating a role for β2-AR in MDSC pro-tumorigenic activity [21].

Interestingly, blockade of β-AR pathways (i.e., beta-blockers) for non-oncological purposes have been observed to improve overall survival and response to treatment in patients with melanoma, lung, breast, and prostate cancer [23–28]. Because patients with melanoma are among those in which retrospective studies demonstrated an improved survival in those taking beta-blockers for non-oncological reasons [23], and because preclinical studies demonstrated a significantly improved tumor control and antitumor immunity through stress reduction or use of nonselective beta-blocker propranolol [17, 21, 29–32], we conducted the first prospective randomized clinical trial observing the effectiveness of adding propranolol to a standard-of-care immune checkpoint inhibitor (ICI), pembrolizumab, in patients with treatment-naïve unresectable or metastatic melanoma (NCT03384836) [33, 34]. Although a small phase I study, this trial resulted in a very encouraging response rate and a phase II trial is underway.

As described below, multiple myeloma (MM) is also a cancer in which there appears to be beneficial effects in patients who were administered propranolol in a randomized controlled biomarker trial [35]. Based on the promising results from this work, we have initiated a new clinical trial, NCT05312255, to prospectively combine propranolol with other therapies in newly diagnosed MM patients. This review will summarize some of the available information regarding the impact of stress in MM that should encourage the development of other new clinical trials that test strategies for stress reduction and/or the use of pharmacological agents to block the biological and immunological impact of nerve-produced stress-hormone signaling in tumor cells and in cells of the immune system.

A Role for Stress in the Progression of Multiple Myeloma

Multiple myeloma (MM) is a disorder of clonal plasma cells with a median age at diagnosis of around 69 years and accounts for ~10% of all newly diagnosed hematological malignancies. Currently, MM remains incurable and a frequently relapsing disease despite incorporation of novel agents into upfront therapy and maintenance therapies that have led to a sustained improvement in the overall survival in these patients [36]. Immune dysregulation is hypothesized to be an important part of the pathogenesis and disease progression in MM [37]. Among the most revolutionary treatments used in this disease include the immune-based therapies targeting the malignant cell clone in a multifaceted approach [38, 39]. At first diagnosis, 80% of MM patients suffer from bone destruction and 90% during the course of their disease, and pain is, along with fatigue and side effects of therapy, one of the most important factors influencing the quality of life of MM patients [40, 41]. Therefore, quality of life in patients with long periods of disease- and treatment-free intervals has become an important concern of these patients and their physicians. Impaired quality of life leads to a great deal of psychological stress in MM as well as other cancer patients [42–44]. Studies have shown that even patients with monoclonal gammopathy of undetermined significance (MGUS) and smoldering multiple myeloma (SMM), precancerous conditions of MM, show high levels of depression, anxiety, and distress [45]. The autonomic nervous system, which includes the sympathetic system that regulates the fight-or-flight response via β-adrenergic signaling, is well known to be activated in both acute and chronic stress states [46]. Bone marrow, the main organ infiltrated by MM cells, is highly innervated by both sympathetic and parasympathetic nerve fibers [47]. In addition to MM cells, mesenchymal stem cells (MSCs) and other cells residing in the hematopoietic bone marrow (BM) are exposed to and regulated by adrenergic (sympathetic) and cholinergic (parasympathetic) signals [47–49]. The adrenergic signals are transmitted by norepinephrine (NE) that binds to the β-ARs [50]. In addition to effects of the autonomic nervous system, other nerve types, particularly sensory nerves, have more recently been identified as having a role in tumor progression and even therapeutic resistance in various solid tumors and MM [51–54].

Multiple Myeloma and Bone Marrow Immune Microenvironment

The BM microenvironment in MM comprises of both hematopoietic and non-hematopoietic cells as well as noncellular elements, such as extracellular matrix and soluble factors [37]. The interactions between myeloma cells and this environment are highly complex and only limitedly understood. MSCs are an important part of the non-hematopoietic BM microenvironment that release cytokines such as interleukin 6 that have an established role in tumorigenesis and progression of MM. Some of the major changes that occur in the BM even during the precursor stages of MM include a selective reduction in cytotoxic memory T cells and increase in immune suppressive populations that include regulatory T cells (Tregs) and MDSCs [55, 56].

Tregs have been recognized as crucial players in the mechanisms of immune escape in various tumors [57]; however, this role is not well understood in MM. It has been elucidated that in mouse models, Tregs accumulate around the MM microenvironment within the BM. Moreover, an apparent correlation has been indicated with minimal residual disease (MRD) status, with increased presence of circulating Tregs in MRD positive patients and decreased frequency in MRD negative patients after treatment [58, 59]. Danzinger et al. studied the BM tumor microenvironment using gene expression-based computational techniques and found that those patients whose BM microenvironment signatures were found to be depleted of granulocytes, adipocytes, osteoclasts, eosinophils, mast cells, neutrophils, monocytes, macrophages, and NK cells demonstrated poor PFS [60]. They also demonstrated a change in the dynamics of the microenvironment with treatment. Patients whose BM microenvironments showed enriched granulocytic signatures also demonstrated increased PFS compared to those who presented with low granulocyte signatures. This raises the question of whether changes in granulocyte signatures in the BM microenvironment could serve as a biomarker to identify those that are at risk for worse outcomes.

MDSCs are immature myeloid cells that share some characteristics with monocytes and neutrophils but have distinct functional differences. Unlike monocytes or neutrophils, MDSCs release soluble immunosuppressive factors that inhibit T cells and promote angiogenesis and metastasis. As such, they are known to positively correlate with tumor progression [61, 62]. Likewise, tumor associated macrophages (TAMs) infiltrate the BM in MM and promote angiogenesis [63]. NK and CD8+ T cells play an important role in anti-myeloma activity by killing myeloma cells via perforins, granzyme, and interferon gamma [64]. They also potentiate the action of important first line therapies in MM such as proteasome inhibitors and immunomodulatory drugs [65]. Dendritic cells – both conventional and plasmacytoid – have been shown to promote growth and survival of healthy plasma and myeloma cells via CD28 and CD80/CD86 pathway and interleukin 6 release [66, 67].

Neural Regulation of the Immune Microenvironment in Multiple Myeloma

The immune, endocrine, and both central and peripheral nervous systems are highly integrated. This integration is greatly emphasized under stress conditions that involve the sympathetic nervous system (SNS) and release of neurotransmitters that act on the immune system [68, 69]. Many in vitro and in vivo studies have demonstrated that catecholamines promote the proliferation and metastasis of numerous cancers including melanoma and cancers of the ovaries, breast, liver, pancreas, and prostate [11, 70–72]. Conversely, inhibition of sympathetic activity decreases tumor growth by inhibiting various pro-tumorigenic signaling pathways.

Interaction between the SNS and the TME is complex and involves regulation of gene expression and cellular function through a variety of mechanisms [73, 74]. The direct effects of the SNS on the TME are mediated by catecholamine neurotransmitters like epinephrine and NE that are released within the TME and engage adrenergic receptors that are expressed on many types of tumor cells and their surrounding stromal components such as TAMs and vascular endothelial cells [49]. Epinephrine is released from the adrenal gland and circulates systemically to the TME through the vasculature, whereas NE is primarily released from sympathetic nerve fibers within the TME, which generally associate with the vasculature and can radiate dendritic fibers into the tumor parenchyma [49]. The indirect effects of SNS on the TME are mediated by the release of catecholamines and neurotransmitters into tissue sites that regulate systemic biological processes and regulate immune cell development (e.g., myelopoiesis in the BM and spleen, lymphocyte differentiation in secondary lymphoid organs such as the spleen and lymph nodes) [75, 76] and trafficking (e.g., monocyte/macrophage recruitment via chemokines such as MCP-1/CCL2 and growth factors such as M-CSF/CSF1), in addition to systemic metabolic and hormonal regulators of tumor growth (e.g., glucose mobilization from the liver and circulating adipokines from white adipose tissue) [77–79]. These multiple regulatory pathways allow the SNS to exert highly pleiotropic effects on tumor progression and metastasis of many solid, epithelial tumors (e.g., breast, prostate, ovary, lung, gastric, pancreas), as well as hematologic malignancies via innervation of lymphoid organs such as the BM, spleen, and lymph nodes [32, 78, 80–82]. MM relies heavily on the BM microenvironment for growth and survival. Recent studies have shown that the autonomic nervous system plays an integral role in the alteration of the BM microenvironment [48, 83].

It has been demonstrated that elevated levels of NE impair dendritic cell maturation. Chronic β2-AR signaling suppresses antitumor CD8+ T cell function and increases populations of MDSCs and Tregs in the spleen and TME. It has been shown in β2-AR-deficient mice that chronic stress-mediated β2-adrenergic signaling, induced by sub-thermoneutral housing temperatures, decreases MDSC-dependent tumor growth. The same group further showed that β2-AR activation increased MDSC accumulation in the TME and tumor vascularization in these models [21]. Alternatively, reducing thermal stress or blocking β-AR signaling in mice significantly improved the efficacy of immunotherapy, chemotherapy, and radiation [22, 29, 30].

Preclinical studies have also shown that MDSCs are metabolically regulated via both β2-AR and fatty acid oxidation (FAO) signaling. It was demonstrated that with increase in tumor burden, there is increase in β2-AR mediated signaling of MDSCs, thereby further upregulating β2-AR expression and in turn creating a positive feedback loop. It has been demonstrated that inhibition of the FAO pathway decreases MDSC populations in tumor-bearing mice. This is of importance because oxidative phosphorylation and FAO are key metabolic pathways that drive the immunosuppressive functions of myeloid cells including TAMs in the TME [21, 84, 85].

Sympathetic signaling also controls the egress of stem cells from BM niches into the peripheral bloodstream. This is prudent to MM, as autologous stem cell transplantation is one of the mainstays of systemic treatment. The adrenergic activation of β3 receptor on bone marrow stromal cells caused reduced expression of chemokine receptor ligand CXCL12 providing a crucial mechanism for retention of these cells in the BM [86]. This ligand has therefore lately been of great interest as a therapeutic target [87, 88].

Sensory nerves have been shown to play a role in solid tumor development and progression, including in pancreatic cancer and melanoma models [54, 89]. However, while MM is not a solid tumor and therefore cannot be directly innervated, its site of origin, the BM, is known to be highly innervated with sensory nerves [51, 90]. Interestingly, Olechnowicz et al. showed that culturing of MM cell lines with murine pre-osteoblasts increased the protein and mRNA expression of nerve growth factor (NGF), suggesting that MM may be able to further promote innervation of the BM. In vivo, the same group showed that treatment of MM tumor-bearing mice resulted in decreased serum levels of NGF upon treatment with a drug known to reduce MM tumor burden, further demonstrating a pro-tumorigenic role for sympathetic and sensory nerves in MM development and progression. MM patients also commonly suffer bone disease and bone pain [91]. While the mechanism is not fully understood, evidence exists that this may, in addition to the pathological observation of osteolytic bone lesions, be a result of the recruitment and activation of sensory nerves by MM cells in the bone microenvironment [51].

Systemic Factors Influencing the Immune Microenvironment

Acute fight-or-flight responses increase heart rate and beat strength by activating β1-adrenergic receptors in the heart muscle, redistribute blood from superficial tissues to long muscles by activating vascular α1- and β2-adrenergic receptors, increase respiratory rate and depth by activating bronchial α1- and β2-adrenergic receptors, mobilize energy by activating β2- and β3-adrenergic receptors in adipose tissue and the liver, and mobilize leukocytes (especially NK cells) into the circulation by activating β2-adrenergic receptors on leukocytes [69].

Unlike other stress-activated neuroendocrine systems such as the hypothalamus-pituitary-adrenal axis, the SNS is easily activated by the mere anticipation of threat [92]. Experimental studies in animal models have shown that chronic social stress can also increase the growth and branching of sympathetic nerve fibers in peripheral tissues (neo-innervation) and thereby upregulate basal activity of target tissue adrenergic receptors and downstream molecular mobility and defense programs [69, 92, 93]. Epinephrine levels in the plasma can spike by >ten-fold during acute fight-or-flight stress responses, leading to rapid physiological changes in cardiovascular, respiratory, muscular, metabolic, neural, immune, and other functions that typically return to baseline within 20–60 min following the abatement of perceived threat [46]. These rapid physiological alterations generally involve posttranslational modifications of protein function that are mediated by activation of two broad classes of adrenergic receptors, α and β, each consisting of multiple receptor subtypes that are differentially expressed across tissue sites with signal transduction pathways that induce distinct molecular effects [69].

Mental health conditions like depression denote a state of chronic stress that has been shown to elicit changes within the TME driven by the SNS [94]. Chronic stress and depression are highly prevalent in cancer patients [95] and have more recently been examined in myeloma patients as well [79]. Patients with MM have improved survival outcomes with current immunological and targeted treatments; however, they still suffer from various skeletal effects of this disease for prolonged periods of time, leading to chronic pain and poor mobility [41]. This significantly reduces their quality of life and inevitably affects their mental health. Lutgendorf et al. first demonstrated a link between stress and tumor progression in patients with ovarian cancer experiencing a high degree of social isolation, who were found to have elevated levels of NE in their tumors [94]. Following this, Cheng et al. showed that prostate cancer samples from depressed patients have increased MDSC differentiated cells recruited to the tumor, as well as increased IL-6 mRNA and protein levels within the tumor tissue, suggesting that adrenergic blockade and neuropeptide-y inhibition could be a helpful strategy in depressed prostate cancer patients [96]. There have been several reports showing beneficial effects of exercise and nutrition on immune cells in both healthy individuals and cancer patients [97]. It is well known that physical activity contributes to a stronger immune environment. Exercise in older adults and patient with hematological disorders was studied by Sitlinger et al. [98] and Duggal et al. [99] They demonstrated that physical activity was inversely correlated with proportion of senescent CD4+ and CD8+ T cells [100]. NK cells that play a critical role in augmentation of treatment of MM as mentioned earlier have shown to be positively influenced by exercise [101]. A similar effect is seen on monocytes and macrophages that help in reducing inflammation [102].

Therapeutic Alteration of the Immune Microenvironment: Role of Beta-Blockers

It was previously shown that the use of beta-blockers (BB) for other medical issues was associated with longer survival in melanoma patients who were receiving immunotherapy including checkpoint blockade [23]. Treatment of tumor-bearing mice housed in standard, cool ambient temperatures with propranolol has been shown to significantly increase the response to checkpoint blockade immunotherapy, indicating that blocking adrenergic receptors may reduce the immunosuppressive effects of adrenergic signaling [103]. Hwa et al. performed a retrospective analysis of the use of BBs in patients with newly diagnosed MM. They found that patients who received BB medication (3-month cutoff) had a statistically significant reduction in disease specific death compared to patients taking non-BB cardiac medications (HR 0.49, 95% CI 0.38–0.63, $P < 0.0001$) and patients not on any cardiac medications (HR 0.53, 95% CI 0.42–0.67,

$P < 0.0001$) [13]. Another retrospective analysis on patients enrolled in a phase II clinical trial assessed the efficacy of pomalidomide in patients with refractory or relapsed MM [104]. In this analysis, the investigators found a statistically significant difference in progression-free survival between patients receiving a BB and patients not on BB therapy (10.9 vs. 6.1 months $p = 0.0168$). Knight et al. performed a phase II trial evaluating the effects of propranolol on biomarker expression in patients with MM. They found that on day 28, patients with propranolol had relatively decreased expression of myeloid progenitor-containing CD33+ cell-associated gene transcripts [35]. The expression of myeloid antigens such as CD33 and CD13 in neoplastic plasma cells has also been associated with shorter progression-free survival [105]. Similarly, Nair et al., utilizing MM patient samples, demonstrated that β-AR signaling regulates hematopoietic stem cell fate and MM oncogene expression. In the same study, in vitro blockade of β-AR with propranolol increased the sensitivity of patient samples to the proteasome inhibitor, bortezomib, and the BCL-2 antagonist, Venetoclax [106]. Further prospective and randomized trials are needed to elucidate the role of BB in improving overall survival in patients with MM.

Conclusions

There is increasing interest in understanding the neuroimmunoregulation of cancers. MM is a cancer of cells in the immune system, and therefore it is prudent to gain deeper knowledge of the specific pathways by which stress and nerve activation regulate the progression of this cancer, as well as the BM microenvironment. Psychosocial stressors affect patients with MM and other cancers in several ways that can impact not only their quality of life but also survival and mortality. Understanding the neural regulation in MM offers a promising scope for the development of agents targeting the SNS and its regulation of the TME. Prospective and randomized clinical trials are needed to establish a clear and beneficial role of β-adrenergic receptor blockade in reducing mortality in patients with MM and to prevent its recurrence after treatment.

References

1. Franklin, M. R.; Platero, S.; Saini, K. S.; Curigliano, G.; Anderson, S. Immuno-oncology trends: preclinical models, biomarkers, and clinical development. *Journal for ImmunoTherapy of Cancer* **2022**, *10* (1).
2. Hiam-Galvez, K. J.; Allen, B. M.; Spitzer, M. H. Systemic immunity in cancer. *Nature reviews cancer* **2021**, *21* (6), 345-359.
3. Dantzer, R.; Kelley, K. W. Stress and immunity: an integrated view of relationships between the brain and the immune system. *Life sciences* **1989**, *44* (26), 1995-2008.
4. Yaman, I.; Çobanoğlu, D. A.; Xie, T.; Ye, Y.; Amit, M. Advances in understanding cancer-associated neurogenesis and its implications on the neuroimmune axis in cancer. *Pharmacology & Therapeutics* **2022**, 108199.
5. Kiecolt-Glaser, J. K.; Glaser, R. Psychoneuroimmunology and cancer: fact or fiction? *European Journal of Cancer* **1999**, *35* (11), 1603-1607.
6. Dhabhar, F. S. Acute stress enhances while chronic stress suppresses skin immunity: the role of stress hormones and leukocyte trafficking. *Annals of the New York Academy of Sciences* **2000**, *917* (1), 876-893.
7. Borniger, J. C. Cancer as a tool for preclinical psychoneuroimmunology. *Brain, Behavior, & Immunity-Health* **2021**, *18*, 100351.
8. McDonald, P. G.; O'Connell, M.; Lutgendorf, S. K. Psychoneuroimmunology and cancer: a decade of discovery, paradigm shifts, and methodological innovations. *Brain, behavior, and immunity* **2013**, *30*, S1-S9.
9. Todd, B. L.; Moskowitz, M. C.; Ottati, A.; Feuerstein, M. Stressors, stress response, and cancer recurrence: a systematic review. *Cancer nursing* **2014**, *37* (2), 114-125.
10. Mravec, B.; Horvathova, L.; Hunakova, L. Neurobiology of Cancer: the Role of beta-Adrenergic Receptor Signaling in Various Tumor Environments. *Int J Mol Sci* **2020**, *21* (21). DOI: https://doi.org/10.3390/ijms21217958.
11. Thaker, P. H.; Lutgendorf, S. K.; Sood, A. K. The neuroendocrine impact of chronic stress on cancer. *Cell Cycle* **2007**, *6* (4), 430-433. DOI: https://doi.org/10.4161/cc.6.4.3829.
12. Moreno-Smith, M.; Lutgendorf, S. K.; Sood, A. K. Impact of stress on cancer metastasis. *Future Oncol* **2010**, *6* (12), 1863-1881. DOI: https://doi.org/10.2217/fon.10.142.
13. Hwa, Y. L.; Shi, Q.; Kumar, S. K.; Lacy, M. Q.; Gertz, M. A.; Kapoor, P.; Buadi, F. K.; Leung, N.; Dingli, D.; Go, R. S.; et al. Beta-blockers improve survival outcomes in patients with multiple myeloma: a retrospective evaluation. *American Journal of Hematology* **2017**, *92* (1), 50–55, Article. DOI: https://doi.org/10.1002/ajh.24582.
14. Servick, K. War of nerves. *Science* **2019**, *365* (6458), 1071–1073. DOI: https://doi.org/10.1126/science.365.6458.1071.
15. Steptoe, A.; Willemsen, G.; Owen, N.; Flower, L.; Mohamed-Ali, V. Acute mental stress elicits delayed increases in circulating inflammatory cytokine levels. *Clin Sci (Lond)* **2001**, *101* (2), 185–192.
16. Bierhaus, A.; Wolf, J.; Andrassy, M.; Rohleder, N.; Humpert, P. M.; Petrov, D.; Ferstl, R.; von Eynatten, M.; Wendt, T.; Rudofsky, G.; et al. A mechanism converting psychosocial stress into mononuclear cell activation. *Proc Natl Acad Sci U S A* **2003**, *100* (4), 1920–1925. DOI: https://doi.org/10.1073/pnas.0438019100.
17. Qiao, G.; Chen, M.; Mohammadpour, H.; MacDonald, C. R.; Bucsek, M. J.; Hylander, B. L.; Barbi, J. J.; Repasky, E. A. Chronic Adrenergic Stress Contributes to Metabolic Dysfunction

and an Exhausted Phenotype in T Cells in the Tumor Microenvironment. *Cancer Immunol Res* **2021**, *9* (6), 651–664. DOI: https://doi.org/10.1158/2326-6066.CIR-20-0445.
18. Faulkner, S.; Jobling, P.; March, B.; Jiang, C. C.; Hondermarck, H. Tumor Neurobiology and the War of Nerves in Cancer. *Cancer Discov* **2019**, *9* (6), 702–710. DOI: https://doi.org/10.1158/2159-8290.CD-18-1398.
19. Zhang, X.; Zhang, Y.; He, Z.; Yin, K.; Li, B.; Zhang, L.; Xu, Z. Chronic stress promotes gastric cancer progression and metastasis: an essential role for ADRB2. *Cell Death Dis* **2019**, *10* (11), 788. DOI: https://doi.org/10.1038/s41419-019-2030-2.
20. Chhatar, S.; Lal, G. Role of adrenergic receptor signalling in neuroimmune communication. *Current Research in Immunology* **2021**, *2*, 202–217.
21. Mohammadpour, H.; MacDonald, C. R.; Qiao, G.; Chen, M.; Dong, B.; Hylander, B. L.; McCarthy, P. L.; Abrams, S. I.; Repasky, E. A. β2 adrenergic receptor-mediated signaling regulates the immunosuppressive potential of myeloid-derived suppressor cells. *J Clin Invest* **2019**, *129* (12), 5537–5552. DOI: https://doi.org/10.1172/jci129502 From NLM.
22. Qiao, G.; Bucsek, M. J.; Winder, N. M.; Chen, M.; Giridharan, T.; Olejniczak, S. H.; Hylander, B. L.; Repasky, E. A. beta-Adrenergic signaling blocks murine CD8(+) T-cell metabolic reprogramming during activation: a mechanism for immunosuppression by adrenergic stress. *Cancer Immunology Immunotherapy* **2019**, *68* (1), 11–22, Article. DOI: https://doi.org/10.1007/s00262-018-2243-8.
23. Kokolus, K. M.; Zhang, Y.; Sivik, J. M.; Schmeck, C.; Zhu, J.; Repasky, E. A.; Drabick, J. J.; Schell, T. D. Beta blocker use correlates with better overall survival in metastatic melanoma patients and improves the efficacy of immunotherapies in mice. *Oncoimmunology* **2018**, *7* (3), e1405205. DOI: https://doi.org/10.1080/2162402X.2017.1405205.
24. Hiller, J. G.; Cole, S. W.; Crone, E. M.; Byrne, D. J.; Shackleford, D. M.; Pang, J. B.; Henderson, M. A.; Nightingale, S. S.; Ho, K. M.; Myles, P. S.; et al. Preoperative beta-Blockade with Propranolol Reduces Biomarkers of Metastasis in Breast Cancer: A Phase II Randomized Trial. *Clin Cancer Res* **2020**, *26* (8), 1803–1811. DOI: https://doi.org/10.1158/1078-0432.CCR-19-2641.
25. Conceicao, F.; Sousa, D. M.; Paredes, J.; Lamghari, M. Sympathetic activity in breast cancer and metastasis: partners in crime. *Bone Res* **2021**, *9* (1), 9. DOI: https://doi.org/10.1038/s41413-021-00137-1.
26. De Giorgi, V.; Grazzini, M.; Benemei, S.; Marchionni, N.; Geppetti, P.; Gandini, S. beta-Blocker use and reduced disease progression in patients with thick melanoma: 8 years of follow-up. *Melanoma Res* **2017**, *27* (3), 268–270. DOI: https://doi.org/10.1097/CMR.0000000000000317.
27. De Giorgi, V.; Grazzini, M.; Benemei, S.; Marchionni, N.; Botteri, E.; Pennacchioli, E.; Geppetti, P.; Gandini, S. Propranolol for Off-label Treatment of Patients With Melanoma: Results From a Cohort Study. *JAMA Oncol* **2018**, *4* (2), e172908. DOI: https://doi.org/10.1001/jamaoncol.2017.2908.
28. Sigorski, D.; Izycka-Swieszewska, E. Sympathetic nervous signaling dictates prostate cancer progression. *Cell Death Discov* **2022**, *8* (1), 109. DOI: https://doi.org/10.1038/s41420-022-00928-3.
29. Bucsek, M. J.; Qiao, G.; MacDonald, C. R.; Giridharan, T.; Evans, L.; Niedzwecki, B.; Liu, H.; Kokolus, K. M.; Eng, J. W. L.; Messmer, M. N.; et al. beta-Adrenergic Signaling in Mice Housed at Standard Temperatures Suppresses an Effector Phenotype in CD8(+) T Cells and Undermines Checkpoint Inhibitor Therapy. *Cancer Research* **2017**, *77* (20), 5639–5651, Article. DOI: https://doi.org/10.1158/0008-5472.Can-17-0546.
30. Eng, J. W. L.; Reed, C. B.; Kokolus, K. M.; Pitoniak, R.; Utley, A.; Bucsek, M. J.; Ma, W. W.; Repasky, E. A.; Hylander, B. L. Housing temperature-induced stress drives therapeutic resistance in murine tumour models through beta(2)-adrenergic receptor activation. *Nature Communications* **2015**, *6*, Article. DOI: https://doi.org/10.1038/ncomms7426.
31. Moretti, S.; Massi, D.; Farini, V.; Baroni, G.; Parri, M.; Innocenti, S.; Cecchi, R.; Chiarugi, P. β-adrenoceptors are upregulated in human melanoma and their activation releases pro-

tumorigenic cytokines and metalloproteases in melanoma cell lines. *Lab Invest* **2013**, *93* (3), 279–290. DOI: https://doi.org/10.1038/labinvest.2012.175 From NLM.
32. Calvani, M.; Pelon, F.; Comito, G.; Taddei, M. L.; Moretti, S.; Innocenti, S.; Nassini, R.; Gerlini, G.; Borgognoni, L.; Bambi, F.; et al. Norepinephrine promotes tumor microenvironment reactivity through beta3-adrenoreceptors during melanoma progression. *Oncotarget* **2015**, *6* (7), 4615–4632. DOI: https://doi.org/10.18632/oncotarget.2652.
33. Gandhi, S.; Pandey, M. R.; Attwood, K.; Ji, W.; Witkiewicz, A. K.; Knudsen, E. S.; Allen, C.; Tario, J. D.; Wallace, P. K.; Cedeno, C. D.; et al. Phase I Clinical Trial of Combination Propranolol and Pembrolizumab in Locally Advanced and Metastatic Melanoma: Safety, Tolerability, and Preliminary Evidence of Antitumor Activity. *Clin Cancer Res* **2020**. DOI: https://doi.org/10.1158/1078-0432.CCR-20-2381.
34. Gandhi, S.; Pandey, M. R.; Attwood, K.; Ji, W.; Witkiewicz, A. K.; Knudsen, E. S.; Allen, C.; Tario, J. D.; Wallace, P. K.; Cedeno, C. D. Phase I Clinical Trial of Combination Propranolol and Pembrolizumab in Locally Advanced and Metastatic Melanoma: Safety, Tolerability, and Preliminary Evidence of Antitumor ActivityPropranolol and Pembrolizumab in Metastatic Melanoma. *Clinical Cancer Research* **2021**, *27* (1), 87–95.
35. Knight, J. M.; Rizzo, J. D.; Hari, P.; Pasquini, M. C.; Giles, K. E.; D'Souza, A.; Logan, B. R.; Hamadani, M.; Chhabra, S.; Dhakal, B.; et al. Propranolol inhibits molecular risk markers in HCT recipients: a phase 2 randomized controlled biomarker trial. *Blood Advances* **2020**, *4* (3), 467–476, Article. DOI: https://doi.org/10.1182/bloodadvances.2019000765.
36. Kumar, S. K.; Dispenzieri, A.; Lacy, M. Q.; Gertz, M. A.; Buadi, F. K.; Pandey, S.; Kapoor, P.; Dingli, D.; Hayman, S. R.; Leung, N.; et al. Continued improvement in survival in multiple myeloma: changes in early mortality and outcomes in older patients. *Leukemia* **2014**, *28* (5), 1122–1128, Article. DOI: https://doi.org/10.1038/leu.2013.313.
37. Kawano, Y.; Roccaro, A. M.; Ghobrial, I. M.; Azzi, J. Multiple Myeloma and the Immune Microenvironment. *Current Cancer Drug Targets* **2017**, *17* (9), 806–818, Review. DOI: https://doi.org/10.2174/1568009617666170214102301.
38. Kawano, Y.; Moschetta, M.; Manier, S.; Glavey, S.; Goerguen, G. T.; Roccaro, A. M.; Anderson, K. C.; Ghobrial, I. M. Targeting the bone marrow microenvironment in multiple myeloma. *Immunological Reviews* **2015**, *263* (1), 160–172, Review. DOI: https://doi.org/10.1111/imr.12233.
39. Kyle, R. A.; Gertz, M. A.; Witzig, T. E.; Lust, J. A.; Lacy, M. Q.; Dispenzieri, A.; Fonseca, R.; Rajkumar, S. V.; Offord, J. R.; Larson, D. R.; et al. Review of 1027 patients with newly diagnosed multiple myeloma. *Mayo Clinic Proceedings* **2003**, *78* (1), 21–33, Article. DOI: https://doi.org/10.4065/78.1.21.
40. Reddy, G. K.; Mughal, T. I.; Lonial, S. Optimizing the management of treatment-related peripheral neuropathy in patients with multiple myeloma. *Supportive cancer therapy* **2006**, *4* (1), 19–22. DOI: https://doi.org/10.3816/SCT.2006.n.027.
41. Rome, S.; Noonan, K.; Bertolotti, P.; Tariman, J. D.; Miceli, T.; Int Myeloma Fdn Nurse Leadership, B. Bone Health, Pain, and Mobility Evidence-based recommendations for patients with multiple myeloma. *Clinical Journal of Oncology Nursing* **2017**, *21* (5), 47–59, Article. DOI: https://doi.org/10.1188/17.Cjon.S5.47-59.
42. Barre, P. V.; Padmaja, G.; Rana, S.; Tiamongla. Stress and quality of life in cancer patients: medical and psychological intervention. *Indian journal of psychological medicine* **2018**, *40* (3), 232–238.
43. Zaleta, A. K.; Miller, M. F.; Olson, J. S.; Yuen, E. Y.; LeBlanc, T. W.; Cole, C. E.; McManus, S.; Buzaglo, J. S. Symptom burden, perceived control, and quality of life among patients living with multiple myeloma. *Journal of the National Comprehensive Cancer Network* **2020**, *18* (8), 1087–1095.
44. Sonneveld, P.; Verelst, S.; Lewis, P.; Gray-Schopfer, V.; Hutchings, A.; Nixon, A.; Petrucci, M. Review of health-related quality of life data in multiple myeloma patients treated with novel agents. *Leukemia* **2013**, *27* (10), 1959–1969.

45. Maatouk, I.; He, S.; Hummel, M.; Hemmer, S.; Hillengass, M.; Goldschmidt, H.; Hartmann, M.; Herzog, W.; Hillengass, J. Patients with precursor disease exhibit similar psychological distress and mental HRQOL as patients with active myeloma. *Blood Cancer Journal* **2019**, *9*, Letter. DOI: https://doi.org/10.1038/s41408-019-0172-1.
46. McEwen, B. S. Perturbing the Organism: The Biology of Stressful Experience. *JAMA* **1993**, *269* (10), 1315–1315. DOI: https://doi.org/10.1001/jama.1993.03500100115046 (acccessed 12/4/2022).
47. Katayama, Y.; Battista, M.; Kao, W. M.; Hidalgo, A.; Peired, A. J.; Thomas, S. A.; Frenette, P. S. Signals from the sympathetic nervous system regulate hematopoietic stem cell egress from bone marrow. *Cell* **2006**, *124* (2), 407–421, Article. DOI: https://doi.org/10.1016/j.cell.2005.10.041.
48. Hanns, P.; Paczulla, A. M.; Medinger, M.; Konantz, M.; Lengerke, C. Stress and catecholamines modulate the bone marrow microenvironment to promote tumorigenesis. *Cell Stress* **2019**, *3* (7), 221–235. DOI: https://doi.org/10.15698/cst2019.07.192.
49. Mattsson, J.; Appelgren, L.; Hamberger, B.; Peterson, H. I. ADRENERGIC-INNERVATION OF TUMOR BLOOD-VESSELS. *Cancer Letters* **1977**, *3* (5–6), 347–351, Article. DOI: https://doi.org/10.1016/s0304-3835(77)97078-1.
50. Calvani, M.; Pelon, F.; Comito, G.; Taddei, M. L.; Moretti, S.; Innocenti, S.; Nassini, R.; Gerlini, G.; Borgognoni, L.; Bambi, F.; et al. Norepinephrine promotes tumor microenvironment reactivity through beta 3-adrenoreceptors during melanoma progression. *Oncotarget* **2015**, *6* (7), 4615–4632, Article. DOI: https://doi.org/10.18632/oncotarget.2652.
51. Hiasa, M.; Okui, T.; Allette, Y. M.; Ripsch, M. S.; Sun-Wada, G.-H.; Wakabayashi, H.; Roodman, G. D.; White, F. A.; Yoneda, T. Bone Pain Induced by Multiple Myeloma Is Reduced by Targeting V-ATPase and ASIC3Bone Pain in Multiple Myeloma. *Cancer research* **2017**, *77* (6), 1283–1295.
52. Kovacs, A.; Vermeer, D. W.; Madeo, M.; Reavis, H. D.; Vermeer, S. J.; Williamson, C. S.; Rickel, A.; Stamp, J.; Lucido, C. T.; Cain, J. Tumor-infiltrating nerves create an electro-physiologically active microenvironment and contribute to treatment resistance. *BioRxiv* **2020**, 2020.2004. 2024.058594.
53. Madeo, M.; Colbert, P. L.; Vermeer, D. W.; Lucido, C. T.; Cain, J. T.; Vichaya, E. G.; Grossberg, A. J.; Muirhead, D.; Rickel, A. P.; Hong, Z. Cancer exosomes induce tumor innervation. *Nature communications* **2018**, *9* (1), 4284.
54. Saloman, J. L.; Albers, K. M.; Li, D.; Hartman, D. J.; Crawford, H. C.; Muha, E. A.; Rhim, A. D.; Davis, B. M. Ablation of sensory neurons in a genetic model of pancreatic ductal adenocarcinoma slows initiation and progression of cancer. *Proceedings of the National Academy of Sciences* **2016**, *113* (11), 3078–3083.
55. Tobias Braga, W. M.; da Silva, B. R.; de Carvalho, A. C.; Maekawa, Y. H.; Bortoluzzo, A. B.; Rizzatti, E. G.; Atanackovic, D.; Braga Colleoni, G. W. FOXP3 and CTLA4 overexpression in multiple myeloma bone marrow as a sign of accumulation of CD4(+) T regulatory cells. *Cancer Immunology Immunotherapy* **2014**, *63* (11), 1189–1197, Article. DOI: https://doi.org/10.1007/s00262-014-1589-9.
56. Xu, Y.; Zhang, X.; Liu, H.; Zhao, P.; Chen, Y.; Luo, Y.; Zhang, Z.; Wang, X. Mesenchymal stromal cells enhance the suppressive effects ofmyeloid-derived suppressor cells of multiple myeloma. *Leukemia & Lymphoma* **2017**, *58* (11), 2668–2676, Article. DOI: https://doi.org/10.1080/10428194.2017.1298753.
57. Najafi M, F. B., Mortezaee K. Contribution of regulatory T cells to cancer: A review. *J Cell Physiol.* **2019 Jun**, *234(6)*, 7983–7993. DOI: doi: https://doi.org/10.1002/jcp.27553.
58. Dahlhoff, J.; Manz, H.; Steinfatt, T.; Delgado-Tascon, J.; Seebacher, E.; Schneider, T.; Wilnit, A.; Mokhtari, Z.; Tabares, P.; Bockle, D.; et al. Transient regulatory T-cell targeting triggers immune control of multiple myeloma and prevents disease progression. *Leukemia* **2022**, *36* (3), 790–800. DOI: https://doi.org/10.1038/s41375-021-01422-y.
59. Papadimitriou, K.; Tsakirakis, N.; Malandrakis, P.; Vitsos, P.; Metousis, A.; Orologas-Stavrou, N.; Ntanasis-Stathopoulos, I.; Kanellias, N.; Eleutherakis-Papaiakovou, E.; Pothos, P.; et al.

Deep Phenotyping Reveals Distinct Immune Signatures Correlating with Prognostication, Treatment Responses, and MRD Status in Multiple Myeloma. *Cancers (Basel)* **2020**, *12* (11). DOI: https://doi.org/10.3390/cancers12113245 From NLM.
60. Danziger, S. A.; McConnell, M.; Gockley, J.; Young, M. H.; Rosenthal, A.; Schmitz, F.; Reiss, D. J.; Farmer, P.; Alapat, D. V.; Singh, A.; et al. Bone marrow microenvironments that contribute to patient outcomes in newly diagnosed multiple myeloma: A cohort study of patients in the Total Therapy clinical trials. *Plos Medicine* **2020**, *17* (11), Article. DOI: https://doi.org/10.1371/journal.pmed.1003323.
61. Nakamura, K.; Smyth, M. J. Myeloid immunosuppression and immune checkpoints in the tumor microenvironment. *Cellular & Molecular Immunology* **2020**, *17* (1), 1–12, Review. DOI: https://doi.org/10.1038/s41423-019-0306-1.
62. Sui, H.; Dongye, S.; Liu, X.; Xu, X.; Wang, L.; Jin, C. Q.; Yao, M.; Gong, Z.; Jiang, D.; Zhang, K.; et al. Immunotherapy of targeting MDSCs in tumor microenvironment. *Frontiers in immunology* **2022**, *13*, 990463–990463,; Review; Research Support, Non-U.S. Gov't. DOI: https://doi.org/10.3389/fimmu.2022.990463.
63. Suyani, E.; Sucak, G. T.; Akyurek, N.; Sahin, S.; Baysal, N. A.; Yagci, M.; Haznedar, R. Tumor-associated macrophages as a prognostic parameter in multiple myeloma. *Annals of Hematology* **2013**, *92* (5), 669–677, Article. DOI: https://doi.org/10.1007/s00277-012-1652-6.
64. Guillerey, C.; de Andrade, L. F.; Vuckovic, S.; Miles, K.; Ngiow, S. F.; Yong, M. C. R.; Teng, M. W. L.; Colonna, M.; Ritchie, D. S.; Chesi, M.; et al. Immunosurveillance and therapy of multiple myeloma are CD226 dependent (vol 125, pg 2077, 2015). *Journal of Clinical Investigation* **2015**, *125* (7), 2904–2904, Correction. DOI: https://doi.org/10.1172/jci82646.
65. Giuliani, M.; Janji, B.; Berchem, G. Activation of NK cells and disruption of PD-L1/PD-1 axis: two different ways for lenalidomide to block myeloma progression. *Oncotarget* **2017**, *8* (14), 24031–24044, Review. DOI: https://doi.org/10.18632/oncotarget.15234.
66. Chauhan, D.; Singh, A. V.; Brahmandam, M.; Carrasco, R.; Bandi, M.; Hideshima, T.; Bianchi, G.; Podar, K.; Tai, Y.-T.; Mitsiades, C.; et al. Functional Interaction of Plasmacytoid Dendritic Cells with Multiple Myeloma Cells: A Therapeutic Target. *Cancer Cell* **2009**, *16* (4), 309–323, Article. DOI: https://doi.org/10.1016/j.ccr.2009.08.019.
67. Murray, M. E.; Gavile, C. M.; Nair, J. R.; Koorella, C.; Carlson, L. M.; Buac, D.; Utley, A.; Chesi, M.; Bergsagel, P. L.; Boise, L. H.; et al. CD28-mediated pro-survival signaling induces chemotherapeutic resistance in multiple myeloma. *Blood* **2014**, *123* (24), 3770–3779, Article. DOI: https://doi.org/10.1182/blood-2013-10-530964.
68. Jobling, P.; Pundavela, J.; Oliveira, S. M. R.; Roselli, S.; Walker, M. M.; Hondermarck, H. Nerve-Cancer Cell Cross-talk: A Novel Promoter of Tumor Progression. *Cancer Research* **2015**, *75* (9), 1777–1781, Review. DOI: https://doi.org/10.1158/0008-5472.Can-14-3180.
69. Steinwerblowsky, R. SYMPATHETIC NERVOUS-SYSTEM AND CANCER. *Experimental Neurology* **1974**, *42* (1), 97–100, Article. DOI: https://doi.org/10.1016/0014-4886(74)90009-0.
70. Boilly, B.; Faulkner, S.; Jobling, P.; Hondermarck, H. Nerve Dependence: From Regeneration to Cancer. *Cancer Cell* **2017**, *31* (3), 342–354, Review. DOI: https://doi.org/10.1016/j.ccell.2017.02.005.
71. Kamiya, A.; Hayama, Y.; Kato, S.; Shimomura, A.; Shimomura, T.; Irie, K.; Kaneko, R.; Yanagawa, Y.; Kobayashi, K.; Ochiya, T. Genetic manipulation of autonomic nerve fiber innervation and activity and its effect on breast cancer progression. *Nature Neuroscience* **2019**, *22* (8), 1289-+, Article. DOI: https://doi.org/10.1038/s41593-019-0430-3.
72. Villers, A.; McNeal, J. E.; Redwine, E. A.; Freiha, F. S.; Stamey, T. A. THE ROLE OF PERINEURAL SPACE INVASION IN THE LOCAL SPREAD OF PROSTATIC ADENOCARCINOMA. *Journal of Urology* **1989**, *142* (3), 763–768, Article. DOI: https://doi.org/10.1016/s0022-5347(17)38881-x.
73. Bussard, K. M.; Mutkus, L.; Stumpf, K.; Gomez-Manzano, C.; Marini, F. C. Tumor-associated stromal cells as key contributors to the tumor microenvironment. *Breast Cancer Research* **2016**, *18*, Review. DOI: https://doi.org/10.1186/s13058-016-0740-2.

74. Mancino, M.; Ametller, E.; Gascon, P.; Almendro, V. The neuronal influence on tumor progression. *Biochimica Et Biophysica Acta-Reviews on Cancer* **2011**, *1816* (2), 105–118, Review. DOI: https://doi.org/10.1016/j.bbcan.2011.04.005.
75. Cole, S. W.; Nagaraja, A. S.; Lutgendorf, S. K.; Green, P. A.; Sood, A. K. Sympathetic nervous system regulation of the tumour microenvironment. *Nature Reviews Cancer* **2015**, *15* (9), 563–572.
76. Katayama, Y.; Battista, M.; Kao, W.-M.; Hidalgo, A.; Peired, A. J.; Thomas, S. A.; Frenette, P. S. Signals from the sympathetic nervous system regulate hematopoietic stem cell egress from bone marrow. *Cell* **2006**, *124* (2), 407–421.
77. Kohm, A. P.; Sanders, V. M. Norepinephrine and beta 2-adrenergic receptor stimulation regulate CD4(+) T and B lymphocyte function in vitro and in vivo. *Pharmacological Reviews* **2001**, *53* (4), 487–525, Review.
78. Nakai, A.; Hayano, Y.; Furuta, F.; Noda, M.; Suzuki, K. Control of lymphocyte egress from lymph nodes through beta(2)-adrenergic receptors. *Journal of Experimental Medicine* **2014**, *211* (13), 2583–2598, Article. DOI: https://doi.org/10.1084/jem.20141132.
79. Powell, N. D.; Sloan, E. K.; Bailey, M. T.; Arevalo, J. M. G.; Miller, G. E.; Chen, E.; Kobor, M. S.; Reader, B. F.; Sheridan, J. F.; Cole, S. W. Social stress up-regulates inflammatory gene expression in the leukocyte transcriptome via beta-adrenergic induction of myelopoiesis. *Proceedings of the National Academy of Sciences of the United States of America* **2013**, *110* (41), 16574–16579, Article. DOI: https://doi.org/10.1073/pnas.1310655110.
80. Bae, G. E.; Kim, H. S.; Won, K. Y.; Kim, G. Y.; Sung, J. Y.; Lim, S. J. Lower Sympathetic Nervous System Density and β-adrenoreceptor Expression Are Involved in Gastric Cancer Progression. *Anticancer Res* **2019**, *39* (1), 231–236. DOI: https://doi.org/10.21873/anticanres.13102 From NLM.
81. Dal Monte, M.; Casini, G.; Filippi, L.; Nicchia, G. P.; Svelto, M.; Bagnoli, P. Functional involvement of beta 3-adrenergic receptors in melanoma growth and vascularization. *Journal of Molecular Medicine-Jmm* **2013**, *91* (12), 1407–1419, Article. DOI: https://doi.org/10.1007/s00109-013-1073-6.
82. Thaker, P. H.; Han, L. Y.; Kamat, A. A.; Arevalo, J. M.; Takahashi, R.; Lu, C.; Jennings, N. B.; Armaiz-Pena, G.; Bankson, J. A.; Ravoori, M.; et al. Chronic stress promotes tumor growth and angiogenesis in a mouse model of ovarian carcinoma (vol 12, pg 939, 2006). *Nature Medicine* **2021**, *27* (12), 2246–2246, Correction. DOI: https://doi.org/10.1038/s41591-021-01566-5.
83. Cheng, Y.; Sun, F.; D'Souza, A.; Dhakal, B.; Pisano, M.; Chhabra, S.; Stolley, M.; Hari, P.; Janz, S. Autonomic nervous system control of multiple myeloma. *Blood Reviews* **2021**, *46*, Review. DOI: https://doi.org/10.1016/j.blre.2020.100741.
84. Hossain, F.; Al-Khami, A. A.; Wyczechowska, D.; Hernandez, C.; Zheng, L.; Reiss, K.; Del Valle, L.; Trillo-Tinoco, J.; Maj, T.; Zou, W.; et al. Inhibition of Fatty Acid Oxidation Modulates Immunosuppressive Functions of Myeloid-Derived Suppressor Cells and Enhances Cancer Therapies. *Cancer Immunology Research* **2015**, *3* (11), 1236–1247, Article. DOI: https://doi.org/10.1158/2326-6066.Cir-15-0036.
85. Xiang, H.; Yang, R.; Tu, J.; Xi, Y.; Yang, S.; Lv, L.; Zhai, X.; Zhu, Y.; Dong, D.; Tao, X. Metabolic reprogramming of immune cells in pancreatic cancer progression. *Biomedicine & pharmacotherapy = Biomedecine & pharmacotherapie* **2022**, *157*, 113992–113992,; Review. DOI: https://doi.org/10.1016/j.biopha.2022.113992.
86. Mendez-Ferrer, S.; Battista, M.; Frenette, P. S. Cooperation of beta(2)- and beta(3)-adrenergic receptors in hematopoietic progenitor cell mobilization. In *Skeletal Biology and Medicine*, Zaidi, M. Ed.; Annals of the New York Academy of Sciences, Vol. 1192; 2010; pp 139–144.
87. Ghobrial, I. M.; Liu, C.-J.; Redd, R. A.; Perez, R. P.; Baz, R.; Zavidij, O.; Sklavenitis-Pistofidis, R.; Richardson, P. G.; Anderson, K. C.; Laubach, J.; et al. A Phase Ib/II Trial of the First-in-Class Anti-CXCR4 Antibody Ulocuplumab in Combination with Lenalidomide or Bortezomib Plus Dexamethasone in Relapsed Multiple Myeloma. *Clinical Cancer Research* **2020**, *26* (2), 344–353, Article. DOI: https://doi.org/10.1158/1078-0432.Ccr-19-0647.

88. Ghobrial, I. M.; Liu, C.-J.; Zavidij, O.; Azab, A. K.; Baz, R.; Laubach, J. P.; Mishima, Y.; Armand, P.; Munshi, N. C.; Basile, F.; et al. Phase I/II trial of the CXCR4 inhibitor plerixafor in combination with bortezomib as a chemosensitization strategy in relapsed/refractory multiple myeloma. *American Journal of Hematology* **2019**, *94* (11), 1244–1253, Article. DOI: https://doi.org/10.1002/ajh.25627.
89. Balood, M.; Ahmadi, M.; Eichwald, T.; Ahmadi, A.; Majdoubi, A.; Roversi, K.; Roversi, K.; Lucido, C. T.; Restaino, A. C.; Huang, S. Nociceptor neurons affect cancer immunosurveillance. *Nature* **2022**, 1–8.
90. Olechnowicz, S. W.; Weivoda, M. M.; Lwin, S. T.; Leung, S. K.; Gooding, S.; Nador, G.; Javaid, M. K.; Ramasamy, K.; Rao, S. R.; Edwards, J. R. Multiple myeloma increases nerve growth factor and other pain-related markers through interactions with the bone microenvironment. *Scientific reports* **2019**, *9* (1), 14189.
91. Diaz-delCastillo, M.; Kamstrup, D.; Olsen, R. B.; Hansen, R. B.; Pembridge, T.; Simanskaite, B.; Jimenez-Andrade, J. M.; Lawson, M. A.; Heegaard, A. M. Differential Pain-Related Behaviors and Bone Disease in Immunocompetent Mouse Models of Myeloma. *JBMR plus* **2020**, *4* (2), e10252.
92. Sloan, E. K.; Capitanio, J. P.; Tarara, R. P.; Mendoza, S. P.; Mason, W. A.; Cole, S. W. Social stress enhances sympathetic innervation of primate lymph nodes: Mechanisms and implications for viral pathogenesis. *Journal of Neuroscience* **2007**, *27* (33), 8857–8865, Article. DOI: https://doi.org/10.1523/jneurosci.1247-07.2007.
93. Sloan, E. K.; Capitanio, J. P.; Tarara, R. P.; Cole, S. W. Social temperament and lymph node innervation. *Brain Behavior and Immunity* **2008**, *22* (5), 717–726, Article. DOI: https://doi.org/10.1016/j.bbi.2007.10.010.
94. Lutgendorf, S. K.; DeGeest, K.; Dahmoush, L.; Farley, D.; Penedo, F.; Bender, D.; Goodheart, M.; Buekers, T. E.; Mendez, L.; Krueger, G.; et al. Social isolation is associated with elevated tumor norepinephrine in ovarian carcinoma patients. *Brain Behavior and Immunity* **2011**, *25* (2), 250–255, Article. DOI: https://doi.org/10.1016/j.bbi.2010.10.012.
95. Chida, Y.; Hamer, M.; Wardle, J.; Steptoe, A. Do stress-related psychosocial factors contribute to cancer incidence and survival? *Nat Clin Pract Oncol* **2008**, *5* (8), 466–475. DOI: https://doi.org/10.1038/ncponc1134 From NLM.
96. Cheng, Y.; Tang, X.-Y.; Li, Y.-X.; Zhao, D.-D.; Cao, Q.-H.; Wu, H.-X.; Yang, H.-B.; Hao, K.; Yang, Y. Depression-induced neuropeptide Y secretion promotes prostate cancer growth by recruiting myeloid cells. *Clinical Cancer Research* **2019**, *25* (8), 2621–2632.
97. Nieman, D. C.; Wentz, L. M. The compelling link between physical activity and the body's defense system. *Journal of Sport and Health Science* **2019**, *8* (3), 201–217, Review. DOI: https://doi.org/10.1016/j.jshs.2018.09.009.
98. Sitlinger, A.; Brander, D. M.; Bartlett, D. B. Impact of exercise on the immune system and outcomes in hematologic malignancies. *Blood Advances* **2020**, *4* (8), 1801–1811, Review. DOI: https://doi.org/10.1182/bloodadvances.2019001317.
99. Duggal, N. A.; Niemiro, G.; Harridge, S. D. R.; Simpson, R. J.; Lord, J. M. Can physical activity ameliorate immunosenescence and thereby reduce age-related multi-morbidity? *Nat Rev Immunol* **2019**, *19* (9), 563–572. DOI: https://doi.org/10.1038/s41577-019-0177-9 From NLM.
100. Spielmann, G.; McFarlin, B. K.; O'Connor, D. P.; Smith, P. J.; Pircher, H.; Simpson, R. J. Aerobic fitness is associated with lower proportions of senescent blood T-cells in man. *Brain Behav Immun* **2011**, *25* (8), 1521–1529. DOI: https://doi.org/10.1016/j.bbi.2011.07.226 From NLM.
101. Bigley, A. B.; Rezvani, K.; Chew, C.; Sekine, T.; Pistillo, M.; Crucian, B.; Bollard, C. M.; Simpson, R. J. Acute exercise preferentially redeploys NK-cells with a highly-differentiated phenotype and augments cytotoxicity against lymphoma and multiple myeloma target cells. *Brain Behav Immun* **2014**, *39*, 160–171. DOI: https://doi.org/10.1016/j.bbi.2013.10.030 From NLM.

102. Timmerman, K. L.; Flynn, M. G.; Coen, P. M.; Markofski, M. M.; Pence, B. D. Exercise training-induced lowering of inflammatory (CD14+CD16+) monocytes: a role in the anti-inflammatory influence of exercise? *J Leukoc Biol* **2008**, *84* (5), 1271–1278. DOI: https://doi.org/10.1189/jlb.0408244 From NLM.
103. Hylander, B. L.; Gordon, C. J.; Repasky, E. A. Manipulation of Ambient Housing Temperature To Study the Impact of Chronic Stress on Immunity and Cancer in Mice. *Journal of Immunology* **2019**, *202* (3), 631–636, Review. DOI: https://doi.org/10.4049/jimmunol.1800621.
104. Hwa, Y. L.; Lacy, M. Q.; Gertz, M. A.; Kumar, S. K.; Muchtar, E.; Buadi, F. K.; Dingli, D.; Leung, N.; Kapoor, P.; Go, R. S.; et al. Use of beta blockers is associated with survival outcome of multiple myeloma patients treated with pomalidomide. *European Journal of Haematology* **2021**, *106* (3), 433–436, Letter. DOI: https://doi.org/10.1111/ejh.13559.
105. Shim, H.; Ha, J. H.; Lee, H.; Sohn, J. Y.; Kim, H. J.; Eom, H.-S.; Kong, S.-Y. Expression of Myeloid Antigen in Neoplastic Plasma Cells Is Related to Adverse Prognosis in Patients with Multiple Myeloma. *Biomed Research International* **2014**, *2014*, Article. DOI: https://doi.org/10.1155/2014/893243.
106. Nair, R.; Subramaniam, V.; Barwick, B. G.; Gupta, V. A.; Matulis, S. M.; Lonial, S.; Boise, L. H.; Nooka, A. K.; Muthumalaiappan, K.; Shanmugam, M. β adrenergic signaling regulates hematopoietic stem and progenitor cell commitment and therapy sensitivity in multiple myeloma. *Haematologica* **2022**, *107* (9), 2226–2231.

Chapter 8
Nervous System Interactions with Nonimmune Elements in Cancer Microenvironment: A Missing Piece?

Kaan Çifcibaşı, Carmen Mota Reyes, Rouzanna Istvanffy, and Ihsan Ekin Demir

Introduction

The tumor microenvironment (TME) is the unique ecosystem in which the cancer cells are able to capitalize on available resources to further its spread. The TME comprises not only cellular components, such as fibroblasts, endothelial cells, immune cells, and nerves, but also the extracellular matrix, a meshwork of cross-linked macromolecules (e.g., collagens, glycoproteins) that form a dynamic scaffold for growth. The TME has an important role for the viability, spread, and therapy resistance of cancer cells [18, 35]. While the tumor develops, cancer cells can alter other resident cells of the tissue and exploit them for their own growth. For instance, cancer cells are able to evoke the formation of new blood vessels, i.e., angiogenesis, providing themselves with nutrients essential for their growth and forming new

K. Çifcibaşı · C. Mota Reyes · R. Istvanffy
Department of Surgery, Klinikum rechts der Isar, Technical University of Munich, School of Medicine, Munich, Germany

German Cancer Consortium (DKTK), Partner Site Munich, Munich, Germany

CRC 1321 Modelling and Targeting Pancreatic Cancer, Munich, Germany

I. E. Demir (✉)
Department of Surgery, Klinikum rechts der Isar, Technical University of Munich, School of Medicine, Munich, Germany

German Cancer Consortium (DKTK), Partner Site Munich, Munich, Germany

CRC 1321 Modelling and Targeting Pancreatic Cancer, Munich, Germany

Department of General Surgery, HPB-Unit, School of Medicine, Acibadem Mehmet Ali Aydinlar University, Istanbul, Turkey

Else Kröner Clinician Scientist Professor for Translational Pancreatic Surgery, Munich, Germany
e-mail: ekin.demir@tum.de

M. Amit, N. N. Scheff (eds.), *Cancer Neuroscience*, https://doi.org/10.1007/978-3-031-32429-1_8

pathways for metastatic spread [44]. Moreover, cancer-associated fibroblasts, activated by cancer cells from resident fibroblasts, have been shown to promote tumor growth and metastasis [29, 69]. Furthermore, nerves, comprising both neuronal axons and Schwann cells, have been shown to have a cancer-promoting role in several cancers [20, 25, 68, 70]. Because of the known importance of the TME for the carcinogenesis, the interaction of cancer cells with other cellular components of the TME has long been in the focus of research, with a particular focus on cancer-immune interactions. Nevertheless, another crucial aspect creating the TME, namely, the interaction between nonimmune cells in tumor stroma, has been neglected. While the field of cancer neuroscience is still in its infancy, recent discoveries have paved a path to many new, exciting aspects widening our perspective on the cancer biology. Discussing the interactions between nerves and nonimmune cells in the TME, our main focus is going to be on the interplay between nerves and fibroblasts and the role nerves play in angiogenesis. Additionally, we will also address the potential part Schwann cells could play in this cross talk.

Neuron-Fibroblast Interactions in Cancer

The body holds several surveillance systems at the periphery to maintain stable conditions, which are essential for its survival. The peripheral nervous system (PNS), comprising of motor, sensory, sympathetic, and parasympathetic nerves, is one of the most important of these surveillance mechanisms. As cancer cells are fairly capable to utilize the preexisting defense mechanisms of the body for their own favor, they are also able to make use of nerves in this manner [12, 14, 60, 74]. In accordance, increases in the neural density and neural size are believed to be not only epiphenomena in many cancers, but they also correlate with a poor prognosis [1, 3, 14].

Fibroblasts are a heterogeneous population of resident cells in many normal tissues [47]. They produce the main components of the extracellular matrix and are essential for the structural integrity of the connective tissue. Moreover, they play an important role in many processes, such as wound healing, aging, and carcinogenesis. During the development of cancer, cancer cells are able to activate resident fibroblasts into cancer-associated fibroblasts (CAFs), which express α-smooth muscle actin (α-SMA) in contrast to inactive fibroblasts resident in normal pancreas [81]. During this activation, cancer cells express Hedgehog ligands, which lead to the paracrine activation of fibroblasts via the Sonic Hedgehog (SHH) signaling pathway [5, 79]. Additionally, evidence for the exosomal microRNA (miRNA) mediated activation of fibroblasts by cancer cells in various cancers such as the pancreatic cancer or melanoma exists [19, 59, 78]. Furthermore, CAFs can also derive from the differentiation of bone marrow derived stem cells, from the transformation of epithelial cells through an epithelial-mesenchymal transformation (EMT)-like process, and from quiescent stellate cells [22, 71]. Classically, two subsets of CAFs have been described: myofibroblastic CAFs (myCAFs) and inflammatory CAFs (iCAFs) [56, 71]. While myCAFs have a higher expression of α-SMA and are localized directly adjacent to neoplastic cell sites, iCAFs are localized more distantly from the neoplastic cells and express tumor-supporting cytokines such as IL-6 [56]. Moreover, single-cell analysis of pancreatic adenocarcinoma (PDAC) has revealed a new subtype, namely, the antigen-presenting CAFs (apCAFs), which activate CD4+ T cells in the TME [21]. This discovery is in support of the hypothesis that there are distinct roles for different subsets of CAFs. While some CAFs produce inflammatory ligands and growth factors, thereby promoting tumor growth [8, 37], or enhance the aggressiveness and drug resistance of cancer cells in an exosome mediated manner [34, 65, 78], some CAFs seem to have tumor-restraining properties, such as the T cell activation and the restraint of the tumor angiogenesis [21, 61]. In addition to the direct influence CAFs might have on cancer cells, there seems to be a close relationship between CAFs and nerves in the tumor environment, which is yet to be elucidated.

A symbiotic relationship between nerves, cancer, and CAFs has been most clearly described in PDAC. PDAC is characterized by a dense, desmoplastic tumor, with extracellular matrix comprising up to 90% of the tumor tissue. Moreover, the neural invasion in PDAC is strongly correlated with the desmoplasia generated by CAFs, suggesting a potential triangular relationship between nerves, cancer cells, and CAFs [14]. For instance, one aspect of this triangular relationship has been described by Li et al. on the axis of SHH paracrine signaling [43]. Pancreatic stellate cells are myofibroblast-like cells located in exocrine regions of the pancreas [57]. Within in vivo and in vitro experiments, Li et al. showed that the SHH, overexpressed by pancreatic cancer cells, activates the hedgehog pathway in pancreatic stellate cells, comparable to the activation of resident fibroblast to CAFs. In return, the activated stellate cells not only assist the migration of cancer cells along nerves but also support the nerve outgrowth toward pancreatic cancer cell colonies. These effects are in line with the upregulated secretion of molecules associated with perineural invasion, such as nerve growth factor (NGF), matrix metalloproteinase 2 (MMP-2), and matrix metalloproteinase 9 (MMP-9). NGF has later been shown to induce pancreatic cancer proliferation and invasion via the PI3K/AKT/GSK signal pathway [33], which is involved in the EMT of cancer cells. Other studies had also shown that activated CAFs can give rise to the EMT of cancer cells, which further enhances their invasiveness [2, 36, 38, 52, 76]. Moreover, Nan et al. have demonstrated that pancreatic stellate cells can promote perineural invasion by secreting hepatocyte growth factor (HGF) via the activation of HGF/c-Met pathway, enhancing the expression of NGF and MMP-9 in vitro [54]. Another described way CAFs can induce the NGF expression in cancer cells is via the secretion of CXCL12. While the activation of the CXCL12/CXCR4 axis promotes the migration and invasion of the cancer cells, the CXCL12 mediated NGF expression results in an enhanced neurite regeneration [58, 77].

However, the interactions between nerves and CAFs are not restricted only to perineural invasion. CAFs also seem to be involved in neural remodeling. For example, CAFs in PDAC secrete leukemia inhibitory factor (LIF), which promotes the differentiation of Schwann cells, induces their migration via JAK/STAT3/AKT pathway, and induces neuronal plasticity. It was also demonstrated that the LIF titer in blood samples from PDAC patients correlated positively with the nerve density in tumor tissue [7]. In addition, pancreatic stellate cells contribute to pancreatic cancer pain. Han et al. demonstrated that the activation of the SHH pathway in pancreatic stellate cells upregulates the NGF and brain-derived neurotrophic factor (BDNF) expression and the transient receptor potential vanilloid 1 protein (TRPV1) sensitization. This finding is also supported by an orthotopic implantation model, in which mice implanted with high SHH expressing stellate or cancer cells showed a more severe, pain-related behavioral phenotype in von Frey filament tests [27]. Based on these results, even though the inhibition of the SHH pathway in PDAC might appear as a potential target to relieve pain, the SHH inhibition has been shown to improve tumor aggressiveness and reduce overall survival, possibly due to the loss of the tumor-restraining properties of apCAFs [21, 41, 61]. An overview of the influence of CAFs on neural invasion and remodeling can be seen in Fig. 8.1.

On the other hand, neural signaling also has a direct influence on fibroblasts (Fig. 8.2). In response to injury of the liver, such as during cirrhosis, hepatic stellate cells become active and fibrogenic. In the injured state, the adrenergic signaling of the liver is also increased. The adrenergic neurotransmitters, norepinephrine and neuropeptide Y, stimulate the proliferation of hepatic stellate cells. Norepinephrine can additionally induce the collagen gene expression of the stellate cells [55]. Likewise, Szpunar et al. demonstrated that the activation of the sympathetic nervous system may alter the collagen structure of the extracellular matrix via the activation

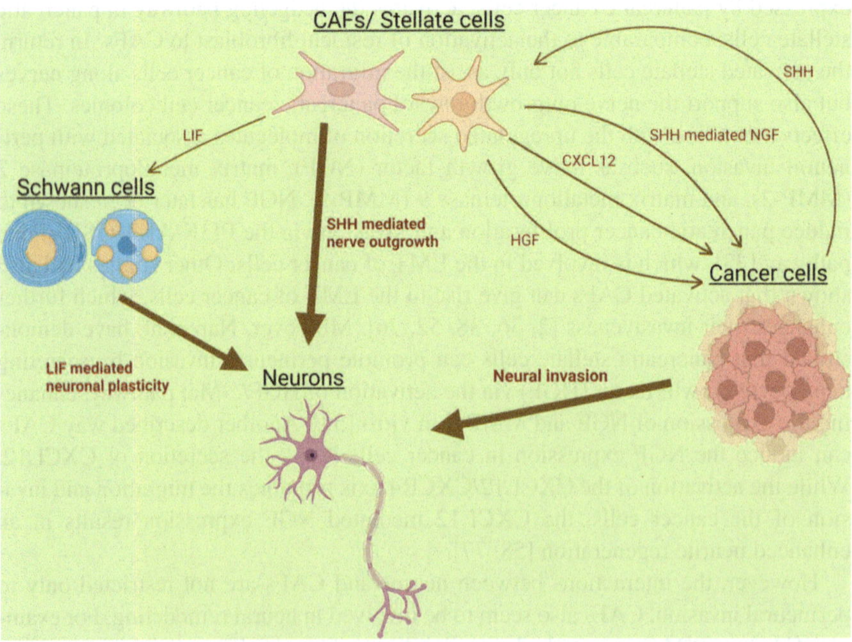

Fig. 8.1 Influence of cancer-associated fibroblasts on neural remodeling and invasion

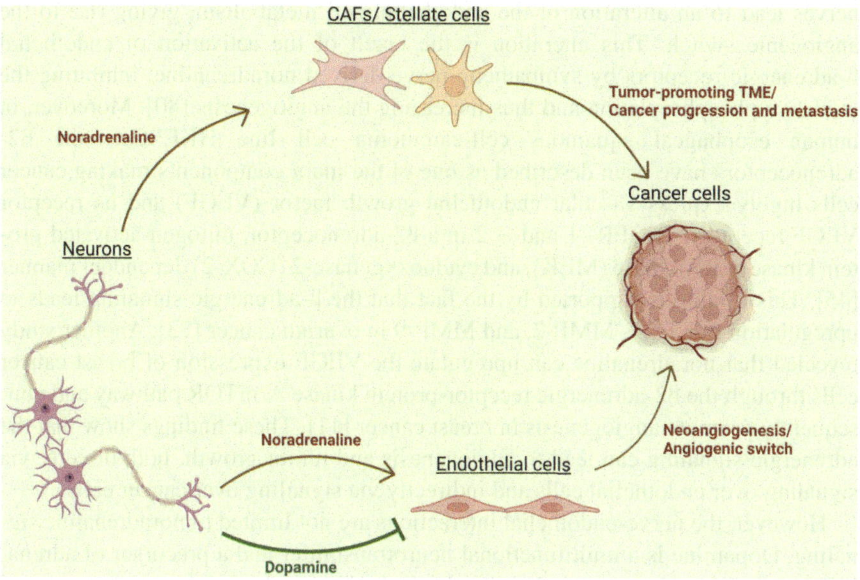

Fig. 8.2 Influence of nerve-nonimmune stroma interactions on cancer progression

of the stromal adrenergic receptors, promoting breast cancer progression and metastasis [72]. Although these findings are for the time being only limited to chronic injuries of liver and to breast cancer, they can set guidance for new discoveries in other types of cancer.

Neuron- Endothelium Interactions in Cancer

Neoangiogenesis is the formation of new blood vessels from already existing vasculature [26]. It is not only necessary for the supply of the tumor with nutrients essential for its growth, but it also facilitates the metastasis formation [23]. In fact, the newly formed vessels are easier to penetrate due to weaker interactions between endothelial cells and not fully formed pericytes [24, 26, 51]. In accordance with the close relationship between nerves and other stromal components, an important link between endothelial cells and nerves exists (Fig. 8.2).

The prognosis and the progression of prostate cancer is in part dependent on the infiltration by autonomic nerves and the resulting innervation [48]. In animal models, it was shown that the adrenergic and cholinergic signaling promotes the prostate cancer progression and metastasis [49]. The "angiogenic switch" can be defined as the transition from the avascularized hyperplasia to growing vascularized tumor [4]. One of the most important contributing mechanisms to the angiogenic switch in prostate cancer was demonstrated by Zahalka et al. They showed that the adrenergic nerves lead to an alteration of the endothelial cell metabolism, giving rise to the angiogenic switch. This alteration is the result of the activation of endothelial ß-adrenergic receptors by sympathetic nerve-derived noradrenaline, inhibiting the oxidative phosphorylation and thus increasing the angiogenesis [80]. Moreover, in human esophageal squamous cell-carcinoma cell line HKESC-1, the ß2-adrenoceptors have been described as one of the main components making cancer cells highly express vascular endothelial growth factor (VEGF) and its receptor VEGF receptor (VEGFR)-1 and − 2 in a ß2-adrenoceptor, mitogen-activated protein kinase/ ERK kinase (MEK), and cyclooxygenase-2 (COX-2) dependent manner [45]. This finding is supported by the fact that the ß-adrenergic signaling leads to upregulation of VEGF, MMP-2, and MMP-9 in ovarian cancer [73]. Another study revealed that noradrenaline can upregulate the VEGF expression of breast cancer cells through the ß2-adrenergic receptor-protein kinase A-mTOR pathway and consequently promote angiogenesis in breast cancer [11]. These findings show that the adrenergic signaling can lead to angiogenesis and tumor growth, both directly via signaling over endothelial cells and indirectly via signaling over cancer cells.

However, the nerve-endothelial interactions are not limited to noradrenaline signaling. Dopamine is a multifunctional neurotransmitter and a precursor of adrenaline and noradrenaline during their synthesis. In the periphery, it is mainly produced in sympathetic neurons, adrenal medulla, and neuroendocrine cells [42, 63, 75]. In contrast to adrenaline and noradrenaline, dopamine seems to have tumor-restraining features [6, 40, 46]. In addition to the direct effects on cancer cells, dopamine can also restrain cancer progression through endothelial cells. Tumor endothelial cells express dopamine receptor 2 [42]. Upon binding to dopamine receptor 2, dopamine prevents the phosphorylation of the VEGF receptor 2 of endothelial cells and consequently restrains the activation of the mitogen-activated protein kinase (MAPK). This results in the inhibition of the proliferation and migration of tumor endothelial cells [66]. This finding is also supported by the inhibitory effect of dopamine along with anticancer therapy on cancer cells in preclinical models of breast and colon cancer [67]. Moreover, dopamine can hinder the endothelial progenitor cells from leaving the bone marrow [10]. Both of these perspectives taken into account, dopamine receptor agonists could have therapeutic value for the treatment of cancer in the future.

Schwann Cell-Stroma Interactions

Schwann cells are a major component of the PNS in addition to neurons. They derive from the embryonic neural crest, and following their development, they form either myelinating cells, or they bundle numerous unmyelinated axons together [39]. They play an essential role in maintaining the homeostasis in the PNS and regulating neuronal repair. These outstandingly plastic cells are also able to dedifferentiate and redifferentiate into different phenotypes to fulfill different tasks, including the demyelination of myelinating Schwann cells while upregulating the expression of L1 cell adhesion molecule (L1CAM), neural cell adhesion molecule (NCAM), p75 neurotrophin receptor (p75NTR), and glial fibrillary acidic protein (GFAP) or the obtainment of a repair phenotype including the upregulation of proteins such as glial cell derived neurotrophic factor (GDNF), artemin, NGF, and VEGF following nerve injury [31, 32]. Schwann cells are observed at neoplastic cell sites even before the onset of cancer [15]. Recent discoveries have revealed that the Schwann cells are key promoters of many cancers, including the pancreatic, cervical, lung, and oral cancer [13, 28, 64, 82, 83]. The mechanisms upon which Schwann cells enhance the aggressiveness of cancer cells include the CCL2/CCR2 mediated promotion of proliferation, migration, and EMT of cancer cells in cervical cancer [28] or the CXCL5/CXCR2 mediated upregulation of Snail and Twist and the resulting increase of the motility and EMT of cancer cells in lung cancer [82]. Moreover, Schwann cells secrete transforming growth factor ß (TGFß) to activate the TGFß-SMAD pathway in pancreatic cancer cells [62], and they have been shown to increase the perineural invasion of cancer cells through the secretion of L1CAM, chemoattracting cancer cells via the activation of MAP kinase signaling [53]. In the same article, Na'ara et al. have also shown L1CAM increases the MMP-2 and MMP-9 expression of cancer cells by activating STAT3, remodeling the extracellular matrix in favor of the invasiveness of cancer cells. Schwann cells have also been shown to suppress cancer-induced pain, delaying the diagnosis [16, 17]. Not only do they have a direct influence on cancer cells, but Schwann cells can also change the TME by modulating its immunological component. This includes their ability to attract myeloid derived suppressor cells and enhance their ability to suppress effector T cells [50] and also to promote the M2 polarization of macrophages, which have been described as tumor promoting due to their increased expression of IL-10, TGFß, and VEGF as well as the increased arginase pathway activity [9, 30, 83].

Even though very little is known about the interactions of Schwann cells with other non-immunological components of the TME, it is not hard to imagine that these cells could have a crucial role supporting other tumor-promoting cells of the TME, such as CAFs or stellate cells. Also, the potential role of Schwann cells in angiogenesis awaits investigation. Further understanding of the interactions of Schwann cells with other components of TME can reveal further potential therapeutic targets.

Conclusion

Nerves, fibroblasts, and endothelial cells are all major components of the TME. They play a crucial role in the development of many cancers and promote carcinogenesis, not only through their direct influence on cancer cells but also through their influence on each other. A closer inspection of the interactions of these cells in the cancer environment may enable us to understand the formation of TME more thoroughly. Moreover, nerves consist not only of neurons but also of Schwann cells. With the rising awareness of the crucial importance of Schwann cells in cancer, the future studies on Schwann cell-stroma interactions might shed light on an even less known perspective of cancer. All these aspects taken together, future studies on nerve-stroma interactions can put together a missing piece in the complex puzzle which cancer is, usher in new therapeutical strategies based on this piece, and improve patient survival.

Sonic hedgehog (SHH) signaling triggers pancreatic stellate cells to secrete nerve growth factor (NGF). NGF promotes the neural invasion of cancer cells [43]. NGF expression of cancer cells is also enhanced by factors such as CXCL12 or hepatocyte growth factor (HGF) [54, 77]. Paracrine SHH signaling also mediates the nerve outgrowth. Moreover, leukemia inhibitory factor (LIF) secreted by CAFs promotes the differentiation of Schwann cells, which induces neuronal plasticity [7].

Sympathetic signaling can change the gene expression profile of cancer-associated fibroblasts and stellate cells in favor of a cancer-promoting tumor microenvironment [55, 72]. Noradrenaline signaling can also trigger the proliferation of endothelial cells, enabling the angiogenic switch in carcinogenesis [80]. Dopamine signaling, on the other hand, inhibits the proliferation and migration of endothelial cells [66]. It also hinders the endothelial progenitor cells from leaving the bone marrow [10].

References

1. Albo, D., Akay, C. L., Marshall, C. L., Wilks, J. A., Verstovsek, G., Liu, H., Agarwal, N., Berger, D. H. & Ayala, G. E. 2011. Neurogenesis in colorectal cancer is a marker of aggressive tumor behavior and poor outcomes. *Cancer,* 117, 4834–45.
2. Asif, P. J., Longobardi, C., Hahne, M. & Medema, J. P. 2021. The Role of Cancer-Associated Fibroblasts in Cancer Invasion and Metastasis. *Cancers (Basel),* 13.
3. Ayala, G. E., Dai, H., Powell, M., Li, R., Ding, Y., Wheeler, T. M., Shine, D., Kadmon, D., Thompson, T., Miles, B. J., Ittmann, M. M. & Rowley, D. 2008. Cancer-related axonogenesis and neurogenesis in prostate cancer. *Clin Cancer Res,* 14, 7593–603.
4. Baeriswyl, V. & Christofori, G. 2009. The angiogenic switch in carcinogenesis. *Semin Cancer Biol,* 19, 329–37.
5. Bailey, J. M., Swanson, B. J., Hamada, T., Eggers, J. P., Singh, P. K., Caffery, T., Ouellette, M. M. & Hollingsworth, M. A. 2008. Sonic hedgehog promotes desmoplasia in pancreatic cancer. *Clin Cancer Res,* 14, 5995–6004.
6. Bakhtou, H., Olfatbakhsh, A., Deezagi, A. & Ahangari, G. 2019. The Expression of Dopamine Receptors Gene and their Potential Role in Targeting Breast Cancer Cells with Selective Agonist and Antagonist Drugs. Could it be the Novel Insight to Therapy? *Curr Drug Discov Technol,* 16, 184–197.
7. Bressy, C., Lac, S., Nigri, J., Leca, J., Roques, J., Lavaut, M. N., Secq, V., Guillaumond, F., Bui, T. T., Pietrasz, D., Granjeaud, S., Bachet, J. B., Ouaissi, M., Iovanna, J., Vasseur, S. & Tomasini, R. 2018. Lif Drives Neural Remodeling in Pancreatic Cancer and Offers a New Candidate Biomarker. *Cancer Res,* 78, 909–921.
8. Calon, A., Espinet, E., Palomo-Ponce, S., Tauriello, D. V., Iglesias, M., Céspedes, M. V., Sevillano, M., Nadal, C., Jung, P., Zhang, X. H., Byrom, D., Riera, A., Rossell, D., Mangues, R., Massagué, J., Sancho, E. & Batlle, E. 2012. Dependency of colorectal cancer on a Tgf-β-driven program in stromal cells for metastasis initiation. *Cancer Cell,* 22, 571–84.

9. Cendrowicz, E., Sas, Z., Bremer, E. & Rygiel, T. P. 2021. The Role of Macrophages in Cancer Development and Therapy. *Cancers (Basel)*, 13.
10. Chakroborty, D., Chowdhury, U. R., Sarkar, C., Baral, R., Dasgupta, P. S. & Basu, S. 2008. Dopamine regulates endothelial progenitor cell mobilization from mouse bone marrow in tumor vascularization. *J Clin Invest*, 118, 1380–9.
11. Chen, H., Liu, D., Yang, Z., Sun, L., Deng, Q., Yang, S., Qian, L., Guo, L., Yu, M., Hu, M., Shi, M. & Guo, N. 2014. Adrenergic signaling promotes angiogenesis through endothelial cell-tumor cell crosstalk. *Endocr Relat Cancer*, 21, 783–95.
12. Colombo, M., Mirandola, L., Chiriva-Internati, M., Basile, A., Locati, M., Lesma, E., Chiaramonte, R. & Platonova, N. 2018. Cancer Cells Exploit Notch Signaling to Redefine a Supportive Cytokine Milieu. *Front Immunol*, 9, 1823.
13. Deborde, S., Omelchenko, T., Lyubchik, A., Zhou, Y., He, S., Mcnamara, W. F., Chernichenko, N., Lee, S. Y., Barajas, F., Chen, C. H., Bakst, R. L., Vakiani, E., He, S., Hall, A. & Wong, R. J. 2016. Schwann cells induce cancer cell dispersion and invasion. *J Clin Invest*, 126, 1538–54.
14. Demir, I. E., Friess, H. & Ceyhan, G. O. 2012. Nerve-cancer interactions in the stromal biology of pancreatic cancer. *Front Physiol*, 3, 97.
15. Demir, I. E., Boldis, A., Pfitzinger, P. L., Teller, S., Brunner, E., Klose, N., Kehl, T., Maak, M., Lesina, M., Laschinger, M., Janssen, K. P., Algül, H., Friess, H. & Ceyhan, G. O. 2014. Investigation of Schwann cells at neoplastic cell sites before the onset of cancer invasion. *J Natl Cancer Inst*, 106.
16. Demir, I. E., Tieftrunk, E., Schorn, S., Saricaoglu Ö. C., Pfitzinger, P. L., Teller, S., Wang, K., Waldbaur, C., Kurkowski, M. U., Wörmann, S. M., Shaw, V. E., Kehl, T., Laschinger, M., Costello, E., Algül, H., Friess, H. & Ceyhan, G. O. 2016. Activated Schwann cells in pancreatic cancer are linked to analgesia via suppression of spinal astroglia and microglia. *Gut*, 65, 1001–14.
17. Demir, I. E., Kujundzic, K., Pfitzinger, P. L., Saricaoglu ÖC, Teller, S., Kehl, T., Reyes, C. M., Ertl, L. S., Miao, Z., Schall, T. J., Tieftrunk, E., Haller, B., Diakopoulos, K. N., Kurkowski, M. U., Lesina, M., Krüger, A., Algül, H., Friess, H. & Ceyhan, G. O. 2017. Early pancreatic cancer lesions suppress pain through Cxcl12-mediated chemoattraction of Schwann cells. *Proc Natl Acad Sci U S A*, 114, E85–E94.
18. Denton, A. E., Roberts, E. W. & Fearon, D. T. 2018. Stromal Cells in the Tumor Microenvironment. *Adv Exp Med Biol*, 1060, 99–114.
19. Dror, S., Sander, L., Schwartz, H., Sheinboim, D., Barzilai, A., Dishon, Y., Apcher, S., Golan, T., Greenberger, S., Barshack, I., Malcov, H., Zilberberg, A., Levin, L., Nessling, M., Friedmann, Y., Igras, V., Barzilay, O., Vaknine, H., Brenner, R., Zinger, A., Schroeder, A., Gonen, P., Khaled, M., Erez, N., Hoheisel, J. D. & Levy, C. 2016. Melanoma mirna trafficking controls tumour primary niche formation. *Nat Cell Biol*, 18, 1006–17.
20. Dwivedi, S., Bautista, M., Shrestha, S., Elhasasna, H., Chaphekar, T., Vizeacoumar, F. S. & Krishnan, A. 2021. Sympathetic signaling facilitates progression of neuroendocrine prostate cancer. *Cell Death Discov*, 7, 364.
21. Elyada, E., Bolisetty, M., Laise, P., Flynn, W. F., Courtois, E. T., Burkhart, R. A., Teinor, J. A., Belleau, P., Biffi, G., Lucito, M. S., Sivajothi, S., Armstrong, T. D., Engle, D. D., Yu, K. H., Hao, Y., Wolfgang, C. L., Park, Y., Preall, J., Jaffee, E. M., Califano, A., Robson, P. & Tuveson, D. A. 2019. Cross-Species Single-Cell Analysis of Pancreatic Ductal Adenocarcinoma Reveals Antigen-Presenting Cancer-Associated Fibroblasts. *Cancer Discov*, 9, 1102–1123.
22. Feig, C., Gopinathan, A., Neesse, A., Chan, D. S., Cook, N. & Tuveson, D. A. 2012. The pancreas cancer microenvironment. *Clin Cancer Res*, 18, 4266–76.
23. Folkman, J. 2002. Role of angiogenesis in tumor growth and metastasis. *Semin Oncol*, 29, 15–8.
24. Folkman, J., Watson, K., Ingber, D. & Hanahan, D. 1989. Induction of angiogenesis during the transition from hyperplasia to neoplasia. *Nature*, 339, 58–61.
25. Fujii-Nishimura, Y., Yamazaki, K., Masugi, Y., Douguchi, J., Kurebayashi, Y., Kubota, N., Ojima, H., Kitago, M., Shinoda, M., Hashiguchi, A. & Sakamoto, M. 2018. Mesenchymal-epithelial

transition of pancreatic cancer cells at perineural invasion sites is induced by Schwann cells. *Pathol Int,* 68, 214–223.
26. Grimm, D., Bauer, J. & Schoenberger, J. 2009. Blockade of neoangiogenesis, a new and promising technique to control the growth of malignant tumors and their metastases. *Curr Vasc Pharmacol,* 7, 347–57.
27. Han, L., Ma, J., Duan, W., Zhang, L., Yu, S., Xu, Q., Lei, J., Li, X., Wang, Z., Wu, Z., Huang, J. H., Wu, E., Ma, Q. & Ma, Z. 2016. Pancreatic stellate cells contribute pancreatic cancer pain via activation of shh signaling pathway. *Oncotarget,* 7, 18146–58.
28. Huang, T., Fan, Q., Wang, Y., Cui, Y., Wang, Z., Yang, L., Sun, X. & Wang, Y. 2020. Schwann Cell-Derived Ccl2 Promotes the Perineural Invasion of Cervical Cancer. *Front Oncol,* 10, 19.
29. Hwang, R. F., Moore, T., Arumugam, T., Ramachandran, V., Amos, K. D., Rivera, A., Ji, B., Evans, D. B. & Logsdon, C. D. 2008. Cancer-associated stromal fibroblasts promote pancreatic tumor progression. *Cancer Res,* 68, 918–26.
30. Italiani, P. & Boraschi, D. 2014. From Monocytes to M1/M2 Macrophages: Phenotypical vs. Functional Differentiation. *Front Immunol,* 5, 514.
31. Jessen, K. R. & Mirsky, R. 2005. The origin and development of glial cells in peripheral nerves. *Nat Rev Neurosci,* 6, 671–82.
32. Jessen, K. R., Mirsky, R. & Lloyd, A. C. 2015. Schwann Cells: Development and Role in Nerve Repair. *Cold Spring Harb Perspect Biol,* 7, a020487.
33. Jiang, J., Bai, J., Qin, T., Wang, Z. & Han, L. 2020. Ngf from pancreatic stellate cells induces pancreatic cancer proliferation and invasion by Pi3K/Akt/Gsk signal pathway. *J Cell Mol Med,* 24, 5901–5910.
34. Josson, S., Gururajan, M., Sung, S. Y., Hu, P., Shao, C., Zhau, H. E., Liu, C., Lichterman, J., Duan, P., Li, Q., Rogatko, A., Posadas, E. M., Haga, C. L. & Chung, L. W. 2015. Stromal fibroblast-derived miR-409 promotes epithelial-to-mesenchymal transition and prostate tumorigenesis. *Oncogene,* 34, 2690–9.
35. Junttila, M. R. & De Sauvage, F. J. 2013. Influence of tumour micro-environment heterogeneity on therapeutic response. *Nature,* 501, 346–54.
36. Jurcak, N. & Zheng, L. 2019. Signaling in the microenvironment of pancreatic cancer: Transmitting along the nerve. *Pharmacol Ther,* 200, 126–134.
37. Kalluri, R. 2016. The biology and function of fibroblasts in cancer. *Nat Rev Cancer,* 16, 582–98.
38. Kalluri, R. & Weinberg, R. A. 2009. The basics of epithelial-mesenchymal transition. *J Clin Invest,* 119, 1420–8.
39. Kidd, G. J., Ohno, N. & Trapp, B. D. 2013. Biology of Schwann cells. *Handb Clin Neurol,* 115, 55–79.
40. Lan, Y. L., Wang, X., Xing, J. S., Lou, J. C., Ma, X. C. & Zhang, B. 2017. The potential roles of dopamine in malignant glioma. *Acta Neurol Belg,* 117, 613–621.
41. Lee, J. J., Perera, R. M., Wang, H., Wu, D. C., Liu, X. S., Han, S., Fitamant, J., Jones, P. D., Ghanta, K. S., Kawano, S., Nagle, J. M., Deshpande, V., Boucher, Y., Kato, T., Chen, J. K., Willmann, J. K., Bardeesy, N. & Beachy, P. A. 2014. Stromal response to Hedgehog signaling restrains pancreatic cancer progression. *Proc Natl Acad Sci U S A,* 111, E3091–E3100.
42. Li, Z. J. & Cho, C. H. 2011. Neurotransmitters, more than meets the eye--neurotransmitters and their perspectives in cancer development and therapy. *Eur J Pharmacol,* 667, 17–22.
43. Li, X., Wang, Z., Ma, Q., Xu, Q., Liu, H., Duan, W., Lei, J., Ma, J., Wang, X., Lv, S., Han, L., Li, W., Guo, J., Guo, K., Zhang, D., Wu, E. & Xie, K. 2014. Sonic hedgehog paracrine signaling activates stromal cells to promote perineural invasion in pancreatic cancer. *Clin Cancer Res,* 20, 4326–38.
44. Li, S., Xu, H. X., Wu, C. T., Wang, W. Q., Jin, W., Gao, H. L., Li, H., Zhang, S. R., Xu, J. Z., Qi, Z. H., Ni, Q. X., Yu, X. J. & Liu, L. 2019. Angiogenesis in pancreatic cancer: current research status and clinical implications. *Angiogenesis,* 22, 15–36.
45. Liu, X., Wu, W. K., Yu, L., Sung, J. J., Srivastava, G., Zhang, S. T. & Cho, C. H. 2008. Epinephrine stimulates esophageal squamous-cell carcinoma cell proliferation via

beta-adrenoceptor-dependent transactivation of extracellular signal-regulated kinase/cyclooxygenase-2 pathway. *J Cell Biochem,* 105, 53–60.
46. Liu, Q., Zhang, R., Zhang, X., Liu, J., Wu, H., Li, Y., Cui, M., Li, T., Song, H., Gao, J., Zhang, Y., Yang, S. & Liao, Q. 2021. Dopamine improves chemotherapeutic efficacy for pancreatic cancer by regulating macrophage-derived inflammations. *Cancer Immunol Immunother,* 70, 2165–2177.
47. Lynch, M. D. & Watt, F. M. 2018. Fibroblast heterogeneity: implications for human disease. *J Clin Invest,* 128, 26–35.
48. Magnon, C., Hall, S. J., Lin, J., Xue, X., Gerber, L., Freedland, S. J. & Frenette, P. S. 2013. Autonomic nerve development contributes to prostate cancer progression. *Science,* 341, 1236361.
49. March, B., Faulkner, S., Jobling, P., Steigler, A., Blatt, A., Denham, J. & Hondermarck, H. 2020. Tumour innervation and neurosignalling in prostate cancer. *Nat Rev Urol,* 17, 119–130.
50. Martyn, G. V., Shurin, G. V., Keskinov, A. A., Bunimovich, Y. L. & Shurin, M. R. 2019. Schwann cells shape the neuro-immune environs and control cancer progression. *Cancer Immunol Immunother,* 68, 1819–1829.
51. Mu, W., Wang, Z. & Zöller, M. 2019. Ping-Pong-Tumor and Host in Pancreatic Cancer Progression. *Front Oncol,* 9, 1359.
52. Mukherjee, A., Ha, P., Wai, K. C. & Naara, S. 2022. The Role of Ecm Remodeling, Emt, and Adhesion Molecules in Cancerous Neural Invasion: Changing Perspectives. *Adv Biol (Weinh),* e2200039.
53. Na'ara, S., Amit, M. & Gil, Z. 2019. L1cam induces perineural invasion of pancreas cancer cells by upregulation of metalloproteinase expression. *Oncogene,* 38, 596–608.
54. Nan, L., Qin, T., Xiao, Y., Qian, W., Li, J., Wang, Z., Ma, J., Ma, Q. & Wu, Z. 2019. Pancreatic Stellate Cells Facilitate Perineural Invasion of Pancreatic Cancer via Hgf/c-Met Pathway. *Cell Transplant,* 28, 1289–1298.
55. Oben, J. A., Yang, S., Lin, H., Ono, M. & Diehl, A. M. 2003. Norepinephrine and neuropeptide Y promote proliferation and collagen gene expression of hepatic myofibroblastic stellate cells. *Biochem Biophys Res Commun,* 302, 685–90.
56. Öhlund, D., Handly-Santana, A., Biffi, G., Elyada, E., Almeida, A. S., Ponz-Sarvise, M., Corbo, V., Oni, T. E., Hearn, S. A., Lee, E. J., Chio, Ii, Hwang, C. I., Tiriac, H., Baker, L. A., Engle, D. D., Feig, C., Kultti, A., Egeblad, M., Fearon, D. T., Crawford, J. M., Clevers, H., Park, Y. & Tuveson, D. A. 2017. Distinct populations of inflammatory fibroblasts and myofibroblasts in pancreatic cancer. *J Exp Med,* 214, 579–596.
57. Omary, M. B., Lugea, A., Lowe, A. W. & Pandol, S. J. 2007. The pancreatic stellate cell: a star on the rise in pancreatic diseases. *J Clin Invest,* 117, 50–9.
58. Orimo, A. & Weinberg, R. A. 2006. Stromal fibroblasts in cancer: a novel tumor-promoting cell type. *Cell Cycle,* 5, 1597–601.
59. Pang, W., Su, J., Wang, Y., Feng, H., Dai, X., Yuan, Y., Chen, X. & Yao, W. 2015. Pancreatic cancer-secreted miR-155 implicates in the conversion from normal fibroblasts to cancer-associated fibroblasts. *Cancer Sci,* 106, 1362–9.
60. Pein, M., Insua-Rodríguez, J., Hongu, T., Riedel, A., Meier, J., Wiedmann, L., Decker, K., Essers, M. A. G., Sinn, H. P., Spaich, S., Sütterlin, M., Schneeweiss, A., Trumpp, A. & Oskarsson, T. 2020. Metastasis-initiating cells induce and exploit a fibroblast niche to fuel malignant colonization of the lungs. *Nat Commun,* 11, 1494.
61. Rhim, A. D., Oberstein, P. E., Thomas, D. H., Mirek, E. T., Palermo, C. F., Sastra, S. A., Dekleva, E. N., Saunders, T., Becerra, C. P., Tattersall, I. W., Westphalen, C. B., Kitajewski, J., Fernandez-Barrena, M. G., Fernandez-Zapico, M. E., Iacobuzio-Donahue, C., Olive, K. P. & Stanger, B. Z. 2014. Stromal elements act to restrain, rather than support, pancreatic ductal adenocarcinoma. *Cancer Cell,* 25, 735–47.
62. Roger, E., Martel, S., Bertrand-Chapel, A., Depollier, A., Chuvin, N., Pommier, R. M., Yacoub, K., Caligaris, C., Cardot-Ruffino, V., Chauvet, V., Aires, S., Mohkam, K., Mabrut, J. Y., Adham, M., Fenouil, T., Hervieu, V., Broutier, L., Castets, M., Neuzillet, C., Cassier, P. A.,

Tomasini, R., Sentis, S. & Bartholin, L. 2019. Schwann cells support oncogenic potential of pancreatic cancer cells through Tgfβ signaling. *Cell Death Dis,* 10, 886.
63. Rubí, B. & Maechler, P. 2010. Minireview: new roles for peripheral dopamine on metabolic control and tumor growth: let's seek the balance. *Endocrinology,* 151, 5570–81.
64. Salvo, E., Tu, N. H., Scheff, N. N., Dubeykovskaya, Z. A., Chavan, S. A., Aouizerat, B. E. & Ye, Y. 2021. Tnfα promotes oral cancer growth, pain, and Schwann cell activation. *Sci Rep,* 11, 1840.
65. Santos, J. C., Ribeiro, M. L., Sarian, L. O., Ortega, M. M. & Derchain, S. F. 2016. Exosomes-mediate micrornas transfer in breast cancer chemoresistance regulation. *Am J Cancer Res,* 6, 2129–2139.
66. Sarkar, C., Chakroborty, D., Mitra, R. B., Banerjee, S., Dasgupta, P. S. & Basu, S. 2004. Dopamine in vivo inhibits Vegf-induced phosphorylation of Vegfr-2, Mapk, and focal adhesion kinase in endothelial cells. *Am J Physiol Heart Circ Physiol,* 287, H1554–H1560.
67. Sarkar, C., Chakroborty, D., Chowdhury, U. R., Dasgupta, P. S. & Basu, S. 2008. Dopamine increases the efficacy of anticancer drugs in breast and colon cancer preclinical models. *Clin Cancer Res,* 14, 2502–10.
68. Schorn, S., Demir, I. E., Haller, B., Scheufele, F., Reyes, C. M., Tieftrunk, E., Sargut, M., Goess, R., Friess, H. & Ceyhan, G. O. 2017. The influence of neural invasion on survival and tumor recurrence in pancreatic ductal adenocarcinoma - A systematic review and meta-analysis. *Surg Oncol,* 26, 105–115.
69. Shan, T., Chen, S., Chen, X., Lin, W. R., Li, W., Ma, J., Wu, T., Cui, X., Ji, H., Li, Y. & Kang, Y. 2017. Cancer-associated fibroblasts enhance pancreatic cancer cell invasion by remodeling the metabolic conversion mechanism. *Oncol Rep,* 37, 1971–1979.
70. Silverman, D. A., Martinez, V. K., Dougherty, P. M., Myers, J. N., Calin, G. A. & Amit, M. 2021. Cancer-Associated Neurogenesis and Nerve-Cancer Cross-talk. *Cancer Res,* 81, 1431–1440.
71. Sun, Q., Zhang, B., Hu, Q., Qin, Y., Xu, W., Liu, W., Yu, X. & Xu, J. 2018. The impact of cancer-associated fibroblasts on major hallmarks of pancreatic cancer. *Theranostics,* 8, 5072–5087.
72. Szpunar, M. J., Burke, K. A., Dawes, R. P., Brown, E. B. & Madden, K. S. 2013. The antidepressant desipramine and α2-adrenergic receptor activation promote breast tumor progression in association with altered collagen structure. *Cancer Prev Res (Phila),* 6, 1262–72.
73. Thaker, P. H., Han, L. Y., Kamat, A. A., Arevalo, J. M., Takahashi, R., Lu, C., Jennings, N. B., Armaiz-Pena, G., Bankson, J. A., Ravoori, M., Merritt, W. M., Lin, Y. G., Mangala, L. S., Kim, T. J., Coleman, R. L., Landen, C. N., Li, Y., Felix, E., Sanguino, A. M., Newman, R. A., Lloyd, M., Gershenson, D. M., Kundra, V., Lopez-Berestein, G., Lutgendorf, S. K., Cole, S. W. & Sood, A. K. 2006. Chronic stress promotes tumor growth and angiogenesis in a mouse model of ovarian carcinoma. *Nat Med,* 12, 939–44.
74. Thompson, C. B. & Palm, W. 2016. Reexamining How Cancer Cells Exploit the Body's Metabolic Resources. *Cold Spring Harb Symp Quant Biol,* 81, 67–72.
75. Tilan, J. & Kitlinska, J. 2010. Sympathetic Neurotransmitters and Tumor Angiogenesis-Link between Stress and Cancer Progression. *J Oncol,* 2010, 539706.
76. Wu, Y. S., Chung, I., Wong, W. F., Masamune, A., Sim, M. S., & Looi, C. Y. 2017. Paracrine IL-6 signaling mediates the effects of pancreatic stellate cells on epithelial-mesenchymal transition via Stat3/Nrf2 pathway in pancreatic cancer cells. *Biochimica et Biophysica Acta (BBA) – General Subjects, 1861*(2), 296–306. https://doi.org/10.1016/j.bbagen.2016.10.006
77. Xu, Q., Wang, Z., Chen, X., Duan, W., Lei, J., Zong, L., Li, X., Sheng, L., Ma, J., Han, L., Li, W., Zhang, L., Guo, K., Ma, Z., Wu, Z., Wu, E. & Ma, Q. 2015. Stromal-derived factor-1α/Cxcl12-Cxcr4 chemotactic pathway promotes perineural invasion in pancreatic cancer. *Oncotarget,* 6, 4717–32.
78. Yang, F., Ning, Z., Ma, L., Liu, W., Shao, C., Shu, Y. & Shen, H. 2017. Exosomal mirnas and mirna dysregulation in cancer-associated fibroblasts. *Mol Cancer,* 16, 148.

79. Yauch, R. L., Gould, S. E., Scales, S. J., Tang, T., Tian, H., Ahn, C. P., Marshall, D., Fu, L., Januario, T., Kallop, D., Nannini-Pepe, M., Kotkow, K., Marsters, J. C., Rubin, L. L. & De Sauvage, F. J. 2008. A paracrine requirement for hedgehog signalling in cancer. *Nature,* 455, 406–10.
80. Zahalka, A. H., Arnal-Estapé, A., Maryanovich, M., Nakahara, F., Cruz, C. D., Finley, L. W. S. & Frenette, P. S. 2017. Adrenergic nerves activate an angio-metabolic switch in prostate cancer. *Science,* 358, 321–326.
81. Zhang, Y., Crawford, H. C. & Pasca Di Magliano, M. 2019. Epithelial-Stromal Interactions in Pancreatic Cancer. *Annu Rev Physiol,* 81, 211–233.
82. Zhou, Y., Shurin, G. V., Zhong, H., Bunimovich, Y. L., Han, B. & Shurin, M. R. 2018. Schwann Cells Augment Cell Spreading and Metastasis of Lung Cancer. *Cancer Res,* 78, 5927–5939.
83. Zhou, Y., Li, J., Han, B., Zhong, R. & Zhong, H. 2020. Schwann cells promote lung cancer proliferation by promoting the M2 polarization of macrophages. *Cell Immunol,* 357, 104211.

Chapter 9
Reprogrammed Schwann Cells in Cancer

Sylvie Deborde and Richard J. Wong

Introduction

Schwann cells (SCs) are glial cells that ensheath axons of neuron in the peripheral nervous system and are necessary for their maintenance and function. They originate from neural crest cells and differentiate in subpopulation of SCs [19, 22, 38] (Fig. 9.1). At the adult stage, there are myelinating and non-myelinating SCs. Myelinating SCs form the insulating myelin sheaths of peripheral axons. They produce myelin, a lipid rich membrane with specific proteins, that allows the fast transmission of neuronal depolarization. Non-myelinating SCs include Remak SCs, which ensheath multiple small-caliber axons in "Remak bundles." Remak SCs play an important role in the development, maintenance, and regeneration of nerves. Other non-myelinating SCs include terminal SCs and nerve repair SCs. Terminal SCs are present at the neuromuscular junction and provide chemical and physical support to the neuromuscular synapse. Nerve repair SCs accumulate at sites of nerve injury after transcriptional reprogramming of myelinating and Remak SCs in response to trauma. The ability of SCs to reprogram (dedifferentiate) and redifferentiate into myelinating SCs at an adult stage reflects the highly plastic nature of these cells. SC reprogramming occurs in cancer among other stress conditions [19, 22, 38]. In this review, we report on the expanding recognition of SCs, and more specifically reprogrammed SCs, as active contributors to cancer progression.

S. Deborde (✉) · R. J. Wong (✉)
Department of Surgery, Memorial Sloan Kettering Cancer Center, New York, NY, USA

David M. Rubenstein Center for Pancreatic Cancer Research, Memorial Sloan Kettering Cancer Center, New York, NY, USA
e-mail: debordes@mskcc.org; wongr@mskcc.org

M. Amit, N. N. Scheff (eds.), *Cancer Neuroscience*,
https://doi.org/10.1007/978-3-031-32429-1_9

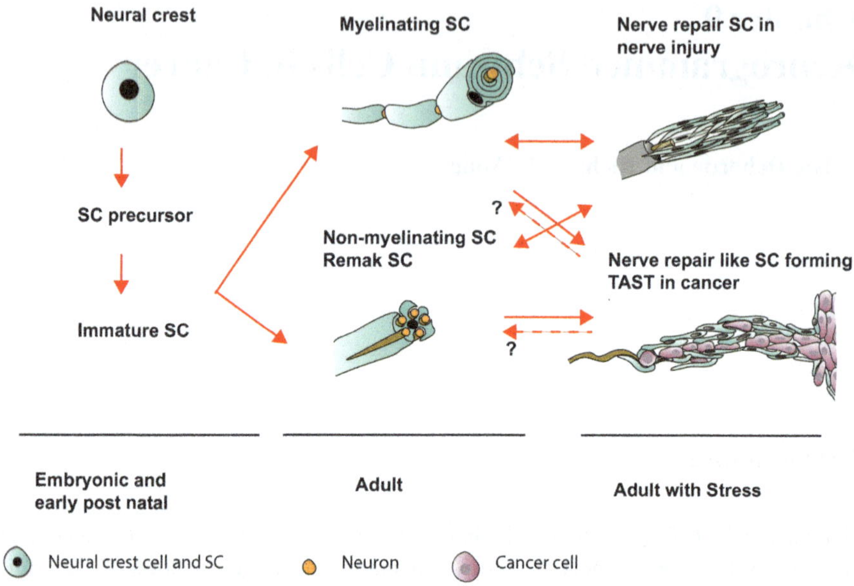

Fig. 9.1 Transitions in the Schwann cell (SC) lineage. Developmental transitions and stress induced transitions at the adult stage. Stress includes nerve injury and cancer. SCs derive from neural crest cells and transition into SC precursors and then into immature SCs. Immature SCs differentiate into myelinating SCs or non-myelinating SCs. At adult stage, both myelinating and non-myelinating SCs can reprogram. They reprogram into nerve repair SCs after nerve injury and nerve repair like SCs that form tumor activated SC tracks (TASTs) in cancer. Redifferentiation into myelinating and Remak SC occurs after nerve repair. This might occur in tumor but remains to be demonstrated

Clinical Significance of the Presence of SCs in Cancer

Recent advances in bioinformatic and single cell analysis have led to the determination of gene signatures of many specific cell types, including SC subtypes [43]. These gene signatures allow for correlations of gene expression with patient outcome. SC signatures, including the non-myelinating SC signature, correlate with patient outcomes in pancreatic adenocarcinoma; high signature scores correlate with the worse survival and progression-free survival in a TCGA cohort of patients with pancreatic adenocarcinoma and in a second cohort of patients with short term (<1 year) survival [6]. In contrast, a cohort of rare Memorial Sloan Kettering Cancer Center long term survival patients (>4 years) showed that elevated SC signatures were associated with improved survival [6].

Pancreatic ductal adenocarcinoma (PDAC) can be sorted into subtypes based on tumor gene expression profiles. One classification presents four subtypes: squamous, pancreatic progenitor, immunogenic, and aberrantly differentiated endocrine exocrine (ADEX) subtypes [2]. SC signature scores vary with the subtypes with the highest scores for the immunogenic subtypes. The overall order of the SC scores in PDAC is immunogenic > ADEX = squamous > progenitor [6]. Further studies are needed to elucidate how exactly these different subtypes of PDAC influence SC presence in cancer.

A correlation between SCs and PDAC patient survival is also observed by detecting glial fibrillary acidic protein (GFAP) expression, an indication of activated or reprogrammed SCs, with immunostaining of surgical specimen sections of PDAC [6]. Patients with high content of GFAP expressing cells have a shorter life survival.

Reprogrammed SCs

Reprogrammed SCs in Nerve Repair and in Mycobacterium leprae *Infection*

The reprogramming of SCs occurs during nerve repair after nerve injury [1, 18], during *Mycobacterium leprae* infection [25], and in several types of cancer [6, 36, 44]. The ability to reprogram adult stage SCs to a more undifferentiated SC type reflects the very plastic and dynamic nature of SCs. During nerve repair, SC reprogramming leads to a change in the expression of about 4000 genes, which is coordinated by c-Jun, Notch, Sox2, and MAPK signaling in addition to other factors [1, 18]. Nerve repair SCs are very motile, release neurotrophic factors and chemokines, recruit immune cells, clear redundant myelin by phagocytosis, and organize into Büngner bands, which are tracks that guide the regenerated axons [21].

SCs also reprogram after bacterial infection. *Mycobacterium leprae*, which causes human leprosy, infects SCs and causes subsequent neurological injury, inducing sensorimotor loss [26, 35, 39]. When infected, SCs downregulate genes associated with myelin differentiation including myelin basic protein (*Mbp*), myelin protein zero (*Mpz*), and early growth response-2 (*Krox20/Egr2*) and upregulate genes associated with mesoderm development including *Snail1*, *Msx2*, *Six1*, *Twist1*, *Hoxb13*, and *Bmp6*. This SC reprogramming changes migratory and immunomodulatory properties, facilitating bacterial dissemination [25].

Reprogrammed SCs in Cancer

Reprogrammed SCs are also detected in tumors in a variety of cancer types including pancreatic cancer [6, 7, 9], thyroid cancer [7], colon cancer [9], lung cancer [37], neuroblastic cancer [44], oral cancer [33], and skin cancer [36]. Transcriptome analyses have revealed information on SC reprogramming in pancreatic cancer [6], lung cancer [37], melanoma [36], oral cancer [33], and ganglioneuroma [44]. Although single SC transcriptome data in cancer are not yet available, analysis using bulk tumor samples have shown genes characteristics of SCs in tumor, implying their presence and activity [6].

In patients with pancreatic adenocarcinoma, the transcriptome analysis of the tumors not only reveals an association of SC gene signatures with patient outcomes, but it also indicates that the gene expression signature of non-myelinating SCs and myelinating SCs associates with different pathways, supporting their roles in different functions. Non-myelinating SC signature positively correlates with axonal guidance and with several pathways related to cancer invasion, including epithelial-mesenchymal transition (EMT), MAPK signaling, PI3-Akt signaling, and extracellular matrix (ECM) organization. In contrast, the myelinating SC signature positively correlates with gene sets related to immune cells and immune function [6]. A human SC line cocultured with pancreatic cancer cells (MiaPaCa-2) has a gene expression signature similar to non-myelinating SCs, and this signature also correlates with poor prognosis in the TCGA dataset. Some of the SC genes upregulated under cancer cell coculture conditions in vitro are involved in axon guidance and MAPK signaling pathways. These pathways are also associated with the transcription factor c-Jun, which is known to coordinate SC reprogramming in nerve repair [1, 18]. c-Jun knockout (KO) SCs cocultured with pancreatic cancer cells have reduced expression of axon guidance genes than control SCs cocultured with cancer cells. These genes include *HRAS*, *CAMK2B*, *WNT4*, *EPHA8*, and *NFATC2*. Furthermore, genes known to be involved in nerve repair and dependent on c-Jun expression are enriched in SCs cultured with cancer cells as compared to SCs grown alone [6].

In lung cancer (lung adenocarcinoma and lung squamous cell carcinoma), genes associated with reprogrammed (dedifferentiated) SCs, including GFAP, NGFR, and GAP43, had higher expression than in normal lung samples. Other genes thought to be associated with SCs include the axon guidance genes Slit2 and Robo2 that are downregulated [37]. The study reports that high GAP43 expression, a marker of undifferentiated SC, is associated with shorter survival. In contrast, low Slit2 and Robo2 expression are associated with longer survival.

In patients with melanoma, high levels of reprogrammed SCs are present in the tissue adjacent to the tumor [36]. Additionally, microdissection of tissue adjacent to the tumor revealed increased expression of genes participating to neuro-regeneration, enriched in axon guidance and immune response pathways, and known to be upregulated after nerve injury. The gene expression profile of SCs cultured in melanoma-conditioned medium is similar to a repair SC profile. They have an increased expression of genes such as Id2, Egr1, c-Jun, and Sox2 and a downregulation of genes promoting myelination such as Sox10, Egr2, and Oct6. Erk and Akt signaling pathways are activated in these cells as in nerve repair SCs. Similar to nerve repair SCs, SCs cultured with melanoma-conditioned medium increase the expression of genes involved in nerve repair, ECM reorganization, immune modulation, chemotaxis, and myelin phagocytosis [36].

In patients with oral cancer, spatial transcriptome analysis of tissue localized either close or far away from the cancer revealed that nerves in close proximity to cancer exhibit increased stress and growth response changes [33]. A lower expression of myelin basic protein (*MBP*) and increased expression of annexin 2 (*ANXA2*) in SCs close to cancer are consistent with SC response to stress. Another SC related gene is *DDX5*, which downregulates *MBP* posttranscriptionally and was found more highly expressed in tissue samples near the tumor. The analysis also reveals a downregulation of protein receptor-type tyrosine-protein phosphatase zeta (*PTPRZ1*) near the tumor, a gene coding for the putative receptor for NCAM [41], a known protein expressed by SC and involved in perineural invasion (PNI) [7]. Nerves near tumor also have genes known to respond to stress and aspects of axonogenesis (e.g., changes in *CFL1* and *MARCKS*) [33].

Transcriptomic and proteomic analysis of human stromal SCs in ganglioneuromas also found increased expression of the nerve repair-associated genes [44] including genes involved in axon guidance, lipid/myelin degradation/metabolism, basement membrane formation or ECM organization, phagocyte attraction, and MHC-II mediated immune regulation. Among the ECM molecules that increase their expression in SCs, epidermal growth factor like protein 8 (EGFL8) induces neurite outgrowth and neuronal differentiation of neuroblastoma cells in culture, acting as a neuritogen that rewires cellular signaling by activating kinases involved in neurogenesis. The activated kinases induced by EGFL8 include HIPK1, p38b/MAPK11, ERK5/MAPK7, SGK1, and TLK2 [44].

In the cancer context, SCs conserve their nerve repair functions that include the ability to promote axon regrowth and to attract immune cells. In addition, SCs can promote cancer invasion. In most cancer types, reprogrammed SCs have a pro-tumorigenic effect by stimulating cancer growth and cancer migration or modifying immune response [8]. In contrast, SCs are present in the less aggressive forms of neuroblastoma and are proposed to provide an anti-tumorigenic effect through their neuroprotective and neuritogenic factors [44].

Mechanism of SC Reprogramming

As in nerve repair, the mechanism of SC reprogramming in cancer involves the transcription factor c-Jun [6]. SC c-Jun is activated in proximity of pancreatic cancer cells in vivo and in vitro [6]. In human specimens of PDAC, SCs close to cancer cells express higher level of activated c-Jun, revealed by phospho-c-Jun staining, as compared to SCs further away from cancer cells as well as SCs in the pancreatic tissue patients without cancer. Similar observations are made in an in vivo model of PNI in which cancer cells are injected under the sciatic nerve sheath and invade the nerve. In this model, SCs at the vicinity of cancer cells express phospho-c-Jun (p-c-Jun) but not SCs further away from the tumor. In an in vitro model, SCs mixed with cancer cells reveals higher SC p-c-Jun as compared to SCs that are grown alone.

The factors that specifically activate c-Jun or induce SC reprogramming following nerve trauma and in cancer invasion remain undefined to date. SC reprogramming may be prompted by cancer cells that induce nerve injury in the tissue adjacent to the tumor based on melanoma cancer models [36]. A recent study shows that oral cancer cells, which express high levels of galanin, treated with palmitic acid are epigenetically modified and highly metastatic and can activate SCs that remodel the ECM [29]. Interestingly galanin is a neuropeptide affecting glial cell myelination

[14], and its expression can be regulated by c-Jun [30]. Furthermore, the highly metastatic cancer cells are characterized by a stable change in the histone mark H3K4me3 and an upregulation of genes involved in neurogenesis and neural remodeling. Some promoters of these genes have a binding signal for early growth response 2 (EGR2), indicating a central role for EGR2 [29]. Interestingly, EGR2 is also a protein involved in SC myelination [28].

Interleukin-6 (IL-6), an inflammatory cytokine, also plays a role in SC activation. IL-6 is released in presence of hypoxia, cancer cells, and neuroinflammation [10]. IL-6 is also produced during nerve repair and is involved in SC ability to recruit macrophages [42]. However, IL-6 may not be critical in the overall process of nerve repair since the loss of IL-6 (i.e., IL-6 global knockout mice) does not prevent the process of nerve repair [17], although compensatory mechanisms could be involved given the diffuseness of the interleukin system.

Another inflammatory cytokine, TNFα, has been shown to activate SCs. TNFα treatments increase levels of c-Jun, GFAP, and p75NTR in cultured SCs, whereas it decreases MBP expression. TNFα stimulates oral cancer proliferation, cytokine production, and nociception in mice with oral cancer. Interestingly, Salvo et al. report that activated SCs contribute to pain behavior in oral cancer and blocking TNFα prevents the nociceptive behaviors induced by the activated SCs [32].

Function of Reprogrammed SCs in Cancer Proliferation, Migration, and Invasion

SCs stimulate cancer cell proliferation, increase cancer migration and invasion, participate in both innate and adaptive immune response, and contribute to pain conduction as previously reviewed [8]. Many of these functions are enabled by SCs that are reprogrammed in response to nerve injury or directly by cancer cells. Here, we focus on describing the role of SCs in promoting cancer invasion.

It is noteworthy that the repair SCs involved in nerve repair represent a transient cell state (Fig. 9.1). These cells redifferentiate into myelinating and Remak SCs in regenerated nerves [20]. Although cancer may be considered a wound that cannot heal [5], it is likely that some reprogrammed SCs in tumor also redifferentiate into myelinating and Remak SCs. This would lead to a heterogenous population of SCs including the myelinating SCs, the non-myelinating Remak SCs, and reprogrammed SCs with several different functions.

SC Secretory Factors and Cancer Progression

SCs can impact cancer progression at the primary tumor and during metastasis. PNI is a form of metastasis defined as the invasion of cancer cells in or around the nerves. PNI allows cancer cells to propagate away from the primary site by following neural pathways and is associated with pain, paralysis, and low patient survival [4]. Several studies implicate the role for SCs in the process of PNI in which cancer cells grow and migrate [8]. In addition to promoting PNI, SCs affect the conventional circulatory metastatic route, as reported in the study showing that SCs enable the progression of highly metastatic palmitate-treated oral cancer cells [29].

Several studies implicate the release of SC-secreted factors in the stimulation of cancer progression, as defined by tumor growth, migration, and invasion (Fig. 9.2). The SC-derived neurotrophic factors, such as nerve growth factor (NGF), brain-derived neurotrophic factor (BDNF), and neurotrophin-3 (NT3), as well as chemokines such as tumor necrosis factor alpha (TNFα) and cytokines such as cc motif chemokine ligand 2 (CCL2) [8] promote cancer growth. The SC factors that stimulate cancer migration in the context of PNI include the glial cell line derived neurotrophic factor (GDNF) [13], the chemokine CCL2 [15, 16], and the cytokine IL-6 [40]. SCs secrete GDNF, and its expression is regulated by c-Jun in the human SC line HEI-286 [6]. Furthermore, GDNF secreted by nerves was shown to be involved

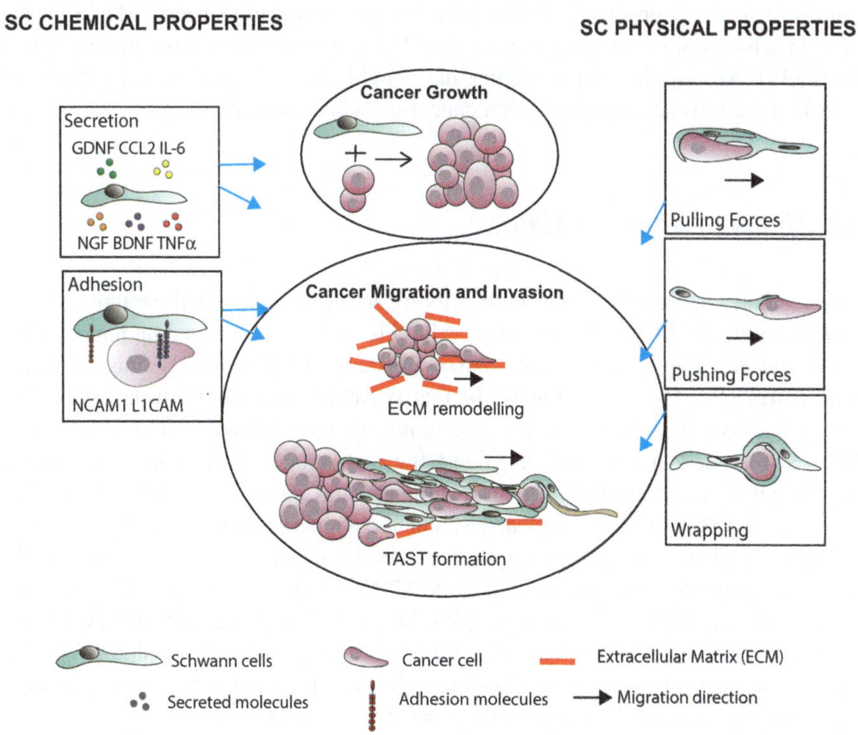

Fig. 9.2 Contribution of chemical and physical properties of Schwann cells (SCs) in cancer growth, migration, and invasion. SCs secrete molecules affecting directly or indirectly cancer growth and regulating cell migration and invasion. SCs express adhesion molecules affecting cancer migration and invasion. SC physical properties affect cancer migration and invasion. SCs wrap, push, and pull cancer cells

in pancreatic cancer cell invasion into nerve bundles [13]. Pancreatic cancer cells (MiaPaCa2) which express RET, the GDNF receptor, associate with and move along neurites from dorsal root ganglia cultured in vitro. This invasion was inhibited in the absence of GDNF or by blocking the RET receptor on cancer cells. SC-derived chemokine CCL2 affects cancer progression directly [16] and indirectly [3, 36, 45]. The direct effect of SC-derived CCL2 is observed on cancer proliferation and migration of the cervical cancer cell lines HeLa and ME-180 via chemokine receptor 2 (CCR2) activation [16]. Indirectly, SC-derived CCL2 has been shown to attract macrophages and induce polarization to the M2 phenotype, thought to contribute to tumor growth via its immunosuppressive function. This has been observed in both clinically and in preclinical models of pancreatic cancer [3], melanoma [36], and lung cancer [45]. The monocytes recruited via SC-derived CCL2 have also been shown to enhance cancer invasion by cathepsin B secretion resulting in ECM degradation [3]. Lastly, transforming growth factor beta (TGF-β) and BDNF have been implicated in SC-induced cancer invasion. Conditioned medium from SCs containing high amounts of TGF-β induced signaling pathway in pancreatic cancer cells related to cell adhesion and motility [31]. In addition, SC-derived BDNF increases cancer invasion by promoting epithelial-to-mesenchymal transition (EMT), the process in which epithelial cells acquire the characteristics of invasive mesenchymal cells [34]. All together, these studies have highlighted the paracrine functions of SCs in the interrelationship between cancer cells and the nerves.

SC Ability to Alter the Matrix

SCs regulate cancer cell migration directly by acting on cancer cells or indirectly by modifying the ECM around the cancer cells (Fig. 9.2). The surrounding matrix influences cancer cell migration. A stiff matrix substrate is usually more permissive than a soft matrix [23]. The remodeling of the matrix can be induced by cancer cells themselves but also by the other cells of the tumor microenvironment (TME), including SCs. SCs seeded in Matrigel can digest the matrix and create tunnels potentially enabling cancer cell invasion in a three-dimensional (3D) invasion assay [7]. Cancer cell activated SCs, similar to activated SCs in nerve repair, produce metalloproteinases (MMPs) which are secreted proteolytic enzymes that can degrade matrices. SC MMPs that have been identified in cancer include MMP2, MMP9, and MMP12 in cervical cancer [16] and MMP1 in melanoma [36]. SCs can also regulate tumoral expression of MMP2 and MMP9 which are thought to enhance perineural invasion [27]. This mechanism depends on the cell adhesion molecule L1CAM that is expressed in both the Schwann cells and the cancer cells in pancreatic cancer [27].

The components of the ECM made by SCs, such as laminin and collagen, can contribute to cancer migration. The transcriptomic analysis of reprogrammed SCs in cancer shows a change in ECM components and factors involved in the matrix reorganization in melanoma [36] and neuroblastoma [44]. In addition to collagen and laminin, the matricellular protein, EGFL8, has been identified as a neuritogenic factor released by stromal SCs in neuroblastoma [44]. Furthermore, analysis of proteins secreted by SCs stimulated by pancreatic cancer cells also identified ECM components and enzymes participating in matrix reorganization including cathepsin-D and MMP2 [12].

SC ECM components also play an important role in transmitting oral cancer metastasis [29]. Cells pretreated with palm oil acquire a metastatic behavior and carry epigenetic memory that allows them to maintain the metastatic behavior. RNA-seq analysis reveals high levels of ECM components related to tumor-associated SCs. In addition, removing SC-extracellular components with a bacterial chondroitinase ABC reduces the metastatic potential of the cancer cells [29].

SC Physical and Mechanical Properties: Formation of TASTs

Cancer cells benefit not only from SC-secreted proteins but also from SC physical and mechanical properties (Fig. 9.2). Both in vivo and in vitro experiments support this concept, but it must be emphasized that in vitro studies have illuminated our knowledge on these SC physical properties. Recent studies led to the discovery of a mechanism of cancer invasion that exploits tracks formed by activated SCs, named tumor activated Schwann cell tracks (TASTs) [6]. TASTs are aligned non-myelinating SCs around cancer cells that permit cancer invasion. The SCs wrap cancer cells and apply forces on the cancer cells to enhance their migration. Several physical aspects of SCs were found to be a determinant in this process including the SC ability to create an intimate contact with the cancer cells, to deform or change their shape allowing cancer cell movement in a confined space, to generate forces on the cancer cells, and to organize in linear manner in 3D [6].

1. *Physical Contact.*

The physical interaction of SCs with cancer cells is needed for SC-dependent cancer invasion. In a 3D in vitro assay, HEI-286 SCs seeded in Matrigel allow MiaPaCa-2 or Panc01 pancreatic cancer cell invasion. This cancer invasion is lost in the absence of SCs or when SCs are replaced by fibroblasts. SC conditioned media, SCs lacking contact with cancer cells, and Matrigel tunnels made by SCs before the SCs are killed with puromycin is each insufficient to induce MiaPaCa-2 cancer invasion [7]. This demonstrates that the physical contact between the two cell types is necessary for enhanced cancer invasion.

Time lapse microscopy shows that SCs induce pancreatic cancer migration by dispersing cancer cell clusters by direct cell contact. SCs have the ability to intercalate between cancer cells via long and dynamic protrusions. Cancer cells align with the SCs, and the shape of the cancer cell clusters deforms from a round to an elongated shape in presence of SCs [7]. The remodeling of cancer cell clusters by SCs is observed in another study on another cancer type. In neuroblastoma, the presence of SCs is associated with cancer cell alignment [44].

The in vitro experiment with pancreatic cancer cells and the SC line reveal that at the site of contact, the cancer cell develops a protrusion in the direction of cell migration. The formation of a protrusion is a critical step for cells initiating cell migration. In this model, the formation of this cancer cell protrusion at site of contact with the SCs depends on the expression of neural cell adhesion molecule 1 (NCAM1). In vitro and in vivo models of PNI show diminished PNI in the absence of NCAM1 in SCs [7].

In the TASTS, SCs make close physical contact with pancreatic cancer cells that align along the SCs [6]. This is found in human specimens and both in vivo and in vitro models of cancer invasion. The association between the two cell types is tight and organized, with the SCs enveloping or wrapping the cancer cells. This wrapping is seen in nerves with PNI in human specimens of PDAC and in the stroma enriched in nerve fibers. The nerves with PNI demonstrate an uneven distribution of the GFAP$^+$ SCs with the GFAP$^+$ SCs localized around the cancer cells and non-GFAP+ SCs further from the cancer cells. The wrapping of cancer cells by SCs is also found in the stroma containing single nerve fibers in which isolated cancer cells are found wrapped by SCs. Similarly, GFAP$^+$ SCs wrap cancer cells injected in the sciatic nerve of mice in a mouse model of PNI [6].

SCs wrapping around cancer cells also occur in vitro in the 3D invasion assay with Matrigel and in microchannels in which cancer cells and SCs cohabit. Time lapse movies on these cells in the microchannels allow the study of dynamic interaction between the two cell types and reveal how HEI-286 SCs wrap pancreatic cancer cells MiaPaCa-2 and Panc01[6]. SCs develop protrusions around cancer cells and deform, so cancer cells can pass through the SCs within the constraint area of the channels.

2. *Stiffness.*

Experiments in microchannels allow the visualization of SCs that deform to allow cancer cell passage in a constraint environment. In microchannels of 10 to 20 μm width, individual SCs occupy the entire channel. In the presence of cancer cells, SCs deform and spread along the walls to let the cancer cells move within the channel filled with SCs. HEI-286 c-Jun KO SCs are less able to wrap the cancer cells and less permissive than the control HEI-286 SCs, suggesting that c-Jun allow the expression of proteins conferring malleable properties [6].

Cell stiffness can be measured using atomic force microscopy (AFM). The determination of the Young's modulus (or elastic modulus) of the HEI-286 SCs reveals that SCs lacking c-Jun (i.e., c-Jun knockout (KO)) are stiffer than the control cells. This finding is consistent with the observation that c-Jun KO SCs block cancer cell passage in microchannels. Cellular stiffness depends on a variety of factors, including the cytoskeleton organization [11]. Indeed, actin organized differently in control and c-Jun KO SCs. SCs organized in stress fibers in control cells and as cortical actin in c-Jun KO SCs. Thus, SC reprogramming alters the mechanical properties of SCs that influence cancer movement in contact with these SCs.

3. *Ability to Generate Forces.*

Another physical property of the SCs that plays a role in cancer cell migration is their ability to exert forces on cancer cells. These types of forces have been recorded using live microscopy of cells in microchannels and particle image velocimetry (PIV) analysis. A recording of a HEI-286 SC pushing a MiaPaCa-2 cell shows a static cancer cell propelled by a HEI-286 SC. Within the SC, an intracellular wave travels from one side of the cell to the other, until it reaches the site of contact with the cancer cell. At this moment, the force is applied to the cancer cell, and this induces the cancer cell movement in the direction of the force, resulting in the SC pushing the cancer cell. SCs also pull or even squeeze cancer cells [6]. Cells from the TME exerting forces on cancer cells are not unique to SCs, as a similar behavior has been reported for fibroblasts [24]. In fibroblast-induced cancer cell movements, force transmission is mediated by a heterophilic adhesion involving N-cadherin present at the fibroblast plasma membrane and E-cadherin at the cancer cell plasma membrane [24].

4. *3D Organization in Bands.*

Unlike fibroblasts, SCs have a unique ability to organize in linear bands in a 3D matrix. These long linear bands allow cancer cells to travel long distances and accelerate the spread of cancer. c-Jun KO SCs lose their ability to organize in such structures and are unable to position themselves in these bands, in contrast to control SCs [6].

Conclusion

In cancer, SCs undergo transcriptional reprogramming that is similar to the dedifferentiation that SCs undergo following nerve injury. These SCs acquire a wide variety of functions which lead to an increase of cancer growth, migration, and invasion. SC secretory and physical factors have been identified as key elements in cancer invasion. Further studies should consider the interplay of these biophysical and biochemical effectors in controlling cancer progression.

References

1. Arthur-Farraj PJ, Latouche M, Wilton DK, Quintes S, Chabrol E, Banerjee A, Woodhoo A, Jenkins B, Rahman M, Turmaine M, Wicher GK, Mitter R, Greensmith L, Behrens A, Raivich G, Mirsky R, Jessen KR (2012) c-Jun reprograms Schwann cells of injured nerves to generate a repair cell essential for regeneration. Neuron 75(4):633–647. https://doi.org/10.1016/j.neuron.2012.06.021
2. Bailey P, Chang DK, Nones K, Johns AL, Patch A-M, Gingras M-C, Miller DK, Christ AN, Bruxner TJC, Quinn MC, Nourse C, Murtaugh LC, Harliwong I, Idrisoglu S, Manning S, Nourbakhsh E, Wani S, Fink L, Holmes O, Chin V, Anderson MJ, Kazakoff S, Leonard C,

Newell F, Waddell N, Wood S, Xu Q, Wilson PJ, Cloonan N, Kassahn KS, Taylor D, Quek K, Robertson A, Pantano L, Mincarelli L, Sanchez LN, Evers L, Wu J, Pinese M, Cowley MJ, Jones MD, Colvin EK, Nagrial AM, Humphrey ES, Chantrill LA, Mawson A, Humphris J, Chou A, Pajic M, Scarlett CJ, Pinho AV, Giry-Laterriere M, Rooman I, Samra JS, Kench JG, Lovell JA, Merrett ND, Toon CW, Epari K, Nguyen NQ, Barbour A, Zeps N, Moran-Jones K, Jamieson NB, Graham JS, Duthie F, Oien K, Hair J, Grützmann R, Maitra A, Iacobuzio-Donahue CA, Wolfgang CL, Morgan RA, Lawlor RT, Corbo V, Bassi C, Rusev B, Capelli P, Salvia R, Tortora G, Mukhopadhyay D, Petersen GM, Initiative APCG, Munzy DM, Fisher WE, Karim SA, Eshleman JR, Hruban RH, Pilarsky C, Morton JP, Sansom OJ, Scarpa A, Musgrove EA, Bailey U-MH, Hofmann O, Sutherland RL, Wheeler DA, Gill AJ, Gibbs RA, Pearson JV, Waddell N, Biankin AV, Grimmond SM (2016) Genomic analyses identify molecular subtypes of pancreatic cancer. Nature 531(7592):47–52. https://doi.org/10.1038/nature16965

3. Bakst RL, Xiong H, Chen C-H, Deborde S, Lyubchik A, Zhou Y, He S, McNamara W, Lee S-Y, Olson OC, Leiner IM, Marcadis AR, Keith JW, Al-Ahmadie HA, Katabi N, Gil Z, Vakiani E, Joyce JA, Pamer E, Wong RJ (2017) Inflammatory Monocytes Promote Perineural Invasion via CCL2-Mediated Recruitment and Cathepsin B Expression. Cancer Research 77(22):6400–6414. https://doi.org/10.1158/0008-5472.can-17-1612
4. Bapat AA, Hostetter G, Hoff DDV, Han H (2011) Perineural invasion and associated pain in pancreatic cancer. Nature Reviews Cancer 11(10):695–707. https://doi.org/10.1038/nrc3131
5. Byun JS, Gardner K (2013) Wounds That Will Not Heal. Am J Pathology 182(4):1055–1064. https://doi.org/10.1016/j.ajpath.2013.01.009
6. Deborde S, Gusain L, Powers A, Marcadis A, Yu Y, Chen C-H, Frants A, Kao E, Tang LH, Vakiani E, Amisaki M, Balachandran VP, Calo A, Omelchenko T, Jessen KR, Reva B, Wong RJ (2022) Reprogrammed Schwann cells organize into dynamic tracks that promote pancreatic cancer invasion. Cancer Discov. https://doi.org/10.1158/2159-8290.cd-21-1690
7. Deborde S, Omelchenko T, Lyubchik A, Zhou Y, He S, McNamara WF, Chernichenko N, Lee S-Y, Barajas F, Chen C-H, Bakst RL, Vakiani E, He S, Hall A, Wong RJ (2016) Schwann cells induce cancer cell dispersion and invasion. The Journal of clinical investigation 126(4):1538–1554. https://doi.org/10.1172/jci82658
8. Deborde S, Wong RJ (2022) The Role of Schwann Cells in Cancer. Adv Biology :2200089. https://doi.org/10.1002/adbi.202200089
9. Demir IE, Boldis A, Pfitzinger PL, Teller S, Brunner E, Klose N, Kehl T, Maak M, Lesina M, Laschinger M, Janssen K-P, Algül H, Friess H, Ceyhan GO (2014) Investigation of Schwann cells at neoplastic cell sites before the onset of cancer invasion. Journal of the National Cancer Institute 106(8):dju184–dju184. https://doi.org/10.1093/jnci/dju184
10. Demir IE, Tieftrunk E, Schorn S, Saricaoglu ÖC, Pfitzinger PL, Teller S, Wang K, Waldbaur C, Kurkowski MU, Wörmann SM, Shaw VE, Kehl T, Laschinger M, Costello E, Algül H, Friess H, Ceyhan GO (2016) Activated Schwann cells in pancreatic cancer are linked to analgesia via suppression of spinal astroglia and microglia. Gut 6:1001–1014. https://doi.org/10.1136/gutjnl-2015-309784
11. Deville SS, Cordes N (2019) The Extracellular, Cellular, and Nuclear Stiffness, a Trinity in the Cancer Resistome—A Review. Frontiers Oncol 9:1376. https://doi.org/10.3389/fonc.2019.01376
12. Ferdoushi A, Li X, Griffin N, Faulkner S, Jamaluddin MFB, Gao F, Jiang CC, Helden DF van, Tanwar PS, Jobling P, Hondermarck H (2020) Schwann Cell Stimulation of Pancreatic Cancer Cells: A Proteomic Analysis. Frontiers in oncology 10:1601. https://doi.org/10.3389/fonc.2020.01601
13. Gil Z, Cavel O, Kelly K, Brader P, Rein A, Gao SP, Carlson DL, Shah JP, Fong Y, Wong RJ (2010) Paracrine regulation of pancreatic cancer cell invasion by peripheral nerves. Journal of the National Cancer Institute 102(2):107–118. https://doi.org/10.1093/jnci/djp456
14. Gresle MM, Butzkueven H, Perreau VM, Jonas A, Xiao J, Thiem S, Holmes FE, Doherty W, Soo P, Binder MD, Akkermann R, Jokubaitis VG, Cate HS, Marriott MP, Gundlach AL, Wynick D, Kilpatrick TJ (2015) Galanin is an autocrine myelin and oligodendrocyte trophic

signal induced by leukemia inhibitory factor. Glia 63(6):1005–1020. https://doi.org/10.1002/glia.22798
15. He S, He S, Chen C-H, Deborde S, Bakst RL, Chernichenko N, McNamara WF, Lee S-Y, Barajas F, Yu Z, Al-Ahmadie HA, Wong RJ (2015) The chemokine (CCL2-CCR2) signaling axis mediates perineural invasion. Molecular Cancer Research 13(2):380–390. https://doi.org/10.1158/1541-7786.mcr-14-0303
16. Huang T, Fan Q, Wang Y, Cui Y, Wang Z, Yang L, Sun X, Wang Y (2020) Schwann Cell-Derived CCL2 Promotes the Perineural Invasion of Cervical Cancer. Frontiers Oncol 10:19. https://doi.org/10.3389/fonc.2020.00019
17. Inserra MM, Yao M, Murray R, Terris DJ (2000) Peripheral Nerve Regeneration in Interleukin 6–Deficient Mice. Archives Otolaryngology Head Neck Surg 126(9):1112–1116. https://doi.org/10.1001/archotol.126.9.1112
18. Jessen KR, Arthur-Farraj P (2019) Repair Schwann cell update: Adaptive reprogramming, EMT, and stemness in regenerating nerves. Glia 67(3):421–437. https://doi.org/10.1002/glia.23532
19. Jessen KR, Mirsky R (2002) Signals that determine Schwann cell identity*. J Anat 200(4):367–376. https://doi.org/10.1046/j.1469-7580.2002.00046.x
20. Jessen KR, Mirsky R (2016) The repair Schwann cell and its function in regenerating nerves. The Journal of physiology 13(594):3521–31. https://doi.org/10.1113/jp270874
21. Jessen KR, Mirsky R (2019) The Success and Failure of the Schwann Cell Response to Nerve Injury. Frontiers in Cellular Neuroscience 13:33. https://doi.org/10.3389/fncel.2019.00033
22. Jessen KR, Mirsky R, Lloyd AC (2015) Schwann Cells: Development and Role in Nerve Repair. Cold Spring Harbor perspectives in biology 7(7):a020487. https://doi.org/10.1101/cshperspect.a020487
23. Kraning-Rush CM, Reinhart-King CA (2012) Controlling matrix stiffness and topography for the study of tumor cell migration. Cell Adhes Migr 6(3):274–279. https://doi.org/10.4161/cam.21076
24. Labernadie A, Kato T, Brugués A, Serra-Picamal X, Derzsi S, Arwert E, Weston A, González-Tarragó V, Elosegui-Artola A, Albertazzi L, Alcaraz J, Roca-Cusachs P, Sahai E, Trepat X (2017) A mechanically active heterotypic E-cadherin/N-cadherin adhesion enables fibroblasts to drive cancer cell invasion. Nature cell biology 19(3):224–237. https://doi.org/10.1038/ncb3478
25. Masaki T, Qu J, Cholewa-Waclaw J, Burr K, Raaum R, Rambukkana A (2013) Reprogramming Adult Schwann Cells to Stem Cell-like Cells by Leprosy Bacilli Promotes Dissemination of Infection. Cell 152(1–2):51–67. https://doi.org/10.1016/j.cell.2012.12.014
26. Mungroo MR, Khan NA, Siddiqui R (2020) Mycobacterium leprae: Pathogenesis, diagnosis, and treatment options. Microb Pathogenesis 149:104475. https://doi.org/10.1016/j.micpath.2020.104475
27. Na'ara S, Amit M, Gil Z (2018) L1CAM induces perineural invasion of pancreas cancer cells by upregulation of metalloproteinase expression. Oncogene 107:219. https://doi.org/10.1038/s41388-018-0458-y
28. Nagarajan R, Svaren J, Le N, Araki T, Watson M, Milbrandt J (2001) EGR2 Mutations in Inherited Neuropathies Dominant-Negatively Inhibit Myelin Gene Expression. Neuron 30(2):355–368. https://doi.org/10.1016/s0896-6273(01)00282-3
29. Pascual G, Domínguez D, Elosúa-Bayes M, Beckedorff F, Laudanna C, Bigas C, Douillet D, Greco C, Symeonidi A, Hernández I, Gil SR, Prats N, Bescós C, Shiekhattar R, Amit M, Heyn H, Shilatifard A, Benitah SA (2021) Dietary palmitic acid promotes a prometastatic memory via Schwann cells. Nature 599(7885):485–490. https://doi.org/10.1038/s41586-021-04075-0
30. Raivich G, Bohatschek M, Costa CD, Iwata O, Galiano M, Hristova M, Nateri AS, Makwana M, Riera-Sans L, Wolfer DP, Lipp H-P, Aguzzi A, Wagner EF, Behrens A (2004) The AP-1 transcription factor c-Jun is required for efficient axonal regeneration. Neuron 43(1):57–67. https://doi.org/10.1016/j.neuron.2004.06.005
31. Roger E, Martel S, Bertrand-Chapel A, Depollier A, Chuvin N, Pommier RM, Yacoub K, Caligaris C, Cardot-Ruffino V, Chauvet V, Aires S, Mohkan K, Mabrut J-Y, Adham M, Fenouil T, Hervieu V, Broutier L, Castets M, Neuzillet C, Cassier PA, Tomasini R, Sentis S, Bartholin

L (2019) Schwann cells support oncogenic potential of pancreatic cancer cells through TGFβ signaling. Cell death & disease 10(12):886–19. https://doi.org/10.1038/s41419-019-2116-x
32. Salvo E, Tu NH, Scheff NN, Dubeykovskaya ZA, Chavan SA, Aouizerat BE, Ye Y (2021) TNFα promotes oral cancer growth, pain, and Schwann cell activation. Scientific reports 11(1):1840–14. https://doi.org/10.1038/s41598-021-81500-4
33. Schmitd LB, Perez-Pacheco C, Bellile EL, Wu W, Casper K, Mierzwa M, Rozek LS, Wolf GT, Taylor JMG, D'Silva NJ (2022) Spatial and Transcriptomic Analysis of Perineural Invasion in Oral Cancer. Clin Cancer Res 28(16):OF1–OF16. https://doi.org/10.1158/1078-0432.ccr-21-4543
34. Shan C, Wei J, Hou R, Wu B, Yang Z, Wang L, Lei D, Yang X (2015) Schwann cells promote EMT and the Schwann-like differentiation of salivary adenoid cystic carcinoma cells via the BDNF/TrkB axis. Oncology reports 35(1):427–435. https://doi.org/10.3892/or.2015.4366
35. Shetty VP, Antia NH, Jacobs JM (1988) The pathology of early leprous neuropathy. J Neurol Sci 88(1–3):115–131. https://doi.org/10.1016/0022-510x(88)90210-9
36. Shurin GV, Kruglov O, Ding F, Lin Y, Hao X, Keskinov AA, You Z, Lokshin AE, LaFramboise WA, Falo LD, Shurin MR, Bunimovich YL (2019) Melanoma-induced reprogramming of Schwann cell signaling aids tumor growth. Cancer Research :canres.3872.2018. https://doi.org/10.1158/0008-5472.can-18-3872
37. Silva VM, Gomes JA, Tenório LPG, Neta GC de O, Paixão K da C, Duarte AKF, Silva GCB da, Ferreira RJS, Koike BDV, Marques C de S, Miguel RD da S, Queiroz AC de, Pereira LX, Fraga CA de C (2019) Schwann cell reprogramming and lung cancer progression: a meta-analysis of transcriptome data. Oncotarget 10(68):7288–7307. https://doi.org/10.18632/oncotarget.27204
38. Stierli S, Imperatore V, Lloyd AC (2019) Schwann cell plasticity-roles in tissue homeostasis, regeneration, and disease. Glia 67(11):2203–2215. https://doi.org/10.1002/glia.23643
39. Stoner GL (1979) Importance of the neural predilection of Mycobacterium leprae in leprosy. Lancet Lond Engl 2(8150):994–6. https://doi.org/10.1016/s0140-6736(79)92564-9
40. Su D, Guo X, Huang L, Ye H, Li Z, Lin L, Chen R, Zhou Q (2020) Tumor-neuroglia interaction promotes pancreatic cancer metastasis. Theranostics 10(11):5029–5047. https://doi.org/10.7150/thno.42440
41. Thomaidou D, Coquillat D, Meintanis S, Noda M, Rougon G, Matsas R (2001) Soluble forms of NCAM and F3 neuronal cell adhesion molecules promote Schwann cell migration: identification of protein tyrosine phosphatases zeta/beta as the putative F3 receptors on Schwann cells. J Neurochem 78(4):767–78. https://doi.org/10.1046/j.1471-4159.2001.00454.x
42. Tofaris GK, Patterson PH, Jessen KR, Mirsky R (2002) Denervated Schwann cells attract macrophages by secretion of leukemia inhibitory factor (LIF) and monocyte chemoattractant protein-1 in a process regulated by interleukin-6 and LIF. Journal of Neuroscience 22(15):6696–6703
43. Wang S, Pisco AO, McGeever A, Brbic M, Zitnik M, Darmanis S, Leskovec J, Karkanias J, Altman RB (2021) Leveraging the Cell Ontology to classify unseen cell types. Nat Commun 12(1):5556. https://doi.org/10.1038/s41467-021-25725-x
44. Weiss T, Taschner-Mandl S, Janker L, Bileck A, Rifatbegovic F, Kromp F, Sorger H, Kauer MO, Frech C, Windhager R, Gerner C, Ambros PF, Ambros IM (2021) Schwann cell plasticity regulates neuroblastic tumor cell differentiation via epidermal growth factor-like protein 8. Nature communications 12(1):1624–19. https://doi.org/10.1038/s41467-021-21859-0
45. Zhou Y, Li J, Han B, Zhong R, Zhong H (2020) Schwann cells promote lung cancer proliferation by promoting the M2 polarization of macrophages. Cellular Immunology 357:104211. https://doi.org/10.1016/j.cellimm.2020.104211

Part III
Perspectives of the Field

Chapter 10
Systemic Interactions Between Cancer and the Nervous System

Yue Wu and Jeremy C. Borniger

Abbreviations

Ach	Acetylcholine
ACTH	Adrenocorticotropic hormone
AGM	Axon guidance molecule
β2-AR	β2-Adrenergic receptor
B2BM	Breast-to-brain metastases
BBB	Blood-brain barrier
BDNF	Brain-derived neurotrophic factor
CAF	Cancer-associated fibroblast
CAM	Cell adhesion molecule
CNS	Central nervous system
CRF	Corticotropin-releasing factor
CRP	C-reactive protein
DeepISTI	Deep intravital subcellular time-lapse imaging
ECM	Extracellular matrix
EGFR	Epidermal growth factor receptor
EV	Extracellular vesicle
GABA	Gamma-aminobutyric acid
GPC3	Glypican 3
HO neuron	Hypocretin/orexin neuron
HPA	Hypothalamic-pituitary-adrenal
LC	Locus coeruleus
MMP	Matrix metalloproteinase
MNT	Macrophage to neuron-like cell transition

Y. Wu · J. C. Borniger (✉)
Cold Spring Harbor Laboratory, Cold Spring Harbor, NY, USA
e-mail: bornige@cshl.edu

M. Amit, N. N. Scheff (eds.), *Cancer Neuroscience*,
https://doi.org/10.1007/978-3-031-32429-1_10

NCAM1	Neural cell adhesion molecule 1
NE	Norepinephrine
NGF	Nerve growth factor
NGLGN3	Neuroligin 3
NSC	Neural stem cell
NSCLC	Non-small cell lung cancer
PD-1	Programmed death receptor-1
PDAC	Pancreatic ductal adenocarcinoma
PNI	Perineural invasion
PSC	Pancreatic stellate cell
S1P1	Sphingosine-1-phosphate receptor 1
SNS	Sympathetic nervous system
SVZ	Subventricular zone
TAM	Tumor-associated macrophage
TCGA	The Cancer Genome Atlas
TM	Tumor microtube
TME	Tumor microenvironment
TNBC	Triple-negative breast cancer
TNF-α	Tumor necrosis factor-alpha
TNT	Tunneling nanotube
TRPA1	Transient receptor potential ankyrin 1
VEGF	Vascular endothelial growth factor
VTA	Ventral tegmental area
xCT	Cystine-glutamate transporter

Introduction

Nervous systems were not classically considered to be actively involved in the process of tumorigenesis and cancer metastasis. However, studies over the last decade have demonstrated the presence of neurons and glial cells in the peritumoral regions of many human tumors, and the density of tumor-associated nerves is often correlated with cancer progression and metastatic spread. In general, it has been demonstrated that neuronal activity is widely pro-cancer in the brain, and blocking neural activity usually confers anticancer benefits [1]. For example, increased neuronal excitability has been observed in preclinical models of both pediatric and adult gliomas [2–4], and nerve ablation can suppress tumor development in various other malignancies [5]. With the growing global cancer burden, research in the nascent field of cancer neuroscience is quickly becoming intense, attracting interdisciplinary efforts from all over the world. This chapter will focus on the reciprocal cross talk between cancer and the nervous system via direct and indirect pathways. Subsequent influences on host behavior and cancer therapeutic resistance will also be discussed.

How Does the Nervous System Control Cancer Development?

Direct Pathways

Cellular Connections

Fifty years ago, tumor cells were found to lack intercellular electrical coupling [6], and many studies support the idea that gap junction proteins act as tumor suppressors [7–9]. However, recent evidence suggests that gap junctions contribute to cancer stem cell renewal [10], cancer microtube-dependent proliferation [11], tumor angiogenesis [12], and brain metastasis [13] in multiple cancer types. A recent study found that potassium (K^+) channel-driven bioelectric signaling regulates metastasis in triple-negative breast cancer [14], which suggests that modulation of cellular resting membrane potential can be used to target nerve-invading cancer cells. Moreover, communication between nerves and brain cancer cells has recently been identified in glioblastomas via bona fide AMPA-mediated chemical synapses [3, 4] or via pseudo-tripartite synapses in breast-to-brain metastases (B2BM) via NMDAR signaling [15]. In addition to synaptic connections, glioma [11, 16] and pancreatic cancer cells [17] can generate long tumor microtubes (TMs) and thinner tunneling nanotubes (TNTs), and these membrane protrusions enable tumor cells to couple with gap junction associated intercellular calcium (Ca^{2+}) wave signaling. The above results demonstrate that neurons can form functional cellular connections with cancer cells to enhance their colonization, proliferation, and migration. Thus far, this cross talk has only been observed in the brain, and it is still unknown if there are direct synaptic interactions between neurons and cancer cells in the periphery.

Paracrine Signaling Pathways

Neurons and glial cells modulate tumor growth and invasion by releasing neurotransmitters, neurotrophins, and axon guidance molecules (AGMs) [18]. Cancer cells express a wide variety of receptors to interact with these molecules and activate downstream signaling cascades important for their growth, metastatic spread, and survival (Fig. 10.1).

Neurotransmitters are chemical messengers secreted by neurons or glial cells to target other cells throughout the body. Neurotransmitter signaling plays a critical role in tumor initiation and progression both in the brain and periphery [19]. In the brain, glutamate acts as an excitatory neurotransmitter to promote cancer cell proliferation and invasion through AMPA receptor signaling in glioma [3, 4] or NMDA receptor in B2BM [15]. Given the cellular context of glioma and the relationship to neural stem cell niches, neurotransmitter signaling can potentially impact glioma growth and recurrence indirectly through the modulation of neural stem cell survival and proliferation. Neural stem cells (NSCs) located in the subventricular zone (SVZ) are regulated by a range of neurotransmitters including glutamate,

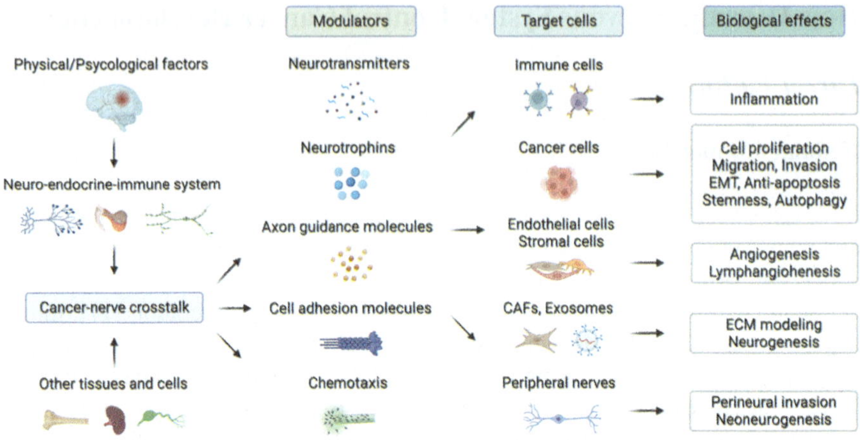

Fig. 10.1 Functional component profile of modulators, target cells, and consequential biological effects involved in tumor-nerve cross talk. This sophisticated interaction is initiated by external and internal cues via the neuroendocrine-immune axis, in which multiple active molecules exert a variety of functions to promote tumor growth and metastasis through targeting distinct effector cells

Gamma-animobutyric acid (GABA), dopamine, serotonin, and acetylcholine (ACh). Exposure to depolarizing (excitatory) neurotransmitters usually promotes stem cell proliferation and neuroblast differentiation [20, 21] (except GABA [22]), making the SVZ a rich environment for glioma growth. In the periphery, autonomic signaling exerts a more diversified function in controlling peripheral tumor development. Three pathways are involved: sympathetic adrenergic signaling, parasympathetic cholinergic signaling, and sensory peptidergic signaling [23]. Postganglionic sympathetic nerve fibers secrete norepinephrine (NE), which binds to β-adrenergic receptors and activates downstream signaling cascades in tumors. Increased adrenergic stimulation in both animal models and humans has been shown to promote tumor development and metastasis [24, 25], as well as impede responses to chemotherapy [26]. Thus, β-blockers were considered as a potential adjuvant treatment to increase cancer patient survival in a variety of cancer types [27–29]. Parasympathetic cholinergic nerves have tissue-dependent effects in many malignancies by releasing Ach and targeting nicotinic or muscarinic receptors. Inhibition of cholinergic signals slowed tumor growth and development in gastric cancer [30, 31] and prevented lymph node metastasis in prostate cancer [32]. Whereas cholinergic signaling via muscarinic receptors directly and indirectly suppresses pancreatic tumorigenesis and cancer stemness, accelerated pancreatic cancer progression was observed when cholinergic signals were inhibited [33]. A similar inhibitory effect was recently demonstrated in a breast cancer mouse model by expressing excitatory bacterial sodium channels in tumor cholinergic nerves through adeno-associated virus (AAV) injections [34]. Some peripheral sensory neurons can secrete glutamate [35] that may target tumor-expressed NMDA receptors to promote tumor development in pancreatic, ovarian, and breast malignancies [36]. *Neuropeptide* roles in cancer have recently been reviewed [37]. Taken together, neurotransmitters in the central nervous system (CNS) regulate primary and metastatic tumors through glutamatergic signaling or modulation of the NSC niche in the brain, whereas autonomic (adrenergic, cholinergic, and sensory peptidergic) signaling influences epithelial-derived tumors in the periphery in diverse ways.

Neurotrophins are growth factors that help neurons survive, proliferate, differentiate, and regulate apoptosis, such as nerve growth factor (NGF), brain-derived neurotrophic factor (BDNF), neurotrophin-3, and neurotrophin-4/5. They interact with Trk or $p75^{NTR}$ receptors to engage downstream signaling pathways (e.g., MAPK, PI3K/Akt, PLC-γ, JNK). Neurotrophic activation has recently been identified in neuroblastoma, medulloblastoma, and many peripheral malignancies to enhance cancer cell proliferation [38]. This indicates the potential anticancer use of neurotrophin/receptor inhibitors [39]. Indeed, anti-NGF/TrkA antibodies, anti-$p75^{NTR}$ siRNA, and pharmacological Trk-modulating interventions have been shown to suppress tumor growth and metastasis in human glioma and triple-negative breast cancer [40, 41].

Axon guidance molecules (AGMs) are major modulators of axonal development in the developing nervous system [42], including proteins such as Slits, Netrin, Ephrins, and Semaphorins. They are extensively expressed outside of the CNS, influencing cancer progression mainly through inflammatory and angiogenesis-related mechanisms [43–45]. For example, disrupting Slit2-ROBO signaling enhances tumor metastasis and perineural invasion (PNI) in pancreatic ductal adenocarcinoma (PDAC) [46]. Ephrins and their receptors also exhibit distinct expression patterns in cancer cells and tumor blood vessels, which influence cancer cell proliferation, migration, and invasion [47].

Cell adhesion molecules (CAMs) are another major class of proteins mediating correct axonal navigation. Besides maintaining cellular attachment, they also serve as signaling molecules for cell proliferation and transcriptional activity [48, 49]. Neuronal activity-dependent release of soluble neuroligin 3 (NLGN3), a postsynaptic adhesion protein, promotes glioma cell proliferation through the PI3K-mTOR pathway [2]. Schwann cell expressed neural cell adhesion molecule 1 (NCAM1) induces reorganization of cancer cell clusters and enhances perineural invasion [50].

These findings imply that neuroactive molecules and their receptors play key roles in tumor progression via direct cellular connections (synapses/membrane protrusions) or paracrine pathways. Antibodies or compounds that regulate the activity of these molecules might be employed as therapeutic anticancer agents. However, many of the molecular pathways and mechanisms are still being investigated.

Indirect Pathways

The tumor microenvironment (TME) is made up of cancer cells, stromal cells (e.g., endothelial cells, fibroblasts), immune cells, and noncellular extracellular matrix (ECM) [51]. Like an ecosystem surrounding a tumor, the TME constantly interacts with tumor cells in complex feedback loops. Tumor cells, as the center of the TME, employ sophisticated signaling networks to reprogram their local environment for their growth, maintenance, resistance to treatment, and metastasis. Alterations in tumor cell gene expression cause a disruption in normal tissue homeostasis, promoting the production of cytokines and growth factors to recruit extratumoral stromal cells, which further supports tumorigenesis, cancer progression, and metastasis [52]. Neurons and nerve fibers have recently been identified as crucial components of the TME that modulate the initiation and progression of a wide range of solid tumors [53, 54]. Various indirect pathways through tumor angiogenesis, immune cells, stromal cells, and the ECM are discussed below (Fig. 10.1).

Angiogenesis is a hallmark of cancer, usually induced by tumor hypoxia, production of the vascular endothelial growth factor (VEGF), and other pro-angiogenic molecules (e.g., matrix metalloproteinases). Zahalka et al. revealed that adrenergic (NE) signals suppressed oxidative phosphorylation in tumor endothelial cells in a prostate cancer mouse model, activating an angiogenic metabolic switch that facilitated rapid tumor growth [55]. Neurotransmitters and neurotrophic factors are further engaged in the process of tumor angiogenesis by binding their receptors on endothelial and tumor cells to induce their migration [56]. This has been thoroughly discussed in recent reviews [56, 57].

Immune cells, as indicated strongly by emerging evidence, in the TME can be directly regulated by the autonomic nervous system. Activation of adrenergic signaling restrains immune cell mobility [58] and enhances cancer metastasis in multiple cancer models [59]. Broadly, adrenergic signaling enhances pancreatic tumor growth and impairs patient survival via recruitment of tumor-associated macrophages (TAMs) [60], whereas cholinergic signaling reduces cancer progression in pancreatic cancer [33] and breast cancer [61]. The cholinergic anticancer property is mediated through widely expressed ACh/ACh receptor (e.g., α7nAChR, α5nAChR) on immune cells, by which the vagus nerve influences the TME and anticancer immunity [62, 63]. Natural killer cells, dendritic cells, mast cells, and other immune cells also have significant influences on the adaptive antitumor immune response by generating chemokines and cytokines, and their regulatory processes and interactions with the nervous systems are currently under investigation [64].

Stromal cells, the most prevalent being cancer-associated fibroblasts (CAFs), act to increase cancer cell viability, proliferation, and mobility [65]. CAF-expressed Slit2 promotes neurite outgrowth from dorsal root ganglia neurons and Schwann cell migration/proliferation in PDAC [66]. CAFs can also produce interleukin-6 (IL-6) under inflammatory conditions [67], where IL-6 is not only a pleiotropic cytokine but may act as a neurotrophic factor involved in the regeneration of peripheral nerves [68]. Similar in another stromal cell-adipocytes, Qing and colleagues revealed a novel role of sympathetic nerves elevating circulatory IL-6 via brown adipocytes following psychological stress [69]. Another study based on The Cancer Genome Atlas (TCGA) database indicates that intratumoral adipocyte-high breast cancers are enriched for inflammation-related gene sets, such as TNF-α signaling via NF-κB and IL-6/JAK/STAT3 signaling pathways. A high amount of intratumoral adipocytes was associated with metastatic pathways, cancer stemness, and a favorable tumor immune microenvironment but less cancer cell proliferation [70]. Whether sympathetic nerves also act within tumors via cell-signaling molecules derived from intratumoral adipocytes remains to be explored [71]. Moreover, activation of pancreatic stellate cells (PSCs) by cytokines, chemokines, or growth factors in pancreatic cancer promotes neurite outgrowth, facilitating cancer cell migration toward nerves [72]. Given the large heterogeneity and plasticity in the TME, investigations on how intratumor nerve fibers interact with stromal cells are sorely needed.

ECM is a crucial component of the TME, and its deposition, cross-linking, and chemical signals are necessary for cancer progression. In cell culture, adhesive proteins present in the ECM and ECM receptors on neurons facilitate nerve cell survival and neurite outgrowth [73, 74]. Further, ECM constituents and their receptors contribute to neural and cancer stem cell behavior and brain tumor progression [75]. However, research into the roles of ECM in neural cancer interactions is still changeling, because widely used external scaffolds (like Matrigel) are nonspecific for different tumor types and lack the intricacy of the in situ protein environment. Recently, several artificial 3D ECMs are developed to mimic the real topographical features of the brain ECM for in vitro investigation of interactions between glioblastoma cells and ECM [76, 77]. With these platforms, pseudo-in situ TME analysis might be realized in a more realistic ternary composite system with neuronal inputs and tumor ECM. Furthermore, ECM can be processed using decellularization procedures to remove cells and donor antigens from tissue or organs while preserving native biological cues essential for cell growth and differentiation. The decellularized brain matrix has been applied for Neuro2a culturing [78] and human cerebral organoid formation [79].

Researchers are attempting to determine how neural signals influence cancer cell behavior via the TME, despite its substantial heterogeneity. This might pave the way for modulating crucial TME components through the indirect pathways to improve cancer treatment, where nerves likely play a critical role.

How Does Cancer Influence Nervous System Function?

Many cancers have nerves in the peritumoral regions, and their presence in the TME is often associated with metastasis. Nerve-cancer interactions include enhanced contact and adhesion between nerves and cancer cells through lengthening and thickening of the nerve fibers [80]. *Perineural invasion* of preexisting nerves and *axonogenesis* of newborn neurites are two primary patterns observed [81] in this interactive process. PNI and axonogenesis are commonly observed in cancer patients and usually associated with poor outcomes, whereas the molecular mechanisms underlying these processes and their significance in cancer progression remain largely unknown. Recruitment of existing nerves is likely sufficient for tumor cell migration, and extension of axons in close proximity to a tumor may further enhance metastatic propensity [82, 83]. Moreover, it is worth noting that cancer cells are also capable of *reprogramming nerve outgrowth or remodeling neuronal phenotypes* [18]. Thus, the presence of intratumoral nerves and/or increased nerve density should be considered as a distinct feature during cancer progression and metastasis [34].

PNI often provides a way for tumor cells to metastasize [23, 84–86], sometimes induces cancer-related pain by causing nerve damage [85], and eventually results in lower survival and a poor prognosis in many cancer patients [84]. Although clear mechanisms, in situ PNI quantification, and PNI-associated risk management strategies are still being investigated, recent studies suggest that the neural tracking hypothesis [87] likely explains major aspects of PNI, indicating an intrinsic capacity of cancer cells to actively migrate along axons. Neural tracking of cancer cells depends on a dynamic balance between nerve injury and repair, in which injured nerves by PNI secrete growth factors to further promote cancer cell proliferation. PNI is a consecutive multistep process as demonstrated by a study from Amit et al. that goes through seven hallmarks: cancer cell survival, nerve homeostasis, inflammatory responses, cancer cell chemotaxis toward the nerves, neurogenesis, cancer cell adhesion to nerve sheaths, and nerve invasion [87]. As emphasized in a recent review, these phases do not occur in a linear fashion and cannot be mechanically separated; rather, it is an interactional, complicated closed-loop system [80]. Cancer cells that escape from autophagy, apoptosis, and immune surveillance gain higher viability, mobility, and invasiveness toward adjacent nerves. Neural injury caused by invading tumor cells triggers inflammatory responses [88, 89] and stimulates the release of soluble signaling molecules in the perineural niche for neuron regeneration [87]. Additionally, Schwann cells, TAMs, T cells, and mast cells [90, 91] are recruited and activated in the TME with enhanced production of neurotrophins, growth factors, and chemokines to induce *axonogenesis* and create a chemoattractant gradient. With the guidance of AGMs and altered expression of adhesion molecules, tumor cells change into a mesenchymal form associated with greater metastasis [50, 91, 92]. Further, matrix metalloproteinases (MMPs) released from cancer and stromal cells in the TME promote the remodeling of ECM, which opens a low-resistance pathway for the chemotaxis of cancer cells along nerve fibers [93–95]. Neurotransmitters, autophagy, IncRNA, and miRNA are all involved in the PNI process, and their roles are discussed in-depth in other reviews [80].

Cancer cells were recently found to influence *nerve outgrowth and remodel neuronal phenotypes* in both the brain and the periphery. For example, upregulation of the cystine-glutamate transporter (xCT) [96] or selective expression of glypican 3 (GPC3) [97] in glioma cells correlates with neuronal hyperexcitability and synaptic remodeling (Fig. 10.2a). Furthermore, Mauffrey et al. revealed that prostate and breast cancers can recruit neural progenitors from the brain's subventricular zone (SVZ), allowing the peripheral tumors to interact with the brain remotely. These neural progenitors are able to infiltrate into the tumor, initiate intratumor neurogenesis, and differentiate into adrenergic neurons after crossing the blood-brain

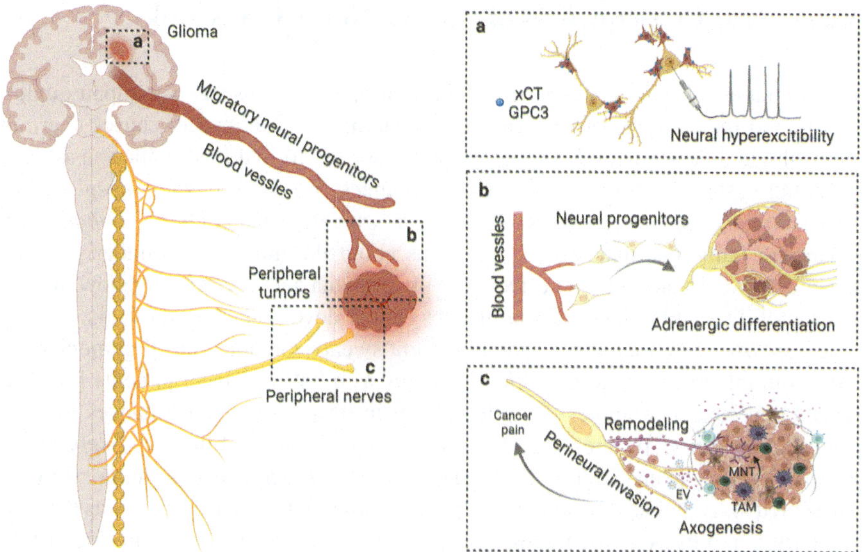

Fig. 10.2 Schematic potential pathways by which cancer influences the nervous system. (**a**) Glioma cells express xCT and GPC3 on their cell membranes to promote neuronal hyperexcitability and synaptic remodeling. (**b**) Neural progenitor cells are able to migrate through the vasculature to infiltrate peripheral tumors and differentiate into adrenergic neurons. (**c**) Tumor cells, in collaboration with other cells (such as CAFs and TAM) and soluble signaling molecules in the TME (e.g., neurotrophins, neurotransmitters, growth factors, AGMs, chemokines, adhesion proteins, MMPs, EVs), participate in neurite outgrowth and remodeling of neural architecture during the PNI process, which is often associated with cancer pain

barrier (BBB) and moving through the bloodstream [82] (Fig. 10.2b). Exciting findings like these highlight the potential breadth of cancer's effect on the nervous system. In head and neck cancer, p53-deficient cancer cells reprogram established sensory nerves into adrenergic phenotypes via cancer-derived extracellular vesicle (EV). As a result, infiltrating sympathetic nerves promote tumor development [98] (Fig. 10.2c). More recently, a study discovered a TAM subset with neuron-like phenotypes through a "macrophage to neuron-like cell transition" (MNT) mechanism to drive de novo tumoral neurogenesis. Single-cell RNA sequencing in lung cancer models indicated Smad3 as the key regulator for promoting the MNT. It is also associated to cancer-related pain due to the expression of pain receptors in this subset (Fig. 10.2c). As a result, MNT might be a precision therapeutic target for tumor innervation and cancer pain [99].

Several major pathways by which cancer influences the form and functions of the nervous system are shown in Fig. 10.2. Understanding the molecular mechanisms underlying cancer's influence on the nervous system would lay a foundation for the development of novel treatments aimed at preventing cancer progression and alleviating neuropathic pain caused by neural tracking of cancer cells.

Behavioral Consequences of Cancer-Nerve Cross Talk

Cancer patients frequently experience debilitating symptoms that can impair quality of life and reduce odds of survival. As a systemic disease, cancer causes aberrant physiologic and behavioral responses by disrupting normal host homeostasis. The potential systemic consequences of nerve-cancer cross talk include *sleep disruption*, *fatigue*, *appetite changes*, *depression*, *anxiety*, and *cognitive impairment*. There are numerous unanswered questions about the mechanisms underlying the cancer-nerve interactions that drive these phenomena and, reciprocally, how manipulation of distinct brain circuits influences cancer development in the body.

Sleep disorders, including insomnia, night sweats, circadian misalignment, daytime somnolence, and fatigue, affect as many as a half of cancer patients [100]. It seems that cancer promotes disrupted sleep; in turn, poor sleep enhances tumorigenesis and cancer progression [101] (Fig. 10.3). In general, tumors likely affect sleep-regulating neurocircuitry through two pathways: a neural and a humoral route. Disrupted sleep induced by cancer, surgery, or other anticancer therapy was thought to be triggered by systemic inflammation acting at the sleep-wake neurocircuitry in the brain [102]. Indeed, tumor- or TME-derived inflammatory cytokines (e.g., tumor necrosis factor-alpha (TNF-α) [103], IL-1 [104], and IFN-α [105]) can enter

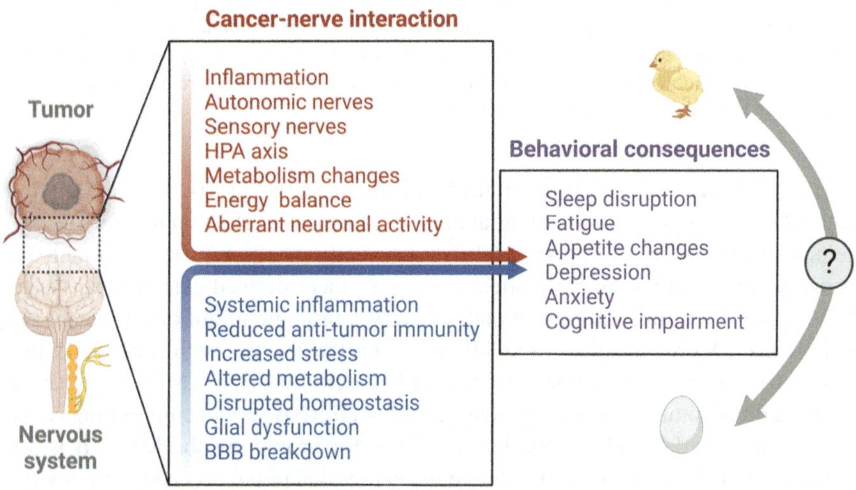

Fig. 10.3 Systemic behavioral consequences of interactions between cancer and the nervous system. Interaction between cancer and the nervous system is a reciprocal process. Tumor-induced inflammation enhances the activity of neuroendocrine (HPA axis, autonomic nervous system) and sensory nerves to change metabolism, eventually causing aberrant neuronal activity in the brain. In turn, the nervous system creates a systemic inflammatory and compromised adaptive immune microenvironment for peripheral tumors. Increased cellular stress and altered metabolism further disrupt whole body homeostasis and promote hazardous permeability of the BBB. This complicated cancer-nerve cross talk often results in various physiological and psychological consequences in cancer patients, such as sleep disruption, depression, anxiety, and cognitive impairment

systemic circulation, eventually reaching the brain parenchyma to cause aberrant neuronal activity in the sleep-related neurocircuitry [106]. For example, IL-1 inhibits wake-promoting neurons in the hypothalamic preoptic area/basal forebrain [107] and serotonergic neurons in the dorsal raphe nucleus [108, 109]. Additionally, tumor-invaded sensory nerves can change their activity patterns in response to noxious stimuli from cancer and TME (e.g., ATP, IL-1, IL-6, NGF, VEGF, TNF-α) to increase pain sensitivity. Pain signals subsequently relayed from the periphery to the lateral hypothalamus [110] or locus coeruleus (LC) can promote arousal [111]. In parallel, tumors can also affect sleep via disrupting host metabolism and energy balance. This includes cancer-induced changes in circulating satiety hormones (e.g., leptin, ghrelin [112, 113]), as well as glucose and insulin release from peripheral organs (such as the stomach, liver, pancreas, adipose tissue). A recent study demonstrated that metabolic and sleep abnormalities in a mouse model of nonmetastatic breast cancer are modulated via wake-promoting lateral-hypothalamic hypocretin/orexin (HO) neurons [114]. Reciprocally, chronic sleep disturbance results in systemic consequences that typically promote cancer progression. Sleep disruption-induced inflammation creates a cancer-promoting environment with increased pro-inflammatory cytokines and growth factors [115, 116]. Additionally, anticancer immunity was recently found to be influenced by sleep, where disrupted sleep-wake states cause alterations in the number of circulating and intratumor lymphocytes and myeloid cells [117, 118]. Cancer elimination is further impaired by reduced differentiation of cytotoxic cells (e.g., natural killer cells and CD8+ T cells) following sleep deprivation [117]. Further, sleep disruption triggers a stress response, activating the hypothalamic-pituitary-adrenal (HPA) axis and the sympathetic nervous system [119]. Glucocorticoids and catecholamines are released to modulate the immune system and energy balance, which are also critical for tumor development. Finally, due to dysfunction of endothelial and glial cells, circulating inflammatory molecules cross the BBB and enter the brain parenchyma, where they impair the function of sleep-wake regulatory systems [120–122].

Unfortunately, physiological changes often coexist with psychological symptoms such as *stress*, *depression*, and *cognitive impairment*, which may contribute to or be the outcome of chronic alterations in physiological status. Indeed, chronic *stress/depression* has been associated with an increased risk of cancer in many solid tumors, such as ovarian, prostate, breast, and pancreatic cancer [123]. Also, it is not surprising that most cancer patients experience psychological stress and/or depression after diagnosis and during anticancer therapy. The links between the psychological and physiological features during cancer progression have been investigated largely through the field of psychoneuroimmunology [124]. Although the exact mechanisms are still unclear, several theories have been proposed to connect cancers with the onset of psychological disorders, including immunologic, neuroendocrine, and metabolic mechanisms [125]. Cancer-induced stress and depression is often coupled with elevated levels of pro-inflammatory cytokines, which activate inflammatory responses by immune cells and the release of C-reactive protein (CRP) and IL-6 [126]. Type 1 helper T and cytotoxic T cell-mediated cellular immunity are selectively suppressed, which results in a compromised immune surveillance and an increased risk of cancer invasion [127]. Systemic inflammation is also thought to induce abnormalities in neurotransmitter metabolism, neural plasticity, and neuroendocrine function. Neuroendocrine pathways constituting the HPA axis and the sympathetic nervous system (SNS) are persistently activated under chronic stress and depression, which likely impairs the immune response and promotes the progression of some types of cancer [128]. Under chronic stress, the hypothalamus is activated to produce the corticotropin-releasing factor (CRF), which stimulates the pituitary gland to release adrenocorticotropic hormone (ACTH) and adrenal cortex to synthesize glucocorticoids (e.g., cortisol, corticosterone). As a result, these stress hormones promote tumor initiation and progression through inducing DNA damage, p53 degradation, and complex regulation of the TME [127]. In chronic stress models, the SNS was also found to be hyperactive, increasing the production of catecholamines (e.g., noradrenaline and adrenaline) [129, 130] to stimulate angiogenesis and cancer cell proliferation by activating β-adrenergic receptors [131]. Therefore, β-blockers might be an applicable strategy in clinical practice for cancer-induced stress/depression, which has been reviewed recently [27].

Thirty to 75% of cancer patients experience *cognitive impairment* before or during therapy [132]. Cancer-related cognitive impairment can be caused by cancer itself, nonspecific (host) factors, comorbid abnormalities, and/or the adverse effects of anticancer therapies [133]. Attention, memory, and information processing impairments are the most common cognitive problems identified in cancer patients [132, 133]. Pretreatment cognitive disorders in non-CNS cancer patients indicate a reciprocal cross talk between the CNS and TME, either directly via paracrine neurotransmitter secretion or indirectly through innate immune system modulation [134]. A recent review summarized potential mechanisms involved in peripheral tumor-induced cognitive impairments in mice model, including central dysregulation of the BBB and important brain regions like the hippocampus, hypothalamus, and ventral tegmental area (VTA) [135]. The integrity of the BBB is crucial for protecting the brain from solutes and cells in blood circulation. Although it was previously considered that the neurotoxic side effects of chemotherapy were the primary cause of BBB dysfunction in cancer patients, it has recently been determined that the tumor-induced inflammatory milieu is the initial source of BBB breakdown [136]. For instance, peripheral tumors release cytokines (e.g., IL-1β, TNF-α) and MMPs to disrupt the basement membrane and tight junctions of the BBB [137]. Moreover, breast cancer-derived EVs are able to cross the BBB via transcytosis by downregulating the brain endothelial expression of Rab7 protein and increasing the efficiency of their transport [138]. Also, brain metastasis-promoting EVs can be internalized by astrocytes to regulate their activity, thereby driving ECM remodeling in vivo [139]. Disrupted integrity of the BBB results in chronic neuronal dysfunction and likely eventually causes cognitive impairment among cancer patients. These cognitive deficits might be regarded as prognostic markers and therapeutic targets in cancer research, although more basic research is needed.

Neural Regulation in Cancer Therapy Resistance

As all tumors are composed of heterogenous malignant cells that are naturally susceptible or resistant to therapy, all types of anticancer treatment may lead to resistance due to the artificial selection of resistant clones. Recent research reveals that neural signaling influences cancer resistance mainly through three categories: chemotherapy resistance, targeted drug resistance, and immunotherapy resistance. Chemotherapy is one of the most widely used treatment options for many cancers. Growing evidence suggests that defective adrenergic signaling affects chemotherapeutic efficacy by upregulating antiapoptotic gene expression in cancer cells [26, 140, 141]. In another study, xenografted breast cancer cells appear to be resistant against chemotherapies by counteracting chemotherapy-induced oxidative stress via Ca^{2+}-dependent antiapoptotic pathways induced by the transient receptor potential ankyrin 1 (TRPA1) [142]. TRPA1 is a Ca^{2+}-influx channel expressed on sensory neurons, which is also overexpressed in a variety of cancers, acting as a sensor for inflammatory cytokines and cytotoxins. Additionally, chronic stress-induced glucocorticoid hypersecretion may confer resistance to chemotherapeutic treatments in tumor cells, where the function of the tumor suppressor p53 is attenuated through the induction of serum- and glucocorticoid-induced protein kinase [143].

Targeted drug resistance represents the second type of resistance that the nervous system influences. Cross talk between cancer and the nervous system generates a complicated intracellular signaling network, promoting cancer development and leading to the failure of targeted therapies. Documented resistances include those from therapies targeting adrenergic signaling and epidermal growth factor receptor (EGFR), as well as neurotrophic factors and their receptors [144]. In the first case, EGFR tyrosine kinase inhibitors – gefitinib and erlotinib – present good therapeutic efficacy among non-small cell lung cancer (NSCLC) patients. However, in vitro and in vivo studies both found that aberrant adrenergic signaling allows NSCLC cells to develop EGFR inhibitor resistance [145]. The mechanism seems to involve stress-mediated activation of β2-adrenergic receptors (β2-ARs) on NSCLC cells, which cooperatively signal with mutant EGFR, resulting in the inactivation of the tumor suppressor, liver kinase B1, and subsequently induce IL-6 expression. These phenomena can be abrogated with β-blockers (e.g., propranolol) or IL-6 inhibition [146]. In the latter case, a neurotransmitter-neurotrophic factor axis plays a role in anticancer drug resistance. For instance, ACh and catecholamines promote β2-AR dependent neurotrophin secretion, increasing sympathetic innervation and local neurotransmitter accumulation during cancer progression [129, 147, 148]. NGF-induced p75 receptor expression causes a decrease in the selectivity of triple-negative breast cancer (TNBC) cells to apoptosis induction [148]. Similarly, the BDNF/TrkB signaling pathway also acts as a universal attenuator of chemotherapy effectiveness by promoting cancer resistance to anoikis (apoptosis in response to loss of sufficient anchorage to epithelial layer or ECM) in multiple cancers. BDNF-induced resistance was also mediated by the downregulation of Bim, a proapoptotic protein that causes mitochondrial-mediated or intrinsic apoptosis [149, 150]. Since neurotrophic factor receptors were found to be overexpressed in many tumors, anticancer therapies using selective inhibitors or antibodies to target neurotrophic factors and their receptors might be used in conjunction with other therapeutics to overcome drug-induced resistance [151].

Immunotherapy resistance is the third major type of therapeutic resistance in cancer. Neural regulation of immune responses is an emerging field, but accumulating data indicates that tumor cells may hijack neural signals to promote immune evasion and suppress antitumor immune responses. Cancer cells often exploit adrenergic signaling-induced T cell suppression as the strategy for immunotherapy resistance [152–154]. For example, activation of adrenergic signaling elevates the response of regulatory (immune suppressive) T cells and suppresses the normal functions of antigen-primed T cells in cancer patients [155]. A recent study found that bone marrow-derived T cells are sequestrated in glioblastoma and other intracranial tumors. This T cell sequestration is associated with tumor-induced loss of sphingosine-1-phosphate receptor 1 (S1P1) from the T cell and hindered T cell trafficking from lymphoid structures to the brain [156]. Immune checkpoints (e.g., PD-1, CTLA-4) act as switches for immune responses, but cancer cells sometimes find ways to use these checkpoints to avoid being attacked by the immune system. Checkpoint targeting monoclonal antibodies, also known as checkpoint inhibitors, help the immune system to better find and attack cancer cells rather than kill them directly [157, 158]. β2-AR signaling was recently found to be a critical modulator for immunological checkpoint inhibition, primarily through direct suppression of CD8+ T cell activation [159]. Using behavioral, pharmacologic, and genetic strategies in mouse models, Bucsek et al. found that adrenergic stress and NE-driven β2-AR signaling not only directly impair T cell function but also influence their trafficking in the TME [159]. Reciprocally, reductions in β2-AR signaling triggers tumors to form an immunologically active TME with more effector CD8+ T cells, less programmed death receptor-1 (PD-1) expression, and higher effector CD8+ T cell to CD4+ regulatory T cell ratio. This conversion substantially boosted the effectiveness of anti-PD-1 checkpoint blockade [159]. These findings emphasize the potential of adrenergic signaling to modulate the immunological state of the TME and support the use of β-blockers in cancer patients to enhance their immunotherapy responses.

Conclusion and Future Directions

Despite the accumulating evidence of neuronal regulation in preclinical cancer models and patients, many of the underlying molecular mechanisms are still waiting to be uncovered. As an ever-moving target, cancer cells do not only divert all of their energy toward growth and spread but also take advantage of other cell types in the tumor microenvironment to facilitate their own survival. This includes a dangerous partnership with the neuroimmune systems, allowing successful clones to adaptively remodel their environment to ensure success. In a manner similar to the vascular system, neurons traverse throughout the body and release neurotransmitters and cytokines, which bind to receptors on cancer cells or stromal cells in the TME, triggering tumor-promoting signaling. In turn, tumor cells recruit adjacent nerves for PNI and/or hijack immune cells to escape immune surveillance and apoptosis. No tumor is an island; rather, they are more like prison breakers, attempting to flee from their initial locations and colonize wherever they can throughout the body. In 2020, the field "cancer neuroscience" was first described independently in a landmark commentary [1]. In this burgeoning research field, what we know so far is only the tip of the iceberg. There are still a number of outstanding questions ahead of us. These include identifying the key molecular mechanisms driving cancer-nerve cross talk, examining what effect their interactions have on host behaviors, and developing targeted therapies aimed at the specific neuronal circuits to arrest tumor initiation and progression.

Recent advances in modern neuroscience techniques allow us to address these questions and expand previous findings in the context of cancer. A variety of distinct approaches have been developed for the investigation of nerve-cancer interactions, such as microfluidic devices for neuron-cancer cell coculture; neurite outgrowth assays to determine cell proliferation and migration; optogenetics in combination with calcium imaging (e.g., GCaMP) or neurotransmitter sensors to modulate neuron activity; high-density microelectrode arrays to detect neuron firing in a high-throughput manner; electroencephalography [160] to record important cancer-related changes in the brain function; metabolic noncanonical amino acid labeling to identify nascent proteome changes in response to neural signaling; as well as cutting-edge neuroimaging techniques including multicolor two-photon light-sheet microscopy with tissue/whole body clearing, cryo-electron tomography, and deep intravital subcellular time-lapse imaging (DeepISTI), among others. Unbiased screening of cancer-induced alterations in neuronal activity and nervous system influences on cancer growth would enable the identification and manipulation of particular neuronal circuits and pathways to regulate cancer progression. Rapidly evolving machine learning and artificial intelligence techniques will further benefit data mining and analysis of currently available data. The following chapter will introduce tools and model systems for studying nerve-cancer interactions in-depth.

To have a comprehensive understanding of cancer-nerve cross talk, attention should be paid not just to local neuron-cancer cell interactions but also to systemic effects of the nervous system on the TME and immune system [1]. The study of cancer-nervous system interactions is appealing to multidisciplinary collaborations from researchers working in neuroscience, developmental biology, immunology, biochemistry, electrophysiology, and cancer biology, and there is much more to be discovered out there. More significantly, it is a challenging task to convert theoretical concepts into solid therapeutic strategies and clinical practices to compensate for current anticancer treatments and improve the quality of life for cancer patients. As Dr. Gustavo Ayala, one of the first to explore interactions between cancer and the nervous system, said, "this is a story to be written by many people over the next 30 years" [161].

References

1. Monje, M., et al., *Roadmap for the Emerging Field of Cancer Neuroscience*. Cell, 2020. **181**(2): p. 219–222.
2. Venkatesh, H.S., et al., *Neuronal Activity Promotes Glioma Growth through Neuroligin-3 Secretion*. Cell, 2015. **161**(4): p. 803–16.
3. Venkatesh, H.S., et al., *Electrical and synaptic integration of glioma into neural circuits*. Nature, 2019. **573**(7775): p. 539–545.
4. Venkataramani, V., et al., *Glutamatergic synaptic input to glioma cells drives brain tumour progression*. Nature, 2019. **573**(7775): p. 532–538.
5. Magnon, C., *Role of the autonomic nervous system in tumorigenesis and metastasis*. Mol Cell Oncol, 2015. **2**(2): p. e975643.
6. Loewenstein, W.R. and Y. Kanno, *Intercellular communication and the control of tissue growth: lack of communication between cancer cells*. Nature, 1966. **209**(5029): p. 1248–9.
7. Mehta, P.P., et al., *Incorporation of the gene for a cell-cell channel protein into transformed cells leads to normalization of growth*. J Membr Biol, 1991. **124**(3): p. 207–25.
8. Zhu, D., et al., *Transfection of C6 glioma cells with connexin 43 cDNA: analysis of expression, intercellular coupling, and cell proliferation*. Proc Natl Acad Sci U S A, 1991. **88**(5): p. 1883–7.
9. Temme, A., et al., *High incidence of spontaneous and chemically induced liver tumors in mice deficient for connexin32*. Curr Biol, 1997. **7**(9): p. 713–6.
10. Hitomi, M., et al., *Differential connexin function enhances self-renewal in glioblastoma*. Cell Rep, 2015. **11**(7): p. 1031–42.
11. Osswald, M., et al., *Brain tumour cells interconnect to a functional and resistant network*. Nature, 2015. **528**(7580): p. 93–8.
12. Alonso, F., et al., *Targeting endothelial connexin40 inhibits tumor growth by reducing angiogenesis and improving vessel perfusion*. Oncotarget, 2016. **7**(12): p. 14015–28.
13. Chen, Q., et al., *Carcinoma-astrocyte gap junctions promote brain metastasis by cGAMP transfer*. Nature, 2016. **533**(7604): p. 493–498.
14. Payne, S.L., et al., *Potassium channel-driven bioelectric signalling regulates metastasis in triple-negative breast cancer*. EBioMedicine, 2021. **75**: p. 103767.
15. Zeng, Q., et al., *Synaptic proximity enables NMDAR signalling to promote brain metastasis*. Nature, 2019. **573**(7775): p. 526–531.
16. Winkler, F. and W. Wick, *Harmful networks in the brain and beyond*. Science, 2018. **359**(6380): p. 1100–1101.
17. Latario, C.J., et al., *Tumor microtubes connect pancreatic cancer cells in an Arp2/3 complex-dependent manner*. Molecular Biology of the Cell, 2020. **31**(12): p. 1259–1272.
18. Pan, C. and F. Winkler, *Insights and opportunities at the crossroads of cancer and neuroscience*. Nature Cell Biology, 2022.
19. Jiang, S.H., et al., *Neurotransmitters: emerging targets in cancer*. Oncogene, 2020. **39**(3): p. 503–515.
20. Platel, J.C., et al., *NMDA receptors activated by subventricular zone astrocytic glutamate are critical for neuroblast survival prior to entering a synaptic network*. Neuron, 2010. **65**(6): p. 859–72.
21. Paez-Gonzalez, P., et al., *Identification of distinct ChAT(+) neurons and activity-dependent control of postnatal SVZ neurogenesis*. Nat Neurosci, 2014. **17**(7): p. 934–42.
22. Liu, X., et al., *Nonsynaptic GABA signaling in postnatal subventricular zone controls proliferation of GFAP-expressing progenitors*. Nat Neurosci, 2005. **8**(9): p. 1179–87.
23. Zahalka, A.H. and P.S. Frenette, *Nerves in cancer*. Nat Rev Cancer, 2020. **20**(3): p. 143–157.
24. Shurin, M.R., et al., *The Neuroimmune Axis in the Tumor Microenvironment*. J Immunol, 2020. **204**(2): p. 280–285.
25. Nagaraja, A.S., et al., *Sustained adrenergic signaling leads to increased metastasis in ovarian cancer via increased PGE2 synthesis*. Oncogene, 2016. **35**(18): p. 2390–7.

26. Kang, Y., et al., *Adrenergic Stimulation of DUSP1 Impairs Chemotherapy Response in Ovarian Cancer.* Clin Cancer Res, 2016. **22**(7): p. 1713–24.
27. Peixoto, R., M.L. Pereira, and M. Oliveira, *Beta-Blockers and Cancer: Where Are We?* Pharmaceuticals (Basel), 2020. **13**(6).
28. Udumyan, R., et al., *Beta-Blocker Drug Use and Survival among Patients with Pancreatic Adenocarcinoma.* Cancer Research, 2017. **77**(13): p. 3700–3707.
29. Le, C.P., et al., *Lymphovascular and neural regulation of metastasis: shared tumour signalling pathways and novel therapeutic approaches.* Best Pract Res Clin Anaesthesiol, 2013. **27**(4): p. 409–25.
30. Zhao, C.M., et al., *Denervation suppresses gastric tumorigenesis.* Sci Transl Med, 2014. **6**(250): p. 250ra115.
31. Hayakawa, Y., et al., *Nerve Growth Factor Promotes Gastric Tumorigenesis through Aberrant Cholinergic Signaling.* Cancer Cell, 2017. **31**(1): p. 21–34.
32. Magnon, C., et al., *Autonomic nerve development contributes to prostate cancer progression.* Science, 2013. **341**(6142): p. 1236361.
33. Renz, B.W., et al., *Cholinergic Signaling via Muscarinic Receptors Directly and Indirectly Suppresses Pancreatic Tumorigenesis and Cancer Stemness.* Cancer Discov, 2018. **8**(11): p. 1458–1473.
34. Kamiya, A., et al., *Genetic manipulation of autonomic nerve fiber innervation and activity and its effect on breast cancer progression.* Nat Neurosci, 2019. **22**(8): p. 1289–1305.
35. Fernández-Montoya, J., C. Avendaño, and P. Negredo, *The Glutamatergic System in Primary Somatosensory Neurons and Its Involvement in Sensory Input-Dependent Plasticity.* International Journal of Molecular Sciences, 2018. **19**(1): p. 69.
36. Li, L. and D. Hanahan, *Hijacking the neuronal NMDAR signaling circuit to promote tumor growth and invasion.* Cell, 2013. **153**(1): p. 86–100.
37. Wu, Y., A. Berisha, and J.C. Borniger, *Neuropeptides in Cancer: Friend and Foe?* Advanced Biology, 2022. **6**(9): p. 2200111.
38. Tan, F., C.J. Thiele, and Z. Li, *Neurotrophin Signaling in Cancer*, in *Handbook of Neurotoxicity*, R.M. Kostrzewa, Editor. 2014, Springer New York: New York, NY. p. 1825–1847.
39. Griffin, N., et al., *Targeting neurotrophin signaling in cancer: The renaissance.* Pharmacol Res, 2018. **135**: p. 12–17.
40. Di Donato, M., et al., *Targeting the Nerve Growth Factor Signaling Impairs the Proliferative and Migratory Phenotype of Triple-Negative Breast Cancer Cells.* Front Cell Dev Biol, 2021. **9**: p. 676568.
41. Forsyth, P.A., et al., *p75 neurotrophin receptor cleavage by alpha- and gamma-secretases is required for neurotrophin-mediated proliferation of brain tumor-initiating cells.* J Biol Chem, 2014. **289**(12): p. 8067–85.
42. Mirakaj, V. and P. Rosenberger, *Immunomodulatory Functions of Neuronal Guidance Proteins.* Trends Immunol, 2017. **38**(6): p. 444–456.
43. Chedotal, A., G. Kerjan, and C. Moreau-Fauvarque, *The brain within the tumor: new roles for axon guidance molecules in cancers.* Cell Death Differ, 2005. **12**(8): p. 1044–56.
44. Biankin, A.V., et al., *Pancreatic cancer genomes reveal aberrations in axon guidance pathway genes.* Nature, 2012. **491**(7424): p. 399–405.
45. Mehlen, P., C. Delloye-Bourgeois, and A. Chedotal, *Novel roles for Slits and netrins: axon guidance cues as anticancer targets?* Nat Rev Cancer, 2011. **11**(3): p. 188–97.
46. Gohrig, A., et al., *Axon guidance factor SLIT2 inhibits neural invasion and metastasis in pancreatic cancer.* Cancer Res, 2014. **74**(5): p. 1529–40.
47. Pasquale, E.B., *Eph receptors and ephrins in cancer: bidirectional signalling and beyond.* Nat Rev Cancer, 2010. **10**(3): p. 165-80.
48. Freemont, A.J. and J.A. Hoyland, *Cell adhesion molecules.* Clin Mol Pathol, 1996. **49**(6): p. M321–30.
49. Windisch, R., et al., *Oncogenic Deregulation of Cell Adhesion Molecules in Leukemia.* Cancers (Basel), 2019. **11**(3).

50. Deborde, S., et al., *Schwann cells induce cancer cell dispersion and invasion.* J Clin Invest, 2016. **126**(4): p. 1538–54.
51. Jahanban-Esfahlan, R., et al., *Combination of nanotechnology with vascular targeting agents for effective cancer therapy.* J Cell Physiol, 2018. **233**(4): p. 2982–2992.
52. Baghban, R., et al., *Tumor microenvironment complexity and therapeutic implications at a glance.* Cell Commun Signal, 2020. **18**(1): p. 59.
53. Wang, W., et al., *Nerves in the Tumor Microenvironment: Origin and Effects.* Front Cell Dev Biol, 2020. **8**: p. 601738.
54. Gysler, S.M. and R. Drapkin, *Tumor innervation: peripheral nerves take control of the tumor microenvironment.* J Clin Invest, 2021. **131**(11).
55. Zahalka, A.H., et al., *Adrenergic nerves activate an angio-metabolic switch in prostate cancer.* Science, 2017. **358**(6361): p. 321–326.
56. Kuol, N., et al., *Role of the Nervous System in Tumor Angiogenesis.* Cancer Microenviron, 2018. **11**(1): p. 1–11.
57. Wang, H., et al., *Role of the nervous system in cancers: a review.* Cell Death Discov, 2021. **7**(1): p. 76.
58. Devi, S., et al., *Adrenergic regulation of the vasculature impairs leukocyte interstitial migration and suppresses immune responses.* Immunity, 2021. **54**(6): p. 1219–1230 e7.
59. Cole, S.W. and A.K. Sood, *Molecular pathways: beta-adrenergic signaling in cancer.* Clin Cancer Res, 2012. **18**(5): p. 1201–6.
60. Partecke, L.I., et al., *Chronic stress increases experimental pancreatic cancer growth, reduces survival and can be antagonised by beta-adrenergic receptor blockade.* Pancreatology, 2016. **16**(3): p. 423–33.
61. Sloan, E.K., et al., *The sympathetic nervous system induces a metastatic switch in primary breast cancer.* Cancer Res, 2010. **70**(18): p. 7042–52.
62. Reijmen, E., et al., *Therapeutic potential of the vagus nerve in cancer.* Immunology Letters, 2018. **202**: p. 38–43.
63. Zhu, P., et al., *Alpha5 nicotinic acetylcholine receptor mediated immune escape of lung adenocarcinoma via STAT3/Jab1-PD-L1 signalling.* Cell Commun Signal, 2022. **20**(1): p. 121.
64. Gonzalez, H., C. Hagerling, and Z. Werb, *Roles of the immune system in cancer: from tumor initiation to metastatic progression.* Genes Dev, 2018. **32**(19–20): p. 1267–1284.
65. Biffi, G. and D.A. Tuveson, *Deciphering cancer fibroblasts.* J Exp Med, 2018. **215**(12): p. 2967–2968.
66. Secq, V., et al., *Stromal SLIT2 impacts on pancreatic cancer-associated neural remodeling.* Cell Death Dis, 2015. **6**: p. e1592.
67. Karakasheva, T.A., et al., *IL-6 Mediates Cross-Talk between Tumor Cells and Activated Fibroblasts in the Tumor Microenvironment.* Cancer Research, 2018. **78**(17): p. 4957–4970.
68. Rothaug, M., C. Becker-Pauly, and S. Rose-John, *The role of interleukin-6 signaling in nervous tissue.* Biochimica et Biophysica Acta (BBA) – Molecular Cell Research, 2016. **1863**(6, Part A): p. 1218–1227.
69. Qing, H., et al., *Origin and Function of Stress-Induced IL-6 in Murine Models.* Cell, 2020. **182**(2): p. 372–387.e14.
70. Tokumaru, Y., et al., *Intratumoral Adipocyte-High Breast Cancer Enrich for Metastatic and Inflammation-Related Pathways but Associated with Less Cancer Cell Proliferation.* Int J Mol Sci, 2020. **21**(16).
71. Santos, G.S.P., et al., *Sympathetic nerve-adipocyte interactions in response to acute stress.* Journal of Molecular Medicine, 2022. **100**(2): p. 151–165.
72. Thomas, D. and P. Radhakrishnan, *Tumor-stromal crosstalk in pancreatic cancer and tissue fibrosis.* Mol Cancer, 2019. **18**(1): p. 14.
73. Tomaselli, K.J., L.F. Reichardt, and J.L. Bixby, *Distinct molecular interactions mediate neuronal process outgrowth on non-neuronal cell surfaces and extracellular matrices.* J Cell Biol, 1986. **103**(6 Pt 2): p. 2659–72.

74. Najafi, M.F., et al., *Which form of collagen is suitable for nerve cell culture?* Neural Regeneration Research, 2013. **8**(23): p. 2165–2170.
75. Reinhard, J., et al., *The extracellular matrix niche microenvironment of neural and cancer stem cells in the brain.* The International Journal of Biochemistry & Cell Biology, 2016. **81**: p. 174–183.
76. Sood, D., et al., *3D extracellular matrix microenvironment in bioengineered tissue models of primary pediatric and adult brain tumors.* Nat Commun, 2019. **10**(1): p. 4529.
77. Norouzi, M., *Recent advances in brain tumor therapy: application of electrospun nanofibers.* Drug Discov Today, 2018. **23**(4): p. 912–919.
78. Granato, A.E.C., et al., *A novel decellularization method to produce brain scaffolds.* Tissue Cell, 2020. **67**: p. 101412.
79. Simsa, R., et al., *Brain organoid formation on decellularized porcine brain ECM hydrogels.* PLoS One, 2021. **16**(1): p. e0245685.
80. Chen, S.H., et al., *Perineural invasion of cancer: a complex crosstalk between cells and molecules in the perineural niche.* Am J Cancer Res, 2019. **9**(1): p. 1–21.
81. Entschladen, F., et al., *Neoneurogenesis: Tumors may initiate their own innervation by the release of neurotrophic factors in analogy to lymphangiogenesis and neoangiogenesis.* Medical Hypotheses, 2006. **67**(1): p. 33–35.
82. Mauffrey, P., et al., *Progenitors from the central nervous system drive neurogenesis in cancer.* Nature, 2019. **569**(7758): p. 672–678.
83. Dobrenis, K., et al., *Granulocyte colony-stimulating factor off-target effect on nerve outgrowth promotes prostate cancer development.* Int J Cancer, 2015. **136**(4): p. 982–8.
84. Liebig, C., et al., *Perineural invasion in cancer: a review of the literature.* Cancer, 2009. **115**(15): p. 3379–91.
85. Bapat, A.A., et al., *Perineural invasion and associated pain in pancreatic cancer.* Nat Rev Cancer, 2011. **11**(10): p. 695–707.
86. Arese, M., et al., *Tumor progression: the neuronal input.* Ann Transl Med, 2018. **6**(5): p. 89.
87. Amit, M., S. Na'ara, and Z. Gil, *Mechanisms of cancer dissemination along nerves.* Nat Rev Cancer, 2016. **16**(6): p. 399–408.
88. Zhang, Y., et al., *Pim-1 kinase as activator of the cell cycle pathway in neuronal death induced by DNA damage.* Journal of Neurochemistry, 2010. **112**(2): p. 497–510.
89. Gasparini, G., et al., *Nerves and Pancreatic Cancer: New Insights into A Dangerous Relationship.* Cancers, 2019. **11**(7): p. 893.
90. Demir, I.E., et al., *Perineural Mast Cells Are Specifically Enriched in Pancreatic Neuritis and Neuropathic Pain in Pancreatic Cancer and Chronic Pancreatitis.* Plos One, 2013. **8**(3).
91. Cavel, O., et al., *Endoneurial Macrophages Induce Perineural Invasion of Pancreatic Cancer Cells by Secretion of GDNF and Activation of RET Tyrosine Kinase Receptor.* Cancer Research, 2012. **72**(22): p. 5733–5743.
92. Shan, C., et al., *Schwann cells promote EMT and the Schwann-like differentiation of salivary adenoid cystic carcinoma cells via the BDNF/TrkB axis.* Oncol Rep, 2016. **35**(1): p. 427–435.
93. Xiang, T., X. Xia, and W. Yan, *Expression of Matrix Metalloproteinases-2/-9 is Associated With Microvessel Density in Pancreatic Cancer.* American Journal of Therapeutics, 2017. **24**(4): p. e431–e434.
94. Klupp, F., et al., *Serum MMP7, MMP10 and MMP12 level as negative prognostic markers in colon cancer patients.* BMC Cancer, 2016. **16**(1): p. 494.
95. Liu, Y., W. Zhou, and D.-W. Zhong, *Meta-analyses of the associations between four common TGF-β1 genetic polymorphisms and risk of colorectal tumor.* Tumor Biology, 2012. **33**(4): p. 1191–1199.
96. Huberfeld, G. and C.J. Vecht, *Seizures and gliomas — towards a single therapeutic approach.* Nature Reviews Neurology, 2016. **12**(4): p. 204–216.
97. Yu, K., et al., *PIK3CA variants selectively initiate brain hyperactivity during gliomagenesis.* Nature, 2020. **578**(7793): p. 166–171.

98. Amit, M., et al., *Loss of p53 drives neuron reprogramming in head and neck cancer.* Nature, 2020. **578**(7795): p. 449–454.
99. Tang, P.C.-T., et al., *Single-cell RNA sequencing uncovers a neuron-like macrophage subset associated with cancer pain.* Science Advances, 2022. **8**(40): p. eabn5535.
100. Davidson, J.R., et al., *Sleep disturbance in cancer patients.* Soc Sci Med, 2002. **54**(9): p. 1309–21.
101. Berisha, A., K. Shutkind, and J.C. Borniger, *Sleep Disruption and Cancer: Chicken or the Egg?* Front Neurosci, 2022. **16**: p. 856235.
102. Walker, W.H., 2nd and J.C. Borniger, *Molecular Mechanisms of Cancer-Induced Sleep Disruption.* Int J Mol Sci, 2019. **20**(11).
103. Kubota, T., et al., *Tumor necrosis factor receptor fragment attenuates interferon-gamma-induced non-REM sleep in rabbits.* J Neuroimmunol, 2001. **119**(2): p. 192–8.
104. Greenberg, D.B., et al., *Treatment-related fatigue and serum interleukin-1 levels in patients during external beam irradiation for prostate cancer.* J Pain Symptom Manage, 1993. **8**(4): p. 196–200.
105. Spath-Schwalbe, E., et al., *Interferon-alpha acutely impairs sleep in healthy humans.* Cytokine, 2000. **12**(5): p. 518–21.
106. Imeri, L. and M.R. Opp, *How (and why) the immune system makes us sleep.* Nature Reviews Neuroscience, 2009. **10**(3): p. 199–210.
107. Alam, M.N., et al., *Interleukin-1beta modulates state-dependent discharge activity of preoptic area and basal forebrain neurons: role in sleep regulation.* Eur J Neurosci, 2004. **20**(1): p. 207–16.
108. Manfridi, A., et al., *Interleukin-1beta enhances non-rapid eye movement sleep when microinjected into the dorsal raphe nucleus and inhibits serotonergic neurons in vitro.* Eur J Neurosci, 2003. **18**(5): p. 1041–9.
109. Brambilla, D., et al., *Interleukin-1 inhibits firing of serotonergic neurons in the dorsal raphe nucleus and enhances GABAergic inhibitory post-synaptic potentials.* Eur J Neurosci, 2007. **26**(7): p. 1862–9.
110. Francis, N. and J.C. Borniger, *Cancer as a homeostatic challenge: the role of the hypothalamus.* Trends Neurosci, 2021. **44**(11): p. 903–914.
111. Alexandre, C., et al., *Decreased alertness due to sleep loss increases pain sensitivity in mice.* Nat Med, 2017. **23**(6): p. 768–774.
112. Au, C.C., J.B. Furness, and K.A. Brown, *Ghrelin and Breast Cancer: Emerging Roles in Obesity, Estrogen Regulation, and Cancer.* Front Oncol, 2016. **6**: p. 265.
113. Garofalo, C. and E. Surmacz, *Leptin and cancer.* J Cell Physiol, 2006. **207**(1): p. 12–22.
114. Borniger, J.C., et al., *A Role for Hypocretin/Orexin in Metabolic and Sleep Abnormalities in a Mouse Model of Non-metastatic Breast Cancer.* Cell Metab, 2018. **28**(1): p. 118–129 e5.
115. Yehuda, S., et al., *REM sleep deprivation in rats results in inflammation and interleukin-17 elevation.* J Interferon Cytokine Res, 2009. **29**(7): p. 393–8.
116. Karin, M. and F.R. Greten, *NF-kappaB: linking inflammation and immunity to cancer development and progression.* Nat Rev Immunol, 2005. **5**(10): p. 749–59.
117. De Lorenzo, B.H.P., et al., *Chronic Sleep Restriction Impairs the Antitumor Immune Response in Mice.* Neuroimmunomodulation, 2018. **25**(2): p. 59–67.
118. Huang, J., et al., *Sleep Deprivation Disturbs Immune Surveillance and Promotes the Progression of Hepatocellular Carcinoma.* Front Immunol, 2021. **12**: p. 727959.
119. Li, S.-B., et al., *Hypothalamic circuitry underlying stress-induced insomnia and peripheral immunosuppression.* Science Advances, 2020. **6**(37): p. eabc2590.
120. He, J., et al., *Sleep restriction impairs blood-brain barrier function.* J Neurosci, 2014. **34**(44): p. 14697–706.
121. Medina-Flores, F., et al., *Sleep loss disrupts pericyte-brain endothelial cell interactions impairing blood-brain barrier function.* Brain Behav Immun, 2020. **89**: p. 118–132.
122. Bellesi, M., et al., *Sleep Loss Promotes Astrocytic Phagocytosis and Microglial Activation in Mouse Cerebral Cortex.* J Neurosci, 2017. **37**(21): p. 5263–5273.

123. Dlamini, Z., et al., *Many Voices in a Choir: Tumor-Induced Neurogenesis and Neuronal Driven Alternative Splicing Sound Like Suspects in Tumor Growth and Dissemination.* Cancers (Basel), 2021. **13**(9).
124. Reiche, E.M., S.O. Nunes, and H.K. Morimoto, *Stress, depression, the immune system, and cancer.* Lancet Oncol, 2004. **5**(10): p. 617–25.
125. Powell, N.D., A.J. Tarr, and J.F. Sheridan, *Psychosocial stress and inflammation in cancer.* Brain Behav Immun, 2013. **30 Suppl**: p. S41–7.
126. Miller, A.H., V. Maletic, and C.L. Raison, *Inflammation and its discontents: the role of cytokines in the pathophysiology of major depression.* Biol Psychiatry, 2009. **65**(9): p. 732–41.
127. Dai, S., et al., *Chronic Stress Promotes Cancer Development.* Front Oncol, 2020. **10**: p. 1492.
128. Smith, S.M. and W.W. Vale, *The role of the hypothalamic-pituitary-adrenal axis in neuroendocrine responses to stress.* Dialogues Clin Neurosci, 2006. **8**(4): p. 383–95.
129. Renz, B.W., et al., *beta2 Adrenergic-Neurotrophin Feedforward Loop Promotes Pancreatic Cancer.* Cancer Cell, 2018. **34**(5): p. 863–867.
130. Lutgendorf, S.K., et al., *Social isolation is associated with elevated tumor norepinephrine in ovarian carcinoma patients.* Brain Behav Immun, 2011. **25**(2): p. 250–5.
131. Thaker, P.H., et al., *Chronic stress promotes tumor growth and angiogenesis in a mouse model of ovarian carcinoma.* Nat Med, 2006. **12**(8): p. 939–44.
132. Janelsins, M.C., et al., *Prevalence, mechanisms, and management of cancer-related cognitive impairment.* Int Rev Psychiatry, 2014. **26**(1): p. 102–13.
133. Pendergrass, J.C., S.D. Targum, and J.E. Harrison, *Cognitive Impairment Associated with Cancer: A Brief Review.* Innov Clin Neurosci, 2018. **15**(1–2): p. 36–44.
134. Olson, B. and D.L. Marks, *Pretreatment Cancer-Related Cognitive Impairment-Mechanisms and Outlook.* Cancers (Basel), 2019. **11**(5).
135. Mampay, M., M.S. Flint, and G.K. Sheridan, *Tumour brain: Pretreatment cognitive and affective disorders caused by peripheral cancers.* Br J Pharmacol, 2021. **178**(19): p. 3977–3996.
136. Kim, J., et al., *Tumor-induced disruption of the blood-brain barrier promotes host death.* Dev Cell, 2021. **56**(19): p. 2712–2721 e4.
137. Kadry, H., B. Noorani, and L. Cucullo, *A blood-brain barrier overview on structure, function, impairment, and biomarkers of integrity.* Fluids Barriers CNS, 2020. **17**(1): p. 69.
138. Morad, G., et al., *Tumor-Derived Extracellular Vesicles Breach the Intact Blood-Brain Barrier via Transcytosis.* ACS Nano, 2019. **13**(12): p. 13853–13865.
139. Morad, G., et al., *Cdc42-Dependent Transfer of mir301 from Breast Cancer-Derived Extracellular Vesicles Regulates the Matrix Modulating Ability of Astrocytes at the Blood-Brain Barrier.* Int J Mol Sci, 2020. **21**(11).
140. Eng, J.W., et al., *Housing temperature-induced stress drives therapeutic resistance in murine tumour models through beta2-adrenergic receptor activation.* Nat Commun, 2015. **6**: p. 6426.
141. Chen, H., et al., *beta2-AR activation induces chemoresistance by modulating p53 acetylation through upregulating Sirt1 in cervical cancer cells.* Cancer Sci, 2017. **108**(7): p. 1310–1317.
142. Takahashi, N., et al., *Cancer Cells Co-opt the Neuronal Redox-Sensing Channel TRPA1 to Promote Oxidative-Stress Tolerance.* Cancer Cell, 2018. **33**(6): p. 985–1003 e7.
143. Feng, Z., et al., *Chronic restraint stress attenuates p53 function and promotes tumorigenesis.* Proc Natl Acad Sci U S A, 2012. **109**(18): p. 7013–8.
144. Liu, D., et al., *Neural regulation of drug resistance in cancer treatment.* Biochim Biophys Acta Rev Cancer, 2019. **1871**(1): p. 20–28.
145. Gridelli, C., et al., *Erlotinib in the treatment of non-small cell lung cancer: current status and future developments.* Anticancer Res, 2010. **30**(4): p. 1301–10.
146. Nilsson, M.B., et al., *Stress hormones promote EGFR inhibitor resistance in NSCLC: Implications for combinations with beta-blockers.* Sci Transl Med, 2017. **9**(415).
147. Renz, B.W., et al., *beta2 Adrenergic-Neurotrophin Feedforward Loop Promotes Pancreatic Cancer.* Cancer Cell, 2018. **33**(1): p. 75–90 e7.

148. Chakravarthy, R., K. Mnich, and A.M. Gorman, *Nerve growth factor (NGF)-mediated regulation of p75(NTR) expression contributes to chemotherapeutic resistance in triple negative breast cancer cells.* Biochem Biophys Res Commun, 2016. **478**(4): p. 1541–7.
149. Jaboin, J., et al., *Brain-derived neurotrophic factor activation of TrkB protects neuroblastoma cells from chemotherapy-induced apoptosis via phosphatidylinositol 3'-kinase pathway.* Cancer Res, 2002. **62**(22): p. 6756–63.
150. Li, Z., et al., *Downregulation of Bim by brain-derived neurotrophic factor activation of TrkB protects neuroblastoma cells from paclitaxel but not etoposide or cisplatin-induced cell death.* Cell Death Differ, 2007. **14**(2): p. 318–26.
151. Demir, I.E., et al., *Nerve growth factor & TrkA as novel therapeutic targets in cancer.* Biochim Biophys Acta, 2016. **1866**(1): p. 37–50.
152. O'Donnell, J.S., et al., *Resistance to PD1/PDL1 checkpoint inhibition.* Cancer Treat Rev, 2017. **52**: p. 71–81.
153. Xia, A., et al., *T Cell Dysfunction in Cancer Immunity and Immunotherapy.* Front Immunol, 2019. **10**: p. 1719.
154. Barrueto, L., et al., *Resistance to Checkpoint Inhibition in Cancer Immunotherapy.* Transl Oncol, 2020. **13**(3): p. 100738.
155. Zhou, L., et al., *Propranolol Attenuates Surgical Stress-Induced Elevation of the Regulatory T Cell Response in Patients Undergoing Radical Mastectomy.* J Immunol, 2016. **196**(8): p. 3460–9.
156. Chongsathidkiet, P., et al., *Sequestration of T cells in bone marrow in the setting of glioblastoma and other intracranial tumors.* Nat Med, 2018. **24**(9): p. 1459–1468.
157. Zappasodi, R., T. Merghoub, and J.D. Wolchok, *Emerging Concepts for Immune Checkpoint Blockade-Based Combination Therapies.* Cancer Cell, 2018. **34**(4): p. 690.
158. Robert, C., *A decade of immune-checkpoint inhibitors in cancer therapy.* Nat Commun, 2020. **11**(1): p. 3801.
159. Bucsek, M.J., et al., *beta-Adrenergic Signaling in Mice Housed at Standard Temperatures Suppresses an Effector Phenotype in CD8(+) T Cells and Undermines Checkpoint Inhibitor Therapy.* Cancer Research, 2017. **77**(20): p. 5639–5651.
160. Cohen, P.S., et al., *Neuropeptide-Y expression in the developing adrenal-gland and in childhood neuroblastoma tumors.* Cancer Research, 1990. **50**(18): p. 6055–6061.
161. Servick, K., *War of nerves.* Science, 2019. **365**(6458): p. 1071–1073.

Chapter 11
Tools and Model Systems to Study Nerve-Cancer Interactions

Peter L. Wang, Nicole A. Lester, Jimmy A. Guo, Jennifer Su, Carina Shiau, and William L. Hwang

Introduction

The study of nerve-cancer interactions relies on a diverse set of tools spanning multiple scientific disciplines. While the connection between nervous system function and cancer can be traced back several centuries to observations made by physicians that associated psychosocial factors with tumor incidence and growth, the anatomical relationship between nerves and cancer came to light in the late nineteenth century following the adoption of specific tissue staining methods for microscopy that enabled the identification of nerves in relation to various types of cancers [1]. These observations led to the appreciation of key histopathological phenomena, including tumor innervation, defined by the presence of nerves within tumors, and perineural invasion (PNI), defined by the surrounding or invasion of nerves by malignant cells [2, 3].

The question of whether nerves could potentiate or inhibit cancer growth led to the adoption of methods for perturbing neuronal signaling to tumors. Early studies utilized surgical denervation in animal models to determine whether tumorigenesis was altered in the absence of nerve signaling [1, 4]. The introduction of beta-adrenergic receptor blocking drugs provided an additional method for eliminating nerve signaling, but with preferred action on the sympathetic nervous system [2, 5]. Methods for stimulating nerve signaling in tumor-bearing mice were also used,

The original version of this chapter was revised. The correction to this chapter is available at https://doi.org/10.1007/978-3-031-32429-1_13

P. L. Wang · N. A. Lester · J. A. Guo · J. Su · C. Shiau · W. L. Hwang (✉)
Center for Systems Biology, Department of Radiation Oncology, and Center for Cancer Research, Massachusetts General Hospital and Harvard Medical School, Boston, MA, USA

Broad Institute of MIT and Harvard, Cambridge, MA, USA
e-mail: whwang1@mgh.harvard.edu

M. Amit, N. N. Scheff (eds.), *Cancer Neuroscience*,
https://doi.org/10.1007/978-3-031-32429-1_11

including models of induced stress, direct electrical stimulation, and beta-adrenergic receptor agonists [6–8]. With the advancement of methods in neuroscience, new techniques, such as optogenetics, chemogenetics, and adeno-associated virus (AAV)-targeted nerve signaling, became available to target specific nerve types with greater spatial and temporal accuracy [9].

Over the past two decades, retrospective studies have demonstrated that patients with increased tumor innervation and PNI across several cancer types, including pancreatic, prostate, head and neck, breast, and colorectal cancers, tend to have worse prognoses [10]. Combined with basic research using modern genomics and bioengineering methods, the picture that has emerged indicates that nerves not only communicate with neoplastic cells directly but also contribute and respond to the greater tumor microenvironment (TME) [2, 11]. As further investigation is required to tease out the complex nature of nerve-cancer interactions, this chapter provides a review of tools and models that are relevant for the cancer neuroscience field as well as emerging perspectives for future studies.

Visualizing Tissue and Cellular Architecture

Since the early observations of neoplastic growths by Rudolf Virchow in the nineteenth century, the microscope has been a pillar for the study of cancer biology [12]. The use of light microscopy similarly opened the door to viewing the nervous system, with seminal observations made by Ramon y Cajal in the late nineteenth century using Camillo Golgi's silver chromate staining method to label individual neurons [13]. In 1897, the first published record of nerve presence in tumors was made by Young using methylene-blue staining of nerves in breast and cervical cancer as well as sarcomas [14]. Over the next century, the wide adoption of hematoxylin and eosin (H&E) staining for medical diagnosis further contributed to the appreciation of nerves in cancer, including the finding that malignant cells could surround and invade nerves, a phenomenon known as perineural invasion, or PNI, in the peripheral nervous system (PNS) and perineuronal satellitosis in the central nervous system (CNS) [15–17].

The invention of immunohistochemistry (IHC), the method by which labeled antibodies are used to identify specific antigens in tissues, revolutionized cancer research by enabling the identification of tumor markers with improved molecular specificity. In the context of PNI, IHC staining of the S100 protein, which is expressed in normal and injured Schwann cells, has become a key identifier of tumor-associated nerves [18]. Indeed, S100 staining has been shown to significantly improve detection of PNI compared to standard H & E alone [19, 20]. Antibodies targeting tyrosine hydroxylase (TH) and vesicular acetylcholine transporter (VAChT), which are expressed on sympathetic and parasympathetic postganglionic nerves, respectively, and transient receptor potential cation channel subfamily V member 1 (TRPV1), which is expressed on nociceptive sensory neurons, further reveal nerve-type specificity in tumors [21, 22]. The presence of PNI is prognostic of clinical outcomes in many cancer types, which has led to its inclusion in the current American Joint Committee on Cancer (AJCC) cancer staging manual (eighth edition) for penile cancer, head and neck cancers, esophageal cancer, and thyroid cancer [11]. IHC-based methods for PNI detection have important implications for the development of potential treatment strategies.

As interest in the biological significance of cancer-nerve interactions has grown, researchers have begun to look more closely at the structural relationship between nerves and cancer cells. Given its high resolution, electron microscopy is an excellent tool for such purposes. Early transmission electron microscopy (TEM) studies of perineural structures in prostate cancer showed that peripheral nerve invasion by cancer cells occurred by direct extension around nerves and not within vascular channels [23]. In a more recent study on breast-to-brain metastasis, TEM revealed pseudo-tripartite synapses between cancer cells and glutamatergic neurons, which fuel metastatic growth [24]. Similar tumor-nerve synapses were likewise visualized by TEM between glutamatergic neurons and gliomas (Fig. 11.1a) [25, 26]. In addition to observing tissue structures, scanning electron microscopy (SEM) has also been used to confirm the presence of exosomes that promote neurogenesis in tumors [27].

High-resolution imaging techniques that utilize fluorophores to distinguish cellular and subcellular structures offer a wide range of applications for modern research. Conventional confocal microscopy, which uses laser scanning through a pinhole that blocks out-of-focus light, enables increased optical resolution and depth of focus [28]. The use of confocal imaging to study intercellular interactions, such as those between cancer cells and Schwann cells, has revealed new insights into how malignant cells invade nerves [29]. In recent years, there has also been increased use of spinning disc confocal microscopy and light-sheet fluorescence microscopy, which facilitate the acquisition of larger sample regions and even live imaging given their faster scanning speed [28, 30]. These developments have occurred alongside the adoption of tissue clearing methods, which utilize solvent-,

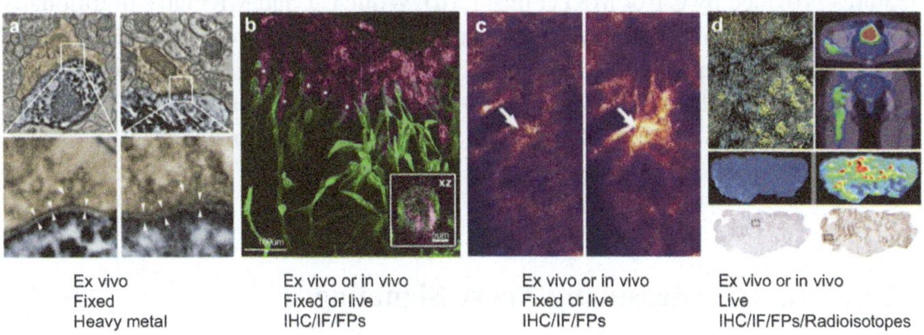

Fig. 11.1 Methods for visualizing tissue and cellular architecture in cancer neuroscience. (**a**) Synapses between presynaptic axons (sepia) and postsynaptic glioma cells (blue) visualized by immuno-electron microscopy [25]. (**b**) Confocal imaging of pancreatic cancer cells (magenta) associated with Schwann cells (green) in Matrigel [29]. (**c**) Intravital imaging of autonomous calcium activity in glioma cells [34]. (**d**) Multiplexed in situ and clinical imaging methods. (Top left) Spatial molecular imaging (SMI) provides mRNA and protein localization at single-cell/subcellular resolution [39]. (Top right) PET/CT detects perineural spread of the sciatic nerve [40]. (Bottom) Raman imaging reveals nanoparticle-stained cancer markers validated by IHC [38]

aqueous-, or hydrogel-based techniques to remove light-obstructive substances from tissues and match the refractive indices of specimens with the imaging media [30]. As a result, "cleared" tissues are optically transparent and enable 3D visualization of complex structures in whole tissues or organs. Using light-sheet fluorescence imaging of optically cleared pancreata, Guillot et al. recently showed that sprouting of sympathetic axons into pancreatic intraepithelial neoplasia (PanIN) lesions occurs early on in a mouse model of pancreatic ductal adenocarcinoma (PDAC) [31].

Live cell imaging in culture systems, tissue explants, or living animals is another important method for studying cancer-nerve interactions. Early studies in 2001 by Ayala et al. utilized phase microscopy to visualize reciprocal tropism between neurons and prostate cancer cell lines in coculture [32]. More recently, live imaging of neural invasion by pancreatic cancer cells using confocal and fluorescent microscopy demonstrated that Schwann cells physically associate with cancer cells and exert forces to enhance their motility (Fig. 11.1b) [29]. Additionally, two-photon microscopy, which has superior imaging depth and low phototoxicity, has been used to identify and map in vivo communication patterns in glioma cell networks (Fig. 11.1c) [25, 33, 34].

The continued progression of imaging techniques has resulted in a variety of multiplexed methods for visualizing multiple molecular and cellular biomarkers in both ex vivo and in vivo conditions. For example, spatial proteotranscriptomics combines optical barcoding approaches and conventional multiplexed immunofluorescence to enable the simultaneous mapping of mRNA and proteins in tissue sections (Fig. 11.1d) (see section "Multimodal Genomic Analysis"). Multiplexed clinical imaging methods are another class of tools that reveal spatial parameters in live patients. For example, the combination of positron emission tomography (PET) with computerized tomography (CT) or magnetic resonance imaging (MRI) enables the detection of metabolically active tumors by PET with respect to anatomical features provided by CT or MRI (Fig. 11.1d). While CT and MRI have traditionally been used for identifying nerve involvement in head and neck cancers, PET/CT and PET/MRI are also now being evaluated [35, 36]. Raman imaging is simultaneously emerging as a technique for localizing nanoparticle-labeled cancer markers in ex vivo and in vivo tissues (Fig. 11.1d) [37, 38]. Combinations of these tools may provide increased opportunities for basic and clinical study.

Modifying and Measuring Nerve Signaling

Following the initial descriptions of tumor innervation in the scientific literature, investigators began to speculate whether nerves could influence tumor growth. Thus, models of surgical denervation followed by tumor induction were employed in animals to examine the effect of eliminating nerve signaling on carcinogenesis. In 1917, Adler and Sittenfield showed that rats inoculated with carcinoma cells into denervated testes rapidly developed large tumors while rats with intact innervation remained tumor-free [41]. A few years later, Cramer demonstrated the opposite effect in mice with denervated skin that developed lower rates of tar painting-induced tumors compared to mice with preserved skin innervation [42]. Batkin and colleagues similarly found in 1970 that tumors transplanted into the gastrocnemius muscle of mice exhibited temporary growth reduction when paired with sciatic denervation [43].

Early studies examining the influence of autonomic nerve signaling on tumor growth utilized animal models of chronic stress to induce sympathetic nerve activity. In 1982, Visintainer and colleagues showed that lack of control over physical stressors reduced tumor rejection and survival in rats receiving implanted tumors [44]. Thaker and colleagues similarly showed that chronic stress promotes tumor growth and angiogenesis in a mouse model of ovarian carcinoma [45]. These studies pointed to the potential role of the sympathetic nervous system and specifically beta-adrenergic signaling activation in cancer progression. Indeed, countering elevated levels of sympathetic activity with propranolol, a nonselective beta-adrenergic receptor antagonist, or with 6-hydroxydopamine (6-OHDA), a chemical sympathectomy agent, inhibits carcinogenesis across several preclinical cancer models [46–48].

Drugs that modify cholinergic signaling have also been used to study subtype-specific neuronal contributions to carcinogenesis. For example, Renz et al. found that activation of muscarinic receptor signaling by the cholinergic agonist bethanechol suppressed tumorigenesis in a mouse model of PDAC [49]. However, stimulation of muscarinic signaling by the nonselective cholinergic agonist carbachol led to increased tumorigenesis in a mouse model of gastric cancer [50]. Similarly, carbachol treatment significantly enhanced lymph node metastasis in a preclinical model of prostate cancer [21]. These apparently conflicting findings across distinct cancer types reflect the complexity of autonomic nerve signaling in cancer and the need to further explore context-specific signaling.

While the role of sensory nerves in cancer has been less explored compared to autonomic nerve signaling, a few studies have examined the relationship between nociceptive fibers and tumor progression. For example, using neonatal capsaicin treatment to ablate sensory nerves, Saloman and colleagues showed that denervation prevented PNI, delayed PanIN formation, and prolonged survival in an autochthonous mouse model of PDAC [51]. In a separate melanoma model, both sensory nerve ablation by resiniferatoxin treatment as well as genetic depletion of Nav1.8-expressing neurons increased melanoma growth [52]. These results similarly demonstrate the potentially divergent responses across distinct cancer types and experimental conditions.

Investigators seeking to induce nerve activation have also adopted methods for direct electrical stimulation. A clinically relevant example of this is vagal nerve stimulation, which is used for the treatment of patients with drug-resistant epilepsy and depression [53]. In mice, vagal nerve stimulation has been shown to dampen endotoxin-induced inflammatory cytokine production by splenic macrophages through a T cell-derived acetylcholine signaling pathway [54, 55]. Interestingly, vagal stimulation has also been shown to enhance antitumor immunity by suppressing the expansion of myeloid-derived suppressor cells (MDSCs) in a mouse model of colorectal cancer [56]. While direct electrical stimulation of other peripheral nerves remains relatively unexplored in the context of cancer, the increasing availability of implantable devices for nerve stimulation offers new opportunities for targeting nerve-cancer pathways.

The use of genetic engineering further enhances the spatial and temporal specificity with which investigators can modulate nerve signaling. Optogenetics, for example, can be used to activate or inhibit specific nerve types by inserting light-sensitive channelrhodopsins into the desired cells [57]. With this approach, Venkataramani et al. showed that stimulation of neurons in a glioma-bearing brain region induced synchronized calcium activity patterns in glioblastoma cells [25]. Using optogenetics to directly target glioma cells, Venkatesh et al. and Venkataramani et al. further observed that depolarization of glioma cell membranes leads to an increase in tumor cell proliferation [25, 26]. Chemogenetic methods, such as designer receptors exclusively activated by designer drugs (DREADDs), and viral-mediated gene delivery are additional techniques for activating and inhibiting nerve signaling [58]. Using nerve-targeted AAV vectors, Kamiya and colleagues recently investigated the differential effects of sympathetic and parasympathetic signaling on breast cancer progression in mice by introducing a constitutively active bacterial voltage-gated sodium channel for nerve stimulation and diphtheria toxin A (DTA) for denervation in the nerve subtype of interest [48]. Importantly, the improved temporal and spatial control of nerve signaling by viral vectors can also be used for optogenetics or chemogenetics by inducing channelrhodopsins or designer receptors in the desired nerve type(s).

In addition to methods for modifying nerve signaling, tools for measuring and monitoring nerve activity are necessary for cancer neuroscience research (Fig. 11.2). Microelectrodes possess the capability to both record and induce nerve activity [59]. Traditionally, single electrodes have been used to record nerve activity in anesthetized animals for the study of peripheral neuropathy or in ex vivo preparations to probe electrical activity in neuronal cells [60]. The introduction of implantable cuff electrodes that wrap around entire nerve bundles and penetrating electrodes that

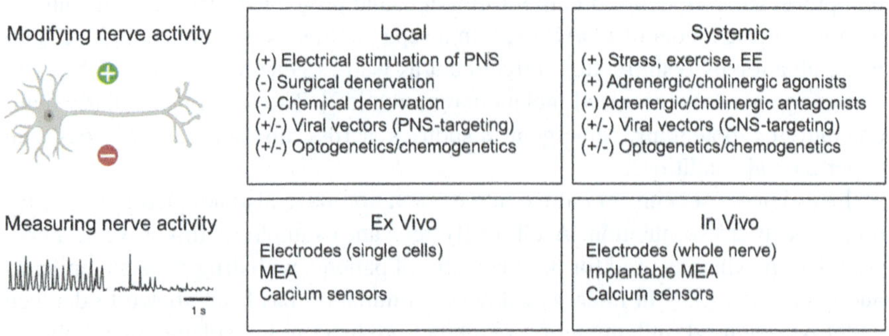

Fig. 11.2 Methods for modifying and measuring nerve signaling. (+) signifies activation, (−) signifies inhibition, and (+/−) signifies both. Abbreviations: *CNS* (central nervous system), *EE* (environmental enrichment), *MEA* (microelectrode array), *PNS* (peripheral nervous system)

insert into nerve fascicles offers increased spatial resolution and temporal control for recording and stimulating nerves in freely moving animals [61]. Microelectrode arrays (MEAs) contain tens to thousands of microelectrodes and can detect signals from all possible sources around every sensor, thus providing a better representation of network function [59]. MEAs can be implanted in vivo or used in vitro to serve as neural interfaces that connect neurons to electronic circuitry [58]. Potential applications for MEAs in cancer neuroscience include stimulating and recording neurons in cocultures with cancer cells or detecting and modulating intratumoral nerve activity with implantable MEAs.

In recent decades, the adoption of fluorescent calcium sensors for detecting neuronal activity has revolutionized the study of in vivo nerve signaling. In particular, GCaMP6 and GCaMP3, two variations of the genetically encoded calcium indicator (GECI) comprised of a fused green fluorescent protein (GFP), calmodulin (Cam), and M13, a peptide sequence from myosin light-chain kinase, have been used to relay nerve activity in both glioma and breast cancer [62]. While calcium sensors report neural activity at timescales much slower than electric signaling, new GECI variants that differ in sensitivity, kinetics, and spectral properties are currently under development [63].

Multimodal Genomic Analysis

Early studies on the molecular basis of cancer focused on tumor suppressor genes and oncogenes, which were found to control abnormal cellular proliferation [64]. For instance, loss-of-function mutations in neurofibromin (*NF1*) were found to cause neurofibromatosis type 1, which is linked to a range of symptoms including cutaneous neurofibromas, optic nerve gliomas, and malignant peripheral nerve sheath tumors [65]. While low-throughput sequencing was used for the initial identification of such genes, the advent of next-generation sequencing (NGS) platforms to perform whole genome, whole exome, and targeted sequencing enabled more systematic and comprehensive identification of genetic mutations, including single nucleotide variations (SNVs) and single nucleotide polymorphisms (SNPs), in patient populations [66, 67]. For example, using whole exome sequencing (WES), Biankin and colleagues identified somatic mutations in SLIT/ROBO signaling genes associated with axon guidance in PDAC patients [68].

The ability to quantify mRNA transcripts allowed further investigation into the cross talk between nerves and cancer cells. While early studies utilized quantitative real-time polymerase chain reaction (qRT-PCR) to identify singular or small groups of differentially expressed genes (DEGs) associated with nerve signaling in tumors, the arrival of microarrays and next-generation RNA sequencing (RNA-seq) enabled unbiased transcriptomic characterization at a much larger scale. To identify potential mediators of PNI, Koide et al. and Abiatari et al. separately compared gene expression profiles between human pancreatic cancer cell lines possessing high versus low potential for neural invasion and found several candidate DEGs, including

CD74 and *KIF14*, which were upregulated and downregulated, respectively, in highly neuroinvasive cells [69, 70]. Gene expression profiling is also used to determine cellular pathways altered by surgical or drug treatments. For example, subdiaphragmatic vagotomy in a mouse model of gastric cancer was associated with inhibition of Wnt signaling and inflammation-related pathways, whereas treatment with the muscarinic agonist pilocarpine suppressed growth-related genes, including *EGFR* and *PI3K*, in human pancreatic cells [49, 50]. RNA-seq can also confer transcriptomic identity in instances of cellular plasticity, such as transdifferentiated adrenergic nerves that originate from sensory nerves in response to *TP53* deletion in cancer cells or CNS-derived neural progenitors that can give rise to new neurons in tumors [71, 72].

Single-cell RNA seq (scRNA-seq) has revolutionized the identification of unique subtypes and transition states of cells. In the context of cancer, scRNA-seq has been especially impactful for profiling immune and stromal cells, identifying subtypes that are associated with treatment efficacy or resistance [73, 74]. However, the application of scRNA-seq to cancer neuroscience is less extensive. One reason is that cells in stromally dense cancers, such as PDAC and sarcomas, cannot be readily dissociated for downstream analysis. Neurons and glial cells have also been traditionally difficult to dissociate [75]. Recently, Hwang and colleagues overcame these barriers by applying single nucleus RNA sequencing (snRNA-seq), a technique initially used in the identification of rare neuronal subtypes in the brain, to identify novel and recurrent malignant cell subtypes in PDAC [76, 77]. This approach also captured intratumoral neurons and Schwann cells, which have been traditionally difficult to dissociate [78]. A further advantage of snRNA-seq is that it can be used to profile frozen or lightly fixed samples such as banked tissues from patients [75].

Spatial proteotranscriptomics has recently emerged as a groundbreaking molecular profiling method that allows researchers to measure and map mRNA transcript and protein abundance in a tissue sample [79]. While various spatial proteotranscriptomic methods have been applied to cancer studies, its application in cancer neuroscience remains limited. Using the GeoMx digital spatial profiler (DSP) instrument developed by Nanostring, which provides a quantitative, high-plex approach for analyzing mRNAs and proteins, Brady et al. categorized prostate tumor phenotypes, including neuroendocrine-related gene expression, from patient samples [80]. Hwang and colleagues used DSP in combination with snRNA-seq to map malignant cell programs, including a neural-like progenitor subtype, to tumor architecture in PDAC [77]. Schmitd et al. likewise applied DSP to oral squamous cell carcinoma and found that nerves closer to cancer cells express genes associated with injury response [81]. The rapidly evolving field of spatial proteotranscriptomics offers researchers the ability to visualize a growing spectrum of genes and proteins involved in cancer-nerve cross talk in unprecedented detail.

In addition to genomic and transcriptomic sequencing, high-throughput analyses of proteins and epigenetic modifications offer novel perspectives for cancer neuroscience research. For example, proteomic analysis of PNI-positive versus PNI-negative microdissected PDAC samples with liquid chromatography-mass spectrometry (LC-MS) revealed signatures consistent with neuronal plasticity and nerve injury response [82]. Epigenomic analysis of low versus high grade prostate

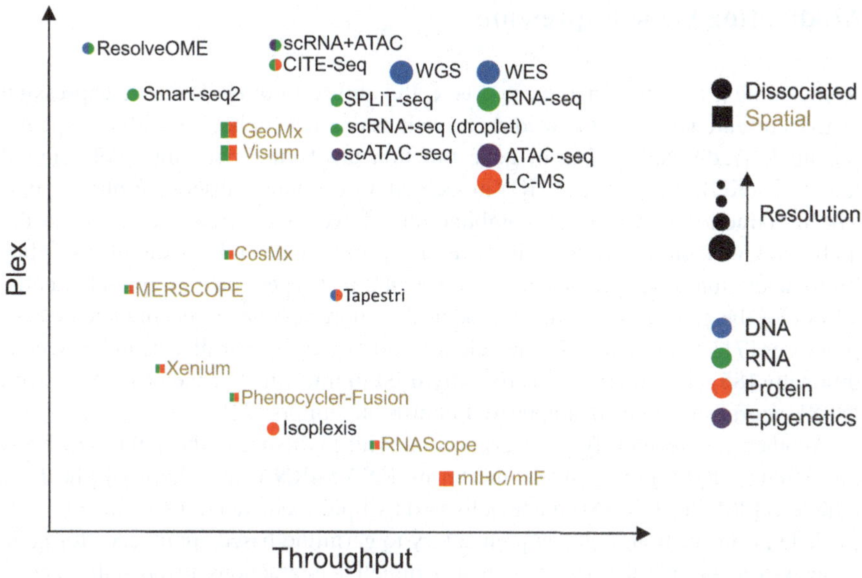

Fig. 11.3 Methods for multimodal genomic, transcriptomic, proteomic, and epigenomic analysis plotted by degree of molecular plex and throughput. Techniques incorporating dissociated cells are specified as circles, while techniques that preserve spatial features are specified as squares. Resolution (e.g., whole tissue < multicellular < cellular < subcellular) is reflected by shape size. Abbreviations: *LC-MS* (liquid chromatography-mass spectrometry), *mIF* (multiplexed immunofluorescence), *mIHC* (multiplexed immunohistochemistry), *WES* (whole exome sequencing), *WGS* (whole genome sequencing)

tumors using a single nucleus assay for transposase-accessible chromatin sequencing (snATAC-seq) further revealed that genomic regions associated with neuronal adhesion genes NRXN1 and NLGN1 are more accessible in high grade tumors [83]. The rapidly growing range of technologies for genomic, transcriptomic, proteomic, and epigenomic analysis provides varying degrees of multiplexing, resolution, and throughput for the study of cancer neuroscience (Fig. 11.3).

With the increased generation and accumulation of large datasets, it is necessary to leverage computational tools and database resources for the interpretation of complex biological information. Indeed, Gene Ontology (GO), Gene Set Enrichment Analysis (GSEA), and related tools are regularly used to match identified genes with their corresponding biological functions, while The Cancer Genome Atlas (TCGA) program is a cancer-specific resource available to the research community. Leveraging data from such large-scale coordinated efforts can lead to more generalizable scientific findings. For example, using data from a custom cohort of 2029 patients with or without PNI, Guo and colleagues recently developed a machine learning classifier that could predict occult PNI missed by initial histopathological review using only bulk transcriptomic data [84]. As artificial intelligence (AI) further integrates into multi-omics studies, researchers and clinicians will be able to access a deeper understanding of nerve-cancer cross talk and translate these findings into diagnostics and treatments.

Modulating Gene Expression

In the context of cancer-nerve cross talk, the ability to modulate gene expression enables researchers to test new ideas as well as engineer solutions with therapeutic potential. Traditional studies of gene function involved generating whole animal knockout (KO) and knock-in (KI) models and observing subsequent phenotypes. The introduction of Cre-lox recombination offered much greater control via the ability to knock out genes in a cell- or tissue-specific manner. For example, the ability to delete tumor suppressor genes in specific cell types, such as epithelial cells, allows for the generation of cancer models that more accurately recapitulate disease processes [71]. Cre-mediated gene deletion and expression can also be induced conditionally [85]. This method was recently used to trace the lineage of CNS-derived DCX+ neural stem cells that appeared in prostate tumors [72].

Another way to modify gene expression with high specificity is through RNA interference (RNAi) using small interfering RNA (siRNA) and short hairpin RNA (shRNA) [86]. Since RNAi can be delivered by lipid-mediated transfection or viral-mediated transduction, it is complementary to germline-based approaches for gene suppression. In an early study of neuroepithelial interactions in prostate cancer, knockdown of semaphorin 4F (SEMA4F) by siRNA in SEMA4F-overexpressing DU145 cancer cells was shown to decrease neurogenesis in vitro [87]. RNAi has also been tested in preclinical models for cancer treatment. For example, Lei et al. recently showed that gold nanocluster-assisted delivery of siRNA targeting nerve growth factor (NGF) effectively inhibited tumor progression in three pancreatic tumor models [88].

The ability to modify genes with viral vectors is also highly applicable to the study of cancer neuroscience. For example, AAV has been used to deliver genes for fluorescent proteins to specific nerve types in tumors, allowing the nerves to be traced to their ganglionic source [48]. Gene delivery by AAV also grants control of nerve signaling by inducing expression of various proteins and channels in neurons, as discussed above. Cao and colleagues further showed that AAV-mediated induction of hypothalamic brain-derived neurotrophic factor (BDNF) recapitulates the antitumor effects of environmental enrichment (EE) in mouse models of melanoma and colorectal cancer [89]. Viral vectors can also be used to deliver complementary DNAs (cDNAs) and open reading frames (ORFs) to overexpress genes in desired cell types [90]. The use of viruses to deliver genes that are lethal to cancer cells has led to the development of a new class of oncolytic therapies, including neurotropic herpes simplex virus (HSV)-based targeting of neuroinvasive cancer cells in peripheral nerve sheath tumors [91, 92]. The ability to exploit nerve-cancer cross talk with virus-mediated gene delivery thus provides many possibilities in both foundational science and potential therapies.

More recently, CRISPR and related derivatives have enabled precise gene editing, modulation of gene expression, and the identification of genes and pathways that can be targeted for specific outcomes [93]. While its application to cancer neuroscience remains relatively unexplored, CRISPR has been applied extensively in

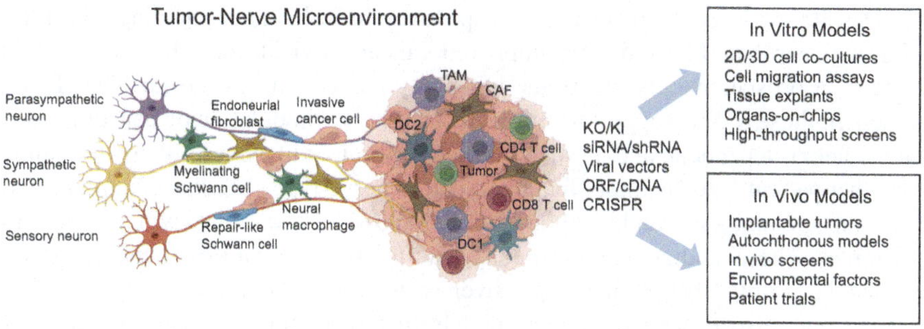

Fig. 11.4 Modulating gene expression in the tumor-nerve microenvironment. A schematic representation of the diverse cell types present in the tumor-nerve microenvironment that can be subjected to genetic modification in the context of in vitro and in vivo experimental models Abbreviations: *CAF* (cancer-associated fibroblast), *DC1* (type 1 dendritic cell), *DC2* (type 2 dendritic cell), *KI* (knock-in), *KO* (knockout), *ORF* (open reading frame), *TAM* (tumor-associated macrophage)

the greater cancer research context [94]. For example, it has been applied in germline editing to develop transgenic mouse models of cancer [95]. Additionally, CRISPR can be used for somatic genome engineering to interrogate gene function and model cancer in specific tissues [96]. CRISPR also enables high-throughput screening to identify genes that are responsible for phenotypes of interest, including targetable cancer drivers or genes that mediate drug resistance [97]. Taken together, methods for modulating gene expression can be applied to all areas of cancer neuroscience to identify new mechanisms and targets (Fig. 11.4).

Future Prospects

As diverse tools from various fields of biology have taken root in cancer neuroscience, new opportunities arise for basic discovery and clinical translation. These necessitate the integration of different experimental modalities to reach maximum potential impact.

While the field of cancer neuroscience has begun to more deeply appreciate the molecular dialog between neuronal and cancer cell subtypes, further studies should include stromal, immune, glial, and endothelial cells, using spatial and single-cell omics approaches to characterize cellular cross talk in various conditions with altered nerve signaling. Indeed, since autonomic nerve signaling is a well-known regulator of immune and vascular function [98], it might follow that targeting it would reduce cancer progression through its actions in part on immune and endothelial cells. Furthermore, its efficacy might be enhanced in combination with immune and/or angiogenic modulation. This should be tested in diverse preclinical models.

Despite evidence from several retrospective clinical studies showing a correlation between beta-blockade and improved cancer survival, there has been mixed results in prospective clinical trials, with only modest success so far [11]. Taken together, these results suggest that the effects of nerve modulation on cancer growth are context-dependent and tissue-specific. Thus, it will be useful to determine whether patient-specific tumor features, such as nerve-type abundance, immune and stromal cell landscape, and vascular endothelial cell state, correlate with response to therapies. These studies may eventually lead to stratification of patients based on molecular profiles that predict responsiveness to certain therapies.

There is also a need for better models that recapitulate human disease. For instance, autochthonous cancer models generated by genetically engineered somatic cells may provide more pathophysiological relevance for examining nerve involvement in tumor initiation and progression. Additional environmental conditions, such as stress, exercise, and microbiome, may also be considered. In vitro models may also benefit by progressing toward a more complete representation of the TME in coculture studies. This includes incorporating additional cell types as well as mimicking physical conditions such as complex extracellular matrix, fluid flow, and mechanical pressures. The increased complexity introduced by these models calls for a mixture of traditional molecular and cell biology tools and unbiased multiomics methods to fully interrogate their biology.

Concluding Remarks

Since the identification of nerve involvement in tumors, there has been a growing desire to better understand the relationship between nerves and cancer cells. By combining tools for visualizing cells and tissue architecture with experimental modification of nerve signaling, early investigators were able to identify specific patterns of nerve-cancer cross talk that correlate with both positive and negative outcomes in cancer progression. With the advent of modern tools for genomic analysis and genetic engineering, the field of cancer neuroscience is set to make more complete sense of these observations while testing new ideas for basic science and clinical therapies. This continued endeavor will synergize even more fields of biology, bringing with it new tools and an evolved understanding of biological systems.

References

1. Mravec, B. Neurobiology of cancer: Definition, historical overview, and clinical implications. *Cancer Med* **11**, 903–921 (2022).
2. Gysler, S. M. & Drapkin, R. Tumor innervation: peripheral nerves take control of the tumor microenvironment. *J Clin Invest* **131**, (2021).
3. Amit, M., Na'ara, S. & Gil, Z. Mechanisms of cancer dissemination along nerves. *Nat Rev Cancer* **16**, 399–408 (2016).

4. Baraldi, J. H., Martyn, G. v, Shurin, G. v & Shurin, M. R. Tumor Innervation: History, Methodologies, and Significance. *Cancers (Basel)* **14**, (2022).
5. Cole, S. W. & Sood, A. K. Molecular pathways: beta-adrenergic signaling in cancer. *Clin Cancer Res* **18**, 1201–6 (2012).
6. Dai, S. *et al.* Chronic Stress Promotes Cancer Development. *Front Oncol* **10**, 1492 (2020).
7. Simon, R. H., Lovett, E. J., Tomaszek, D. & Lundy, J. Electrical Stimulation of the Midbrain Mediates Metastatic Tumor Growth. *Science (1979)* **209**, 1132–1133 (1980).
8. Kim-Fuchs, C. *et al.* Chronic stress accelerates pancreatic cancer growth and invasion: a critical role for beta-adrenergic signaling in the pancreatic microenvironment. *Brain Behav Immun* **40**, 40–7 (2014).
9. Hui, Y. *et al.* Strategies for Targeting Neural Circuits: How to Manipulate Neurons Using Virus Vehicles. *Front Neural Circuits* **16**, 882366 (2022).
10. Shi, D. D. *et al.* Therapeutic avenues for cancer neuroscience: translational frontiers and clinical opportunities. *Lancet Oncol* **23**, e62–e74 (2022).
11. Shi, D. D. *et al.* Therapeutic avenues for cancer neuroscience: translational frontiers and clinical opportunities. *Lancet Oncol* **23**, e62–e74 (2022).
12. Loison, L. The Microscope against Cell Theory: Cancer Research in Nineteenth-Century Parisian Anatomical Pathology. *J Hist Med Allied Sci* **71**, 271–92 (2016).
13. Garcia-Lopez, P., Garcia-Marin, V. & Freire, M. The histological slides and drawings of cajal. *Front Neuroanat* **4**, 9 (2010).
14. Young, H. H. On the presence of nerves in tumors and of other structures in them as revealed by a modification if Ehrlich's method of 'vital staining' with methylene blue. *J Exp Med* **2**, 1–12 (1897).
15. Batsakis, J. G. Nerves and neurotropic carcinomas. *Ann Otol Rhinol Laryngol* **94**, 426–7 (1985).
16. Liebig, C., Ayala, G., Wilks, J. A., Berger, D. H. & Albo, D. Perineural invasion in cancer: a review of the literature. *Cancer* **115**, 3379–91 (2009).
17. Civita, P., Valerio, O., Naccarato, A. G., Gumbleton, M. & Pilkington, G. J. Satellitosis, a Crosstalk between Neurons, Vascular Structures and Neoplastic Cells in Brain Tumours; Early Manifestation of Invasive Behaviour. *Cancers (Basel)* **12**, (2020).
18. Hernandez, S., Serrano, A. G. & Solis Soto, L. M. The Role of Nerve Fibers in the Tumor Immune Microenvironment of Solid Tumors. *Adv Biol* **6**, e2200046 (2022).
19. Conte, G. A. *et al.* S100 Staining Adds to the Prognostic Significance of the Combination of Perineural Invasion and Lymphovascular Invasion in Colorectal Cancer. *Appl Immunohistochem Mol Morphol* **28**, 354–359 (2020).
20. Berlingeri-Ramos, A. C., Detweiler, C. J., Wagner, R. F. & Kelly, B. C. Dual S-100-AE1/3 Immunohistochemistry to Detect Perineural Invasion in Nonmelanoma Skin Cancers. *J Skin Cancer* **2015**, 620235 (2015).
21. Magnon, C. *et al.* Autonomic nerve development contributes to prostate cancer progression. *Science* **341**, 1236361 (2013).
22. Le, T. T. *et al.* Sensory nerves enhance triple-negative breast cancer invasion and metastasis via the axon guidance molecule PlexinB3. *NPJ Breast Cancer* **8**, 116 (2022).
23. Hassan, M. O. & Maksem, J. The prostatic perineural space and its relation to tumor spread: an ultrastructural study. *Am J Surg Pathol* **4**, 143–8 (1980).
24. Zeng, Q. *et al.* Synaptic proximity enables NMDAR signalling to promote brain metastasis. *Nature* **573**, 526–531 (2019).
25. Venkataramani, V. *et al.* Glutamatergic synaptic input to glioma cells drives brain tumour progression. *Nature* **573**, 532–538 (2019).
26. Venkatesh, H. S. *et al.* Electrical and synaptic integration of glioma into neural circuits. *Nature* **573**, 539–545 (2019).
27. Madeo, M. *et al.* Cancer exosomes induce tumor innervation. *Nat Commun* **9**, 4284 (2018).
28. Elliott, A. D. Confocal Microscopy: Principles and Modern Practices. *Curr Protoc Cytom* **92**, e68 (2020).

29. Deborde, S. *et al.* Reprogrammed Schwann Cells Organize into Dynamic Tracks that Promote Pancreatic Cancer Invasion. *Cancer Discov* **12**, 2454–2473 (2022).
30. Ueda, H. R. *et al.* Tissue clearing and its applications in neuroscience. *Nat Rev Neurosci* **21**, 61–79 (2020).
31. Guillot, J. *et al.* Sympathetic axonal sprouting induces changes in macrophage populations and protects against pancreatic cancer. *Nat Commun* **13**, 1985 (2022).
32. Ayala, G. E. *et al.* In vitro dorsal root ganglia and human prostate cell line interaction: redefining perineural invasion in prostate cancer. *Prostate* **49**, 213–23 (2001).
33. Lecoq, J., Orlova, N. & Grewe, B. F. Wide. Fast. Deep: Recent Advances in Multiphoton Microscopy of In Vivo Neuronal Activity. *J Neurosci* **39**, 9042–9052 (2019).
34. Hausmann, D. *et al.* Autonomous rhythmic activity in glioma networks drives brain tumour growth. *Nature* (2022) doi:https://doi.org/10.1038/s41586-022-05520-4.
35. Dercle, L. *et al.* Diagnostic and prognostic value of 18F-FDG PET, CT, and MRI in perineural spread of head and neck malignancies. *Eur Radiol* **28**, 1761–1770 (2018).
36. Nie, X. *et al.* Does PET scan have any role in the diagnosis of perineural spread associated with the head and neck tumors? *Adv Clin Exp Med* **31**, 827–835 (2022).
37. Wang, Y. W. *et al.* Raman-Encoded Molecular Imaging with Topically Applied SERS Nanoparticles for Intraoperative Guidance of Lumpectomy. *Cancer Res* **77**, 4506–4516 (2017).
38. Andreou, C., Weissleder, R. & Kircher, M. F. Multiplexed imaging in oncology. *Nat Biomed Eng* **6**, 527–540 (2022).
39. He, S. *et al.* High-plex imaging of RNA and proteins at subcellular resolution in fixed tissue by spatial molecular imaging. *Nat Biotechnol* (2022) doi:https://doi.org/10.1038/s41587-022-01483-z.
40. Kimura, K. *et al.* Perineural spread of gastric cancer to the sciatic nerve incidentally detected by 18F-FDG PET/CT. *Eur J Nucl Med Mol Imaging* **48**, 940–941 (2021).
41. Adler, I. & Sittenfield, M. J. *Preliminary note on the possible effects of the nervous system upon the growth and development of tumors.* http://aacrjournals.org/jcancerres/article-pdf/2/2/239/2169658/239.pdf (1917).
42. Cramer, W. *Innervation as a factor in the experimental production of cancer.* (1925).
43. Batkin, S., Piette, L. H. & Wildman, E. Effect of muscle denervation on growth of transplanted tumor in mice. *Proc Natl Acad Sci U S A* **67**, 1521–7 (1970).
44. Thaker, P. H. *et al.* Chronic stress promotes tumor growth and angiogenesis in a mouse model of ovarian carcinoma. *Nat Med* **12**, 939–44 (2006).
45. Visintainer, M. A., Volpicelli, J. R. & Seligman, M. E. Tumor rejection in rats after inescapable or escapable shock. *Science* **216**, 437–9 (1982).
46. Eng, J. W.-L. *et al.* Housing temperature-induced stress drives therapeutic resistance in murine tumour models through β2-adrenergic receptor activation. *Nat Commun* **6**, 6426 (2015).
47. Brohée, L. *et al.* Propranolol sensitizes prostate cancer cells to glucose metabolism inhibition and prevents cancer progression. *Sci Rep* **8**, 7050 (2018).
48. Kamiya, A. *et al.* Genetic manipulation of autonomic nerve fiber innervation and activity and its effect on breast cancer progression. *Nat Neurosci* **22**, 1289–1305 (2019).
49. Renz, B. W. *et al.* Cholinergic Signaling via Muscarinic Receptors Directly and Indirectly Suppresses Pancreatic Tumorigenesis and Cancer Stemness. *Cancer Discov* **8**, 1458–1473 (2018).
50. Zhao, C.-M. *et al.* Denervation suppresses gastric tumorigenesis. *Sci Transl Med* **6**, 250ra115 (2014).
51. Saloman, J. L. *et al.* Ablation of sensory neurons in a genetic model of pancreatic ductal adenocarcinoma slows initiation and progression of cancer. *Proc Natl Acad Sci U S A* **113**, 3078–83 (2016).
52. Prazeres, P. H. D. M. *et al.* Ablation of sensory nerves favours melanoma progression. *J Cell Mol Med* **24**, 9574–9589 (2020).
53. Johnson, R. L. & Wilson, C. G. A review of vagus nerve stimulation as a therapeutic intervention. *J Inflamm Res* **11**, 203–213 (2018).

54. Borovikova, L. v *et al.* Vagus nerve stimulation attenuates the systemic inflammatory response to endotoxin. *Nature* **405**, 458–62 (2000).
55. Rosas-Ballina, M. *et al.* Acetylcholine-synthesizing T cells relay neural signals in a vagus nerve circuit. *Science* **334**, 98–101 (2011).
56. Dubeykovskaya, Z. *et al.* Neural innervation stimulates splenic TFF2 to arrest myeloid cell expansion and cancer. *Nat Commun* **7**, 10517 (2016).
57. Zhang, F. *et al.* Multimodal fast optical interrogation of neural circuitry. *Nature* **446**, 633–639 (2007).
58. Roth, B. L. DREADDs for Neuroscientists. *Neuron* **89**, 683–694 (2016).
59. Obien, M. E. J., Deligkaris, K., Bullmann, T., Bakkum, D. J. & Frey, U. Revealing neuronal function through microelectrode array recordings. *Front Neurosci* **8**, 423 (2014).
60. Walsh, M. E. *et al.* Use of Nerve Conduction Velocity to Assess Peripheral Nerve Health in Aging Mice. *J Gerontol A Biol Sci Med Sci* **70**, 1312–1319 (2015).
61. Rijnbeek, E. H., Eleveld, N. & Olthuis, W. Update on Peripheral Nerve Electrodes for Closed-Loop Neuroprosthetics. *Front Neurosci* **12**, 350 (2018).
62. Chen, T.-W. *et al.* Ultrasensitive fluorescent proteins for imaging neuronal activity. *Nature* **499**, 295–300 (2013).
63. Lin, M. Z. & Schnitzer, M. J. Genetically encoded indicators of neuronal activity. *Nat Neurosci* **19**, 1142–1153 (2016).
64. Weinberg, R. A. Oncogenes and tumor suppressor genes. *CA Cancer J Clin* **44**, 160–70 (1994).
65. Gutmann, D. H. *et al.* Neurofibromatosis type 1. *Nat Rev Dis Primers* **3**, 17004 (2017).
66. Goodwin, S., McPherson, J. D. & McCombie, W. R. Coming of age: ten years of next-generation sequencing technologies. *Nat Rev Genet* **17**, 333–351 (2016).
67. Suwinski, P. *et al.* Advancing Personalized Medicine Through the Application of Whole Exome Sequencing and Big Data Analytics. *Front Genet* **10**, 49 (2019).
68. Biankin, A. v *et al.* Pancreatic cancer genomes reveal aberrations in axon guidance pathway genes. *Nature* **491**, 399–405 (2012).
69. Koide, N. *et al.* Establishment of perineural invasion models and analysis of gene expression revealed an invariant chain (CD74) as a possible molecule involved in perineural invasion in pancreatic cancer. *Clin Cancer Res* **12**, 2419–26 (2006).
70. Abiatari, I. *et al.* Consensus transcriptome signature of perineural invasion in pancreatic carcinoma. *Mol Cancer Ther* **8**, 1494–504 (2009).
71. Amit, M. *et al.* Loss of p53 drives neuron reprogramming in head and neck cancer. *Nature* **578**, 449–454 (2020).
72. Mauffrey, P. *et al.* Progenitors from the central nervous system drive neurogenesis in cancer. *Nature* **569**, 672–678 (2019).
73. Guruprasad, P., Lee, Y. G., Kim, K. H. & Ruella, M. The current landscape of single-cell transcriptomics for cancer immunotherapy. *J Exp Med* **218**, (2021).
74. Lavie, D., Ben-Shmuel, A., Erez, N. & Scherz-Shouval, R. Cancer-associated fibroblasts in the single-cell era. *Nat Cancer* **3**, 793–807 (2022).
75. Habib, N. *et al.* Massively parallel single-nucleus RNA-seq with DroNc-seq. *Nat Methods* **14**, 955–958 (2017).
76. Habib, N. *et al.* Div-Seq: Single-nucleus RNA-Seq reveals dynamics of rare adult newborn neurons. *Science* **353**, 925–8 (2016).
77. Hwang, W. L. *et al.* Single-nucleus and spatial transcriptome profiling of pancreatic cancer identifies multicellular dynamics associated with neoadjuvant treatment. *Nat Genet* **54**, 1178–1191 (2022).
78. Yim, A. K. Y. *et al.* Disentangling glial diversity in peripheral nerves at single-nuclei resolution. *Nat Neurosci* **25**, 238–251 (2022).
79. Merritt, C. R. *et al.* Multiplex digital spatial profiling of proteins and RNA in fixed tissue. *Nat Biotechnol* **38**, 586–599 (2020).
80. Brady, L. *et al.* Inter- and intra-tumor heterogeneity of metastatic prostate cancer determined by digital spatial gene expression profiling. *Nat Commun* **12**, 1426 (2021).

81. Schmitd, L. B. *et al.* Spatial and Transcriptomic Analysis of Perineural Invasion in Oral Cancer. *Clin Cancer Res* **28**, 3557–3572 (2022).
82. Alrawashdeh, W. *et al.* Perineural invasion in pancreatic cancer: proteomic analysis and in vitro modelling. *Mol Oncol* **13**, 1075–1091 (2019).
83. Eksi, S. E. *et al.* Epigenetic loss of heterogeneity from low to high grade localized prostate tumours. *Nat Commun* **12**, 7292 (2021).
84. Guo, J. A. *et al.* Pan-cancer Transcriptomic Predictors of Perineural Invasion Improve Occult Histopathologic Detection. *Clin Cancer Res* **27**, 2807–2815 (2021).
85. Schmidt-Supprian, M. & Rajewsky, K. Vagaries of conditional gene targeting. *Nat Immunol* **8**, 665–668 (2007).
86. Setten, R. L., Rossi, J. J. & Han, S. The current state and future directions of RNAi-based therapeutics. *Nat Rev Drug Discov* **18**, 421–446 (2019).
87. Ayala, G. E. *et al.* Cancer-related axonogenesis and neurogenesis in prostate cancer. *Clin Cancer Res* **14**, 7593–603 (2008).
88. Lei, Y. *et al.* Gold nanoclusters-assisted delivery of NGF siRNA for effective treatment of pancreatic cancer. *Nat Commun* **8**, 15130 (2017).
89. Cao, L. *et al.* Environmental and genetic activation of a brain-adipocyte BDNF/leptin axis causes cancer remission and inhibition. *Cell* **142**, 52–64 (2010).
90. Yang, X. *et al.* A public genome-scale lentiviral expression library of human ORFs. *Nat Methods* **8**, 659–61 (2011).
91. Kelly, K. *et al.* Attenuated multimutated herpes simplex virus-1 effectively treats prostate carcinomas with neural invasion while preserving nerve function. *FASEB J* **22**, 1839–48 (2008).
92. Gil, Z. *et al.* Nerve-sparing therapy with oncolytic herpes virus for cancers with neural invasion. *Clin Cancer Res* **13**, 6479–85 (2007).
93. Adli, M. The CRISPR tool kit for genome editing and beyond. *Nat Commun* **9**, 1911 (2018).
94. Katti, A., Diaz, B. J., Caragine, C. M., Sanjana, N. E. & Dow, L. E. CRISPR in cancer biology and therapy. *Nat Rev Cancer* **22**, 259–279 (2022).
95. Mou, H., Kennedy, Z., Anderson, D. G., Yin, H. & Xue, W. Precision cancer mouse models through genome editing with CRISPR-Cas9. *Genome Med* **7**, 53 (2015).
96. Kaltenbacher, T. *et al.* CRISPR somatic genome engineering and cancer modeling in the mouse pancreas and liver. *Nat Protoc* **17**, 1142–1188 (2022).
97. Dubrot, J. *et al.* In vivo CRISPR screens reveal the landscape of immune evasion pathways across cancer. *Nat Immunol* **23**, 1495–1506 (2022).
98. Devi, S. *et al.* Adrenergic regulation of the vasculature impairs leukocyte interstitial migration and suppresses immune responses. *Immunity* **54**, 1219–1230.e7 (2021)

Chapter 12
Future Direction of Cancer Neuroscience

Jami L. Saloman, Nicole N. Scheff, and Brian M. Davis

Why Do Nerves Make Tumors Feel Welcome?

One of the most robust findings, validated in multiple labs, across multiple cancer models of solid tumors that occur outside of the CNS, is that ablation of peripheral neurons (sensory and/or autonomic) slows or halts cancer progression [1–3]. A common feature of these studies is that the earlier the ablation occurs during disease progression, the more effective the intervention [4–7]. In fact, the most efficacious procedures involve denervation before cancer can be detected (in the case of genetic models) [5, 6] or prior to inoculation (in the case of transplant models [8]). This begs the question: how does the presence of nerves contribute to the nascent tumor microenvironment (TME)? The parsimonious explanation is that nerves provide a hospitable environment for developing tumors, a line of thinking that we are calling "the safe harbor hypothesis."

This "safe harbor" potentially begins with the stability of the peripheral nervous system (PNS) relative to incipient cancer cells. In virtually all cases, cancer cells arise from cell types that normally exhibit some level of replication. However, due to either inherited or acquired mutations, these cell types reproduce in a manner that ignores normal physiological processes and the replication restrictions that prevent one cell type from overrunning their neighbors. All multicellular

J. L. Saloman
Department of Medicine, Division of Gastroenterology, University of Pittsburgh, Pittsburgh, PA, USA

N. N. Scheff
Department of Neurobiology, Hillman Cancer Center, University of Pittsburgh, Pittsburgh, PA, USA

B. M. Davis (✉)
Department of Neurobiology, University of Pittsburgh, Pittsburgh, PA, USA
e-mail: bmd1@pitt.ed

M. Amit, N. N. Scheff (eds.), *Cancer Neuroscience*,
https://doi.org/10.1007/978-3-031-32429-1_12

organisms have mechanisms designed to eradicate rogue mutated cells either through spontaneous apoptosis or recognition by other cell types for directed phagocytosis (e.g., via detection of antigen expression). Thus, for a cell that is undergoing transformation into malignancy, it must be able to compensate for the demands of unregulated replication and escape internal (i.e., apoptosis) and external assaults (e.g., from the immune system) on its survival. The safe harbor hypothesis suggests that peripheral nerves serve as a supportive niche providing relative stability, including fixed anatomy, supportive intercellular interactions, and suppression of the immune system.

Anatomical Considerations

The PNS is comprised of the enteric nervous system, sensory neurons, and second order motoneurons of the sympathetic and parasympathetic nervous systems. The neuronal processes associated with these structures course through every tissue in the body. In the case of the autonomic and enteric motoneurons, tissue innervating processes are axons conveying top-down communication. In the case of sensory neurons, their processes combine feature of axons and dendrites and are referred to as afferent fibers. The density of both motor and sensory processes in most tissues is underappreciated due to the limitations of light microscopy for visualization. These processes are found in every tissue type, accompanying blood, and lymph vessels, innervating all organ capsules and coursing through the parenchyma of most organs (Fig. 12.1); even leukemias have the potential to interact with nerves in the bone marrow. Thus, any potential cancer cell can count on being only a few cell diameters away from a neural process. Moreover, peripheral nerves have three layers of connective tissue wrappings (i.e., peri-, epi-, and endoneurium) that act as a barrier to immune cell infiltration. Cancer cells that attain access to any one of these layers will achieve a level of isolation from the immune system lacking in most peripheral tissues and will be able to divide and migrate in the absence of immunosurveillance. Moreover, recent studies show that nerves can go beyond providing a stable, protected environment by actively attracting cancers via chemical and physical recruitment [9, 10].

While it is accepted that the terminal endings of these axons and fibers can exhibit anatomical plasticity in response to disease, the great majority of the anatomical units that make up the peripheral nervous system, especially the named nerves and their macroscopic branches, are static; barring pathology, these structures are present for the life of the animal. In addition, the neuronal component is postmitotic (however, there is significant disagreement about neurogenesis in the myenteric nervous plexus [11–13]), and the glial components of the PNS (satellite and Schwann cells) exhibit limited cell division. This overall stability could be critical for those tissues that have a propensity for producing potential cancer cells over the life of the individual. When given a choice of

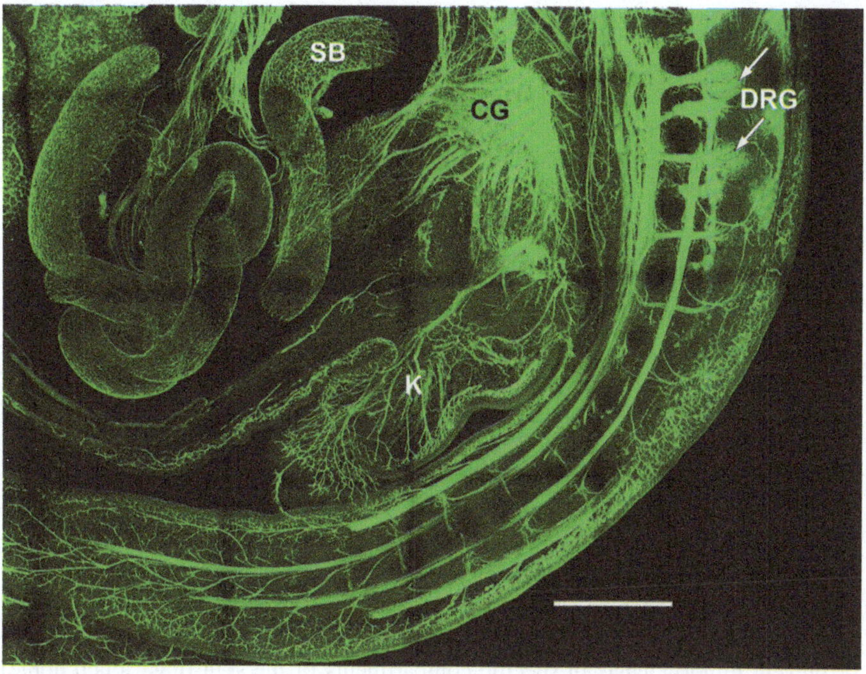

Fig. 12.1 Peripheral nervous system distribution in developing embryo. E16.5 rat embryo was stained with Tuj1 (anti-beta tubulin III) antibody to visualize all processes arising from the peripheral nervous system. Multiple images were taken on a scanning confocal microscope and stitched together to provide a macroscopic, sagittal cross-sectional image of the posterior half of the embryo. Processes are easily visualized at this age, revealing the extensive nature of innervation of every organ and tissue. The small bowel (SB) is one of the most densely innervated organs, likely due to the presence of sensory, sympathetic, and enteric neurons and their processes. *CG* celiac ganglion, *K* kidney, *DRG* dorsal root ganglion. Calibration bar = 1.0 mm. (with permission from Andrea Kalinoski and Marthe Howard, University of Toledo)

niches in which to expand, the overall stability of peripheral nerve architecture could provide a selective advantage to founder cancer stem cells that find their way into proximity to neural structures. This idea dovetails with the concept of "leader" cancer cells that colonize sites distant from the primary tumor because of their capacity to invade through the surrounding basement membrane, travel through the bloodstream or lymphatic system, and ultimately expand to establish colonies at the metastatic sites. Leader cells are hypothesized to be tumor cells that lose epithelial features, such as expression of E-cadherin, and gain mesenchymal features, including expression of vimentin and N-cadherin, whereas follower cells appear to be highly proliferative but have poor invasion capacity. Interestingly, one of the first papers to characterize the differences between leader and follower cells found that neuropilin 1 (NRP1), a checkpoint protein expressed by tumors and neurons, was one of the genes upregulated significantly in putative leader cancer cells [14].

Intercellular Interactions

At the cellular level, peripheral neurons and their associated glia cells also maintain a stable extracellular environment that is required for optimal neural transmission. This includes expression of numerous ion transporters/exchangers that are designed to buffer perturbations in the extracellular environment that could also stabilize the changes caused by tumorigenesis. Moreover, developing tumors are known to express genes normally associated with neural transmission (e.g., Ramp1, Nkr1, Adrb2) [3, 15–20] and might therefore be able to optimize their interaction with the neuronal environment. There are also reports of tumor cells using neurotransmitters released in the periphery as either signaling molecules that regulate replication and differentiation [21] or even as metabolites for energy production [22, 23]. Some of the best evidence for these types of interactions come from the studies of glioma. Using optogenetics, Monje and coworkers showed that increased neural activity promoted tumor growth in a glioblastoma xenograft model, potentially via release of neuroligin-3 (NLGN3). Soluble NLGN3 was shown to induce activation of the PI3K-mTOR pathway that included glioma production of NLGN3 [24]. A similar mitogenic role has been proposed for cutaneous sensory neurons in basal cell carcinoma (BCC) [5]. Peterson et al. (2015) found that in a genetic model of BCC (that utilized a deletion of the hedgehog inhibitory gene *Pitch1*), denervation could slow or prevent tumorigenesis in specific compartments of the skin (e.g., touch dome) [5]. Homeostatic hedgehog (Hh) signaling was found to be correlated with susceptibility to tumor formation, and the authors concluded that sensory neurons release Hh ligands required for initiation of BCC. In mammals, there are three different ligands that activate hedgehog signaling (sonic Hh, Indian Hh, and Desert Hh), all of which were shown to be expressed in the DRG neuron in subsequent single cell RNAseq studies [25].

Immunosuppression

Interest in neuroimmune interactions has waxed and waned over the decades. With the advent of transcriptomic analysis, we now have the ability to conduct unbiased analysis of gene expression. These studies have revealed that peripheral neurons express a wide range of genes normally thought to be restricted to immune cells. For example, sensory neurons are known to express mRNA for receptors for interleukins 1, 4, 18, and 33, interferon α receptor1, interferon receptor γ 1 and 2, and tumor necrosis factor (TNF) receptors 1α and 1β [25–27]. Binding of cytokines to these receptors directly affects neuronal excitability and synaptic plasticity, including ion channel assembly and conductance [28–30]. The impact of cytokines on neurons has been widely studied in the context of nociceptive sensitization, pain, and neurogenic inflammation [29, 31], but how cytokine-dependent changes in neural activity impact tumor immunity is an open question.

Previously, the neural contribution to initiating immune responses has been attributed to the release of chemoattractants such as CCL2. However, there is now a body of evidence that neuropeptides (small peptides released from neurons along with classical neurotransmitters) also possess chemotactic activity, recruiting both monocytes and lymphocytes in vitro and in vivo under nonmalignant conditions [32–34]. Neuron-released vasoactive intestinal peptide (VIP) has been shown to recruit innate lymphoid cells in response to intestinal pathogens [35], and calcitonin gene-related peptide (CGRP) is chemotactic for human T lymphocytes [33], whereas substance P recruits both myeloid and lymphoid cells [36–38]. In the context of cancer, some of these neuron-derived signals have been shown to induce tumor cell migration [39, 40], but it remains unclear whether their chemotactic activity extends to tumor infiltrating leukocytes. Some data suggest that neuropeptides may attract both monocytes and tumor cells in vitro [41, 42].

Once leukocytes are in the vicinity, canonical chemical neurotransmission can regulate their activity as they express a range of receptors and membrane proteins that can respond to signals derived from sympathetic, parasympathetic, and sensory neurons. Sympathetic nerves, which can be activated by physical and psychological stress, release catecholamines and VIP. VIP induces the differentiation of monocytes into myeloid derived suppressor cells (MDSCs) in gastric cancer [43]. The catecholamine norepinephrine induces expression of arginase 1 and programmed death ligand 1 (PDL1) on MDSCs enhancing their immunosuppressive capabilities [18]. Inhibiting VIP signaling in T cells improves T cell activity, as well as the efficacy of checkpoint inhibitors in both leukemia and pancreatic ductal adenocarcinoma (PDAC) [44, 45]. The role of the parasympathetic neurotransmitter acetylcholine in PDAC is controversial. One report, using a genetic PDAC model, found that vagotomy accelerated tumor growth and suggested that this occurred via loss of cholinergic efferents (although the role of vagal afferents was not considered). The authors proposed that vagotomy removed cholinergic signaling that normally inhibits MDSCs [20], resulting in immune disinhibition [20]. In contrast, Yang et al. [46] found that vagotomy increased CD8+ cytotoxic T cell infiltration and decreased tumor size. While the role of cholinergic signaling in PDAC remains unclear, unpublished studies by Saloman and colleagues suggest that sensory denervation also increases CD8+ T cell infiltration and this was associated with smaller tumor size in an inducible pancreatic tumor model. Consistent with this, sensory nerve-derived CGRP (released by the majority of visceral afferents and all classes of cutaneous afferents) has been directly linked to T cell suppression in both head and neck cancer and melanoma [17, 47].

Human T cells express receptors for several neurotransmitters including dopamine, neuropeptide Y, CGRP, and glutamate [48]. Treating isolated T cells from head and neck cancer patients with dopamine or glutamate increases T cell migration [49]. Moreover, dopamine, glutamate, CGRP, and NPY increase T cell proliferation, reduced PD1 expression, and improved T cell killing of hepatocellular carcinoma cells [48]. These types of results have led Levite and colleagues to suggest that a "personalized adoptive neuro-immunotherapy" may represent a novel therapeutic approach to combat immunosuppression [48].

In addition to chemical transmission, another mechanism through which neurons could contribute to immunosuppression is via expression of checkpoint proteins. Innate and adaptive immune cells express checkpoint proteins that form intercellular complexes, defined as complex structures consisting of transmembrane and cytoplasmic protein components. These physical cell-cell interactions serve as a primary mode of regulation for the immune system [50]. Tumor cells hijack this regulatory system to suppress antitumor immunity [51, 52]. Of significant interest is the discovery of expression of these canonical checkpoint proteins in enteric, sympathetic, and sensory neurons, including PDL1 [53, 54]. Of further importance, canonical axon guidance molecules including members of the semaphorin and neuropilin families have been shown to also act as immune checkpoint proteins [55, 56]. In the context of cancer, neuropilin-1 signaling (that is maintained in adult neurons) enhances the suppressive capacity of regulatory T cells [57]. Near peripheral neurons, especially at exposed terminal endings, the formation of intercellular complexes between neurons, and immune cells expressing checkpoint proteins could play a major role in limiting immune activity.

Exploration of the "Safe Harbor Hypothesis": Future Directions

How Nervy Are Tumors?

The previous paragraphs suggest some of the general features that make the nervous system an optimal coconspirator in tumorigenesis. But to date, most nerve-cancer studies that explore the neurotrophic/neurotropic behavior of cancer cells have focused on the usual suspects, those genes known to be implicated in neuronal development or plasticity. For example, pancreatic tumors in both humans and animal models express neuronal growth factors in both the NGF (e.g., NGF, BDNF) and GDNF (e.g., GDNF, artemin, and neurturin) families as well as the receptors for these cytokines (NTRK1, P75, Ret, NTRK1, GFRa2) [58]. Moreover, the expression of these genes changes during the course of disease in a manner that is correlated with tumorigenesis [59]. While informative, these studies do not provide a complete picture of how cancer cells evolve in their relationship with the nervous system. Thus, one potential future direction for the cancer neuroscience field would be unbiased transcriptomic (or proteomic) analyses of gene expression (e.g., scRNAseq or ATACseq) of developing cancer cells (including analysis of precancerous progenitor cells) over the course of tumorigenesis in the presence and absence of nerves. The goal would be to identify new genes that could be participating in neuron-like interactions between individual cancer cells and between cancer cells and neurons. Interrogation of the resulting cancer cell databases could initially focus on pathways known to regulate neuronal development as expression of these genes appears to be a common feature of tumors. Another feature of tumors that that could be addressed with unbiased transcriptomic studies is identification of genes that could be the basis of intercellular communication between neurons and tumor cells. For example, cancer cells are known to express virtually every class of voltage-gated channels including sodium-, potassium-, and chloride-gated ion channels as well as proteins required for responses to neurotransmitters/neuromodulators including norepinephrine, acetylcholine, glutamate, GABA, CGRP, and substance P [60–62]. But we are lacking data that indicates the context in which these genes are expressed. Ultimately the function of any channel/receptor can only be understood if we know what interacting proteins are expressed, or in the case of channels and receptors that are made up of multiple subunits (e.g., glutamate, GABA receptors), the subunit composition of a given receptor/channel.

The proposed "omic" studies would only be the first step identifying the next generation of important tumor-expressed "neurocentric" genes. The next step would be to use existing neuroscience techniques to determine the extent to which cancer cells can mimic neuron behaviors. For example, unpublished studies by Scheff and coworkers indicate that mechanical deformation of cultured cancer cells can result in spreading calcium transients from one cell to the next, similar to what is seen in the myenteric plexus following mechanical deformation of the colon [63]. The timescale of current spread is relatively slow (on the order of seconds, also like the myenteric plexus) and changes as the calcium transients move inward from the applied mechanical stimulus. These initial studies were done via standard imagining techniques (Fura-2 calcium imaging), typically used to study populations of neurons, and are limited by the relatively slow time course of calcium transients and the inability to reveal membrane properties. These enticing preliminary results argue for follow-up studies using techniques with better temporal resolution and the ability to measure membrane properties. For example, patch-clamp recordings combined with pharmacology would reveal the extent to which the different channel types expressed in cancer cells behave in a manner similar to that seen in neurons, including their response to pharmacological manipulations that have been proposed as potential therapeutic strategies [64–71]. Examination of the functional characteristics of neural genes expressed in cancers cell lines could be the foundation of preclinical studies that examine cancer cells in situ by electrophysiological investigation of slice preparations of tumors removed from patients and animal models. "Tumoroids" that include different combinations of tumor cells, neurons, and other cell types that potentially influence the trajectory of cancer cells (e.g., immune cells) could also be explored using neurophysiological techniques. These approaches would not only identify the best animal models for mimicking the phenotype of the disease in humans but also shed light on how interactions between different cell types affect the trajectory of cancer cells in terms of expression of nerve-cancer relevant molecules.

Dissecting Neuronal Contribution to the TME

This review started by noting three different features of peripheral nerves that could be contributing to tumor cell survival and expansion. The most dramatic demonstration of the potential role of peripheral nerves has been consistently seen in experiments that ablate the neuronal component. This manipulation removes all three features of nerves listed above: structure, intercellular signaling, and immune suppression. There is little doubt the role of nerves will be complex and will probably be different for different types of cancers. Fortunately, there are multiple techniques that could be employed to begin to determine the relative importance of neurons to tumorigenesis.

Alteration of Anatomical Elements Surgical ablation studies remove all the anatomical elements of peripheral nerves. Every peripheral nerve is composed of multiple types of neuronal processes, whether it be a mix of different types of sensory fibers as in a pure cutaneous nerve or a combination of sensory and motor fibers (arising from autonomic and/or voluntary motoneurons). Transgenic models are now available that allow selective ablation of many different types of neurons such that it should be possible to determine the relative contribution of different populations of neurons to the production of a tumor-supportive environment. For example, peptidergic sensory fibers (those that release neuropeptides substance P and CGRP) make up less than half of all cutaneous sensory neurons. These could be selectively ablated using genetic strategies without compromising the non-peptidergic populations. Combining these mice with xenograft/allografts or transgenic models of cancer could be used to directly test the contribution of the different afferent populations to tumor development. Unfortunately, no similar approach currently exists for deleting Schwann cells that would not produce a cascade of off-target effects that would make interpretation impossible. However, there are specific promoters that can target expression to Schwann cells that could be used to alter gene expression in these cells.

Given the emphasis on sprouting that has been seen in the close proximity to peripheral tumors, it is somewhat surprising that little work has been done to actually examine the ultrastructure of nerves in the vicinity of both tumor and immune cells. With the advent of imaging techniques including three-dimensional electron microscopy (EM) reconstruction [72], it is now possible to examine anatomic specializations that might occur between tumor, nerve, glia, and immune cells at different stages of tumorigenesis. This type of analysis could be very powerful in that it could reveal the extent to which direct contact between nerves, tumor, and immunes is occurring, whether these connections are reciprocal and provided indications of their strengths; e.g., is there evidence for synapse-like connections or gap junctions that could be contributing to tumor viability?

Altering Intercellular Interactions Altering intercellular interactions between neurons and cancer cells can be accomplished by modulating neuronal activity of neurons during tumorigenesis. This has already been done in the context of glioma

[73–75], but no work has been reported yet in the context of peripheral tumors. There is a wide range of transgenic models that employ either optogenetics (e.g., channelrhodopsin) or chemogenetics (e.g., DREADDs) that make it possible to target select population of neurons. Implantable LEDs combined with transgenic mice expressing novel opsins make it possible to target modulation of neural activity to individual nerves or structures. In addition, electrical stimulators (e.g., vagal stimulators) that are highly tunable with respect to stimulation frequency and duration could be used, although these devices do not currently allow for activation of specific populations of axons within a nerve (although it is possible to favor activation of different sizes of axons) [76, 77]. Inhibition or activation of select populations of peripheral neurons would allow investigators to determine if neuronal activity is making significant contributions to tumor growth or whether it is the other features of peripheral nerves (e.g., expression of checkpoint proteins, physical exclusion of immune cells) that are the critical factor. In some cases, activity modulation could be combined with disruption of transmitter release. For example, transgenic mice are available that allow deletion of *Calcα*, the gene required for production of CGRPα that has been shown to play a role in immune suppression [17]. Importantly, the neuronal features contributing to tumorigenesis is likely to change as tumors develop, requiring time course studies.

Modulation of Expression of Immune Modulatory Genes Because of the large number of available cre-lines that target peripheral neurons, it is possible to delete immune modulatory genes known to be expressed in sensory neurons. Mice containing a floxed version of checkpoint proteins including PD1, PDL1, and neuropilin, as well as other checkpoint proteins and immune cytokines are available, making it possible to produce mouse lines lacking any of these genes in specific neuronal populations. A major caveat with this approach, however, is the likely redundancy of mechanisms underlying the role of neuroimmune interactions in tumorigenesis. Deletion of a single gene may produce false negative results that may end up excluding a gene whose deletion in combination with other genes could produce a significant result. One solution may be to combine single gene deletion with pharmacological approaches in which gene deletion is seen as an adjuvant therapy for a drug treatment that has been validated in a particular cancer model.

Integration of Cancer Neuroscience into Clinical Oncology Given the striking discoveries made in the study of cancer neuroscience thus far, it is not surprising that actionable strategies are already being suggested and implemented in clinical oncology. Clinical disruptions of the neural niche attempted so far have focused mainly on the anatomical and intercellular signaling components of our proposed "safe harbor hypothesis." There are key anatomic considerations for each organ; complex innervation from multiple extrinsic and intrinsic sources in visceral cancers will likely make it difficult to identify the primary driver for nerve-cancer interactions. However, celiac ganglionectomy and neurolysis are used to treat intractable visceral cancer pain [78–82]. This selective denervation is a promising option to block progression and metastasis if executed earlier in the disease process. Similarly, denervation in tumors of the head and neck region can be associated with severe organ dysfunction given the need for sensory feedback to drive fine motor movements. An alternative to removing the neural anatomy is reducing expansion of the neural component (i.e., axonogenesis) into the tumor. Growth factor inhibitors have been explored extensively in the pain field. The monoclonal antibody against NGF, tanezumab, initially showed promising efficacy for osteoarthritis pain, but failed to get FDA approval because of negative off target effects. However, tanezumab trial is still ongoing for schwannomatosis pain, and a newer NGF antibody, EP-001A, is in a phase I trial. Modulation of the intercellular signaling between nerves and the environment is now also actively being explored as adjuvant therapy using autonomic neuropharmacology (e.g., adrenergic and cholinergic inhibitors) with the primary endpoints assessing cytokine biomarkers and tumor progression in both nonmetastatic (perioperative use) and metastatic patients. Unfortunately, large randomized controlled trials are still lacking and will be needed to demonstrate clinical benefit. The final targetable component of the neural niche is neuroimmune communication, and the impact of the nervous system on antitumor immunity is only just beginning to be considered. An upsurge of small phase II clinical trials is expected over the next several years investigating adjuvant potential of FDA-approved neuromodulators (e.g., anti-CGRP, propranolol) for the improvement of antitumor immunity when used alone and when used in combination with existing immunotherapeutics. These short-term study designs will begin to lay the foundation for subsequent large-scale, long-term oncological studies.

Concluding Remarks: Confirmation Bias

Although perineural invasion, the oldest known interaction between nerves and cancer, was first documented in the nineteenth century [83], the idea that the nervous system plays a more direct role in cancer development only gained traction in the previous decade [76, 84, 85]. Since that time, many high-profile papers often highlight the role of single molecules. While most of these studies have layers of supporting data, the one thing that is clear about the biology of cancer is that individual molecules are unlikely to regulate multicellular interactions in a significant way in the patient. It is with this concern in mind that we propose that investigations of single molecules and specific canonical cancer pathways be augmented by various unbiased strategies (e.g., −omic assays) with the potential to reveal the broader context in which nerves and cancers interact in a way that regulates tumorigenesis. As this field continues to evolve, there will be increasing pressure to emphasize positive results that support current notions of how nerves support cancer development. Because of this pressure, we must be on guard for confirmation bias that will result in provocative results that are either not biologically relevant (i.e., not translatable outside of the artificial nature of a given experimental model) or not reproducible by other labs. While targeting the nervous system is unlikely to be a silver bullet for treating all cancer, it is possible that disruption of the neural niche could result in enough instability in the tumor that efficacy of current cancer therapeutic strategies could be significantly increased.

References

1. Demir IE, Mota Reyes C, Alrawashdeh W, Ceyhan GO, Deborde S, Friess H, Gorgulu K, Istvanffy R, Jungwirth D, Kuner R, Maryanovich M, Na'ara S, Renders S, Saloman JL, Scheff NN, Steenfadt H, Stupakov P, Thiel V, Verma D, Yilmaz BS, White RA, Wang TC, Wong RJ, Frenette PS, Gil Z, Neural Influences in Cancer International Research C and Davis BM. Future directions in preclinical and translational cancer neuroscience research. *Nat Cancer*. 2021;1:1027–1031.
2. Saloman JL, Albers KM, Rhim AD and Davis BM. Can Stopping Nerves, Stop Cancer? *Trends Neurosci*. 2016;39:880–889.
3. Zahalka AH, Arnal-Estape A, Maryanovich M, Nakahara F, Cruz CD, Finley LWS and Frenette PS. Adrenergic nerves activate an angio-metabolic switch in prostate cancer. *Science*. 2017;358:321–326.
4. Magnon C, Hall SJ, Lin J, Xue X, Gerber L, Freedland SJ and Frenette PS. Autonomic nerve development contributes to prostate cancer progression. *Science*. 2013;341:1236361.
5. Peterson SC, Eberl M, Vagnozzi AN, Belkadi A, Veniaminova NA, Verhaegen ME, Bichakjian CK, Ward NL, Dlugosz AA and Wong SY. Basal cell carcinoma preferentially arises from stem cells within hair follicle and mechanosensory niches. *Cell Stem Cell*. 2015;16:400–12.
6. Saloman JL, Albers KM, Li D, Hartman DJ, Crawford HC, Muha EA, Rhim AD and Davis BM. Ablation of sensory neurons in a genetic model of pancreatic ductal adenocarcinoma slows initiation and progression of cancer. *Proc Natl Acad Sci U S A*. 2016;113:3078–83.
7. Zhao CM, Hayakawa Y, Kodama Y, Muthupalani S, Westphalen CB, Andersen GT, Flatberg A, Johannessen H, Friedman RA, Renz BW, Sandvik AK, Beisvag V, Tomita H, Hara A, Quante M, Li Z, Gershon MD, Kaneko K, Fox JG, Wang TC and Chen D. Denervation suppresses gastric tumorigenesis. *Sci Transl Med*. 2014;6:250ra115.
8. Vats K, Kruglov O, Sahoo B, Soman V, Zhang J, Shurin GV, Chandran UR, Skums P, Shurin MR, Zelikovsky A, Storkus WJ and Bunimovich YL. Sensory Nerves Impede the Formation of Tertiary Lymphoid Structures and Development of Protective Antimelanoma Immune Responses. *Cancer Immunol Res*. 2022;10:1141–1154.
9. Deborde S, Gusain L, Powers A, Marcadis A, Yu Y, Chen CH, Frants A, Kao E, Tang LH, Vakiani E, Amisaki M, Balachandran VP, Calo A, Omelchenko T, Jessen KR, Reva B and Wong RJ. Reprogrammed Schwann Cells Organize into Dynamic Tracks that Promote Pancreatic Cancer Invasion. *Cancer Discov*. 2022;12:2454–2473.
10. Deborde S, Omelchenko T, Lyubchik A, Zhou Y, He S, McNamara WF, Chernichenko N, Lee SY, Barajas F, Chen CH, Bakst RL, Vakiani E, He S, Hall A and Wong RJ. Schwann cells induce cancer cell dispersion and invasion. *J Clin Invest*. 2016;126:1538-54.
11. Kulkarni S, Micci MA, Leser J, Shin C, Tang SC, Fu YY, Liu L, Li Q, Saha M, Li C, Enikolopov G, Becker L, Rakhilin N, Anderson M, Shen X, Dong X, Butte MJ, Song H, Southard-Smith EM, Kapur RP, Bogunovic M and Pasricha PJ. Adult enteric nervous system in health is maintained by a dynamic balance between neuronal apoptosis and neurogenesis. *Proc Natl Acad Sci U S A*. 2017;114:E3709-E3718.
12. Virtanen H, Garton D and Andressoo JO. Reply. *Cell Mol Gastroenterol Hepatol*. 2022;14:968–969.
13. Virtanen H, Garton DR and Andressoo JO. Myenteric Neurons Do Not Replicate in Small Intestine Under Normal Physiological Conditions in Adult Mouse. *Cell Mol Gastroenterol Hepatol*. 2022;14:27–34.

14. Konen J, Summerbell E, Dwivedi B, Galior K, Hou Y, Rusnak L, Chen A, Saltz J, Zhou W, Boise LH, Vertino P, Cooper L, Salaita K, Kowalski J and Marcus AI. Image-guided genomics of phenotypically heterogeneous populations reveals vascular signalling during symbiotic collective cancer invasion. *Nat Commun.* 2017;8:15078.
15. Atherton MA, Park S, Horan NL, Nicholson S, Dolan JC, Schmidt BL and Scheff NN. Sympathetic modulation of tumor necrosis factor alpha-induced nociception in the presence of oral squamous cell carcinoma. *Pain.* 2023;164:27–42.
16. Dalaklioglu S and Erin N. Substance P Prevents Vascular Endothelial Dysfunction in Metastatic Breast Carcinoma. *Protein Pept Lett.* 2016;23:952–957.
17. McIlvried LA, Atherton MA, Horan NL, Goch TN and Scheff NN. Sensory Neurotransmitter Calcitonin Gene-Related Peptide Modulates Tumor Growth and Lymphocyte Infiltration in Oral Squamous Cell Carcinoma. *Adv Biol (Weinh).* 2022;6:e2200019.
18. Mohammadpour H, MacDonald CR, Qiao G, Chen M, Dong B, Hylander BL, McCarthy PL, Abrams SI and Repasky EA. beta2 adrenergic receptor-mediated signaling regulates the immunosuppressive potential of myeloid-derived suppressor cells. *J Clin Invest.* 2019;129:5537–5552.
19. Renz BW, Takahashi R, Tanaka T, Macchini M, Hayakawa Y, Dantes Z, Maurer HC, Chen X, Jiang Z, Westphalen CB, Ilmer M, Valenti G, Mohanta SK, Habenicht AJR, Middelhoff M, Chu T, Nagar K, Tailor Y, Casadei R, Di Marco M, Kleespies A, Friedman RA, Remotti H, Reichert M, Worthley DL, Neumann J, Werner J, Iuga AC, Olive KP and Wang TC. beta2 Adrenergic-Neurotrophin Feedforward Loop Promotes Pancreatic Cancer. *Cancer Cell.* 2018;33:75–90 e7.
20. Renz BW, Tanaka T, Sunagawa M, Takahashi R, Jiang Z, Macchini M, Dantes Z, Valenti G, White RA, Middelhoff MA, Ilmer M, Oberstein PE, Angele MK, Deng H, Hayakawa Y, Westphalen CB, Werner J, Remotti H, Reichert M, Tailor YH, Nagar K, Friedman RA, Iuga AC, Olive KP and Wang TC. Cholinergic Signaling via Muscarinic Receptors Directly and Indirectly Suppresses Pancreatic Tumorigenesis and Cancer Stemness. *Cancer Discov.* 2018;8:1458–1473.
21. Zhang Y, Lin C, Liu Z, Sun Y, Chen M, Guo Y, Liu W, Zhang C, Chen W, Sun J, Xia R, Hu Y, Yang X, Li J, Zhang Z, Cao W, Sun S, Wang X and Ji T. Cancer cells co-opt nociceptive nerves to thrive in nutrient-poor environments and upon nutrient-starvation therapies. *Cell Metab.* 2022;34:1999–2017 e10.
22. Gu I, Gregory E, Atwood C, Lee SO and Song YH. Exploring the Role of Metabolites in Cancer and the Associated Nerve Crosstalk. *Nutrients.* 2022;14.
23. Yi H, Talmon G and Wang J. Glutamate in cancers: from metabolism to signaling. *J Biomed Res.* 2019;34:260–270.
24. Venkatesh HS, Johung TB, Caretti V, Noll A, Tang Y, Nagaraja S, Gibson EM, Mount CW, Polepalli J, Mitra SS, Woo PJ, Malenka RC, Vogel H, Bredel M, Mallick P and Monje M. Neuronal Activity Promotes Glioma Growth through Neuroligin-3 Secretion. *Cell.* 2015;161:803–16.
25. Usoskin D, Furlan A, Islam S, Abdo H, Lonnerberg P, Lou D, Hjerling-Leffler J, Haeggstrom J, Kharchenko O, Kharchenko PV, Linnarsson S and Ernfors P. Unbiased classification of sensory neuron types by large-scale single-cell RNA sequencing. *Nat Neurosci.* 2015;18:145–53.
26. Hockley JRF, Taylor TS, Callejo G, Wilbrey AL, Gutteridge A, Bach K, Winchester WJ, Bulmer DC, McMurray G and Smith ESJ. Single-cell RNAseq reveals seven classes of colonic sensory neuron. *Gut.* 2019;68:633–644.
27. Meerschaert KA, Adelman PC, Friedman RL, Albers KM, Koerber HR and Davis BM. Unique Molecular Characteristics of Visceral Afferents Arising from Different Levels of the Neuraxis: Location of Afferent Somata Predicts Function and Stimulus Detection Modalities. *J Neurosci.* 2020;40:7216–7228.
28. Gahring LC, Days EL, Kaasch T, Gonzalez de Mendoza M, Owen L, Persiyanov K and Rogers SW. Pro-inflammatory cytokines modify neuronal nicotinic acetylcholine receptor assembly. *J Neuroimmunol.* 2005;166:88–101.

29. Czeschik JC, Hagenacker T, Schafers M and Busselberg D. TNF-alpha differentially modulates ion channels of nociceptive neurons. *Neurosci Lett*. 2008;434:293–8.
30. Vezzani A and Viviani B. Neuromodulatory properties of inflammatory cytokines and their impact on neuronal excitability. *Neuropharmacology*. 2015;96:70–82.
31. Schafers M and Sorkin L. Effect of cytokines on neuronal excitability. *Neurosci Lett*. 2008;437:188–93.
32. Dunzendorfer S, Schratzberger P, Reinisch N, Kahler CM and Wiedermann CJ. Secretoneurin, a novel neuropeptide, is a potent chemoattractant for human eosinophils. *Blood*. 1998;91:1527–32.
33. Foster CA, Mandak B, Kromer E and Rot A. Calcitonin gene-related peptide is chemotactic for human T lymphocytes. *Ann N Y Acad Sci*. 1992;657:397–404.
34. Schratzberger P, Reinisch N, Prodinger WM, Kahler CM, Sitte BA, Bellmann R, Fischer-Colbrie R, Winkler H and Wiedermann CJ. Differential chemotactic activities of sensory neuropeptides for human peripheral blood mononuclear cells. *J Immunol*. 1997;158:3895–901.
35. Yu HB, Yang H, Allaire JM, Ma C, Graef FA, Mortha A, Liang Q, Bosman ES, Reid GS, Waschek JA, Osborne LC, Sokol H, Vallance BA and Jacobson K. Vasoactive intestinal peptide promotes host defense against enteric pathogens by modulating the recruitment of group 3 innate lymphoid cells. *Proc Natl Acad Sci U S A*. 2021;118.
36. Cima K, Vogelsinger H and Kahler CM. Sensory neuropeptides are potent chemoattractants for human basophils in vitro. *Regul Pept*. 2010;160:42–8.
37. Hood VC, Cruwys SC, Urban L and Kidd BL. Differential role of neurokinin receptors in human lymphocyte and monocyte chemotaxis. *Regul Pept*. 2000;96:17–21.
38. Sloniecka M, Le Roux S, Zhou Q and Danielson P. Substance P Enhances Keratocyte Migration and Neutrophil Recruitment through Interleukin-8. *Mol Pharmacol*. 2016;89:215–25.
39. Drell TLt, Joseph J, Lang K, Niggemann B, Zaenker KS and Entschladen F. Effects of neurotransmitters on the chemokinesis and chemotaxis of MDA-MB-468 human breast carcinoma cells. *Breast Cancer Res Treat*. 2003;80:63–70.
40. Medeiros PJ, Al-Khazraji BK, Novielli NM, Postovit LM, Chambers AF and Jackson DN. Neuropeptide Y stimulates proliferation and migration in the 4T1 breast cancer cell line. *Int J Cancer*. 2012;131:276–86.
41. Ruff M, Schiffmann E, Terranova V and Pert CB. Neuropeptides are chemoattractants for human tumor cells and monocytes: a possible mechanism for metastasis. *Clin Immunol Immunopathol*. 1985;37:387–96.
42. Wu Y, Berisha A and Borniger JC. Neuropeptides in Cancer: Friend and Foe? *Adv Biol (Weinh)*. 2022;6:e2200111.
43. Li G, Wu K, Tao K, Lu X, Ma J, Mao Z, Li H, Shi L, Li J, Niu Y, Xiang F and Wang G. Vasoactive intestinal peptide induces CD14+HLA-DR-/low myeloid-derived suppressor cells in gastric cancer. *Mol Med Rep*. 2015;12:760–8.
44. Ogawa T and Wada J. [Electron microscopic and peroxidase cytochemical analysis of acute leukemia]. *Rinsho Ketsueki*. 1982;23:1009–18.
45. Petersen MBK, Azad A, Ingvorsen C, Hess K, Hansson M, Grapin-Botton A and Honore C. Single-Cell Gene Expression Analysis of a Human ESC Model of Pancreatic Endocrine Development Reveals Different Paths to beta-Cell Differentiation. *Stem Cell Reports*. 2017;9:1246–1261.
46. Yang MW, Tao LY, Jiang YS, Yang JY, Huo YM, Liu DJ, Li J, Fu XL, He R, Lin C, Liu W, Zhang JF, Hua R, Li Q, Jiang SH, Hu LP, Tian GA, Zhang XX, Niu N, Lu P, Shi J, Xiao GG, Wang LW, Xue J, Zhang ZG and Sun YW. Perineural Invasion Reprograms the Immune Microenvironment through Cholinergic Signaling in Pancreatic Ductal Adenocarcinoma. *Cancer Res*. 2020;80:1991–2003.
47. Balood M, Ahmadi M, Eichwald T, Ahmadi A, Majdoubi A, Roversi K, Roversi K, Lucido CT, Restaino AC, Huang S, Ji L, Huang KC, Semerena E, Thomas SC, Trevino AE, Merrison H, Parrin A, Doyle B, Vermeer DW, Spanos WC, Williamson CS, Seehus CR, Foster SL, Dai H, Shu CJ, Rangachari M, Thibodeau J, S VDR, Drapkin R, Rafei M, Ghasemlou N, Vermeer

PD, Woolf CJ and Talbot S. Nociceptor neurons affect cancer immunosurveillance. *Nature*. 2022;611:405–412.
48. Levite M, Safadi R, Milgrom Y, Massarwa M and Galun E. Neurotransmitters and Neuropeptides decrease PD-1 in T cells of healthy subjects and patients with hepatocellular carcinoma (HCC), and increase their proliferation and eradication of HCC cells. *Neuropeptides*. 2021;89:102159.
49. Saussez S, Laumbacher B, Chantrain G, Rodriguez A, Gu S, Wank R and Levite M. Towards neuroimmunotherapy for cancer: the neurotransmitters glutamate, dopamine and GnRH-II augment substantially the ability of T cells of few head and neck cancer patients to perform spontaneous migration, chemotactic migration and migration towards the autologous tumor, and also elevate markedly the expression of CD3zeta and CD3epsilon TCR-associated chains. *J Neural Transm (Vienna)*. 2014;121:1007–27.
50. He X and Xu C. Immune checkpoint signaling and cancer immunotherapy. *Cell Res*. 2020;30:660–669.
51. Dunn GP, Old LJ and Schreiber RD. The three Es of cancer immunoediting. *Annu Rev Immunol*. 2004;22:329–60.
52. Kim R, Emi M and Tanabe K. Cancer immunoediting from immune surveillance to immune escape. *Immunology*. 2007;121:1–14.
53. Meerschaert KA, Edwards BS, Epouhe AY, Jefferson B, Friedman R, Babyok OL, Moy JK, Kehinde F, Liu C, Workman CJ, Vignali DAA, Albers KM, Koerber HR, Gold MS, Davis BM, Scheff NN and Saloman JL. Neuronally expressed PDL1, not PD1, suppresses acute nociception. *Brain Behav Immun*. 2022;106:233–246.
54. Zeisel A, Hochgerner H, Lonnerberg P, Johnsson A, Memic F, van der Zwan J, Haring M, Braun E, Borm LE, La Manno G, Codeluppi S, Furlan A, Lee K, Skene N, Harris KD, Hjerling-Leffler J, Arenas E, Ernfors P, Marklund U and Linnarsson S. Molecular Architecture of the Mouse Nervous System. *Cell*. 2018;174:999–1014 e22.
55. Casazza A, Laoui D, Wenes M, Rizzolio S, Bassani N, Mambretti M, Deschoemaeker S, Van Ginderachter JA, Tamagnone L and Mazzone M. Impeding macrophage entry into hypoxic tumor areas by Sema3A/Nrp1 signaling blockade inhibits angiogenesis and restores antitumor immunity. *Cancer Cell*. 2013;24:695–709.
56. Tordjman R, Lepelletier Y, Lemarchandel V, Cambot M, Gaulard P, Hermine O and Romeo PH. A neuronal receptor, neuropilin-1, is essential for the initiation of the primary immune response. *Nat Immunol*. 2002;3:477–82.
57. Delgoffe GM, Woo SR, Turnis ME, Gravano DM, Guy C, Overacre AE, Bettini ML, Vogel P, Finkelstein D, Bonnevier J, Workman CJ and Vignali DA. Stability and function of regulatory T cells is maintained by a neuropilin-1-semaphorin-4a axis. *Nature*. 2013;501:252–6.
58. Ozawa F, Friess H, Tempia-Caliera A, Kleeff J and Buchler MW. Growth factors and their receptors in pancreatic cancer. *Teratog Carcinog Mutagen*. 2001;21:27–44.
59. Stopczynski RE, Normolle DP, Hartman DJ, Ying H, DeBerry JJ, Bielefeldt K, Rhim AD, DePinho RA, Albers KM and Davis BM. Neuroplastic changes occur early in the development of pancreatic ductal adenocarcinoma. *Cancer Res*. 2014;74:1718–27.
60. Liang Y, Li H, Gan Y and Tu H. Shedding Light on the Role of Neurotransmitters in the Microenvironment of Pancreatic Cancer. *Front Cell Dev Biol*. 2021;9:688953.
61. Litan A and Langhans SA. Cancer as a channelopathy: ion channels and pumps in tumor development and progression. *Front Cell Neurosci*. 2015;9:86.
62. Peng J, Sun BF, Chen CY, Zhou JY, Chen YS, Chen H, Liu L, Huang D, Jiang J, Cui GS, Yang Y, Wang W, Guo D, Dai M, Guo J, Zhang T, Liao Q, Liu Y, Zhao YL, Han DL, Zhao Y, Yang YG and Wu W. Single-cell RNA-seq highlights intra-tumoral heterogeneity and malignant progression in pancreatic ductal adenocarcinoma. *Cell Res*. 2019;29:725–738.
63. Kugler EM, Michel K, Zeller F, Demir IE, Ceyhan GO, Schemann M and Mazzuoli-Weber G. Mechanical stress activates neurites and somata of myenteric neurons. *Front Cell Neurosci*. 2015;9:342.
64. Arese M, Bussolino F, Pergolizzi M and Bizzozero L. An Overview of the Molecular Cues and Their Intracellular Signaling Shared by Cancer and the Nervous System: From

Neurotransmitters to Synaptic Proteins, Anatomy of an All-Inclusive Cooperation. *Int J Mol Sci*. 2022;23.

65. Falcinelli M, Al-Hity G, Baron S, Mampay M, Allen MC, Samuels M, Jones W, Cilibrasi C, Flaherty RL, Giamas G, Thaker PH and Flint MS. Propranolol reduces IFN-gamma driven PD-L1 immunosuppression and improves anti-tumour immunity in ovarian cancer. *Brain Behav Immun*. 2023.
66. Gallo S, Vitacolonna A and Crepaldi T. NMDA Receptor and Its Emerging Role in Cancer. *Int J Mol Sci*. 2023;24.
67. Jayachandran P, Battaglin F, Strelez C, Lenz A, Algaze S, Soni S, Lo JH, Yang Y, Millstein J, Zhang W, Shih JC, Lu J, Mumenthaler SM, Spicer D, Neman J, Roussos Torres ET and Lenz HJ. Breast cancer and neurotransmitters: emerging insights on mechanisms and therapeutic directions. *Oncogene*. 2023.
68. Kaur J and Dora S. Purinergic signaling: Diverse effects and therapeutic potential in cancer. *Front Oncol*. 2023;13:1058371.
69. Schuller HM and Al-Wadei HA. Neurotransmitter receptors as central regulators of pancreatic cancer. *Future Oncol*. 2010;6:221–8.
70. Wang C, Shen Y, Ni J, Hu W and Yang Y. Effect of chronic stress on tumorigenesis and development. *Cell Mol Life Sci*. 2022;79:485.
71. Wang X, Li Y, Shi Y, Luo J, Zhang Y, Pan Z, Wu F, Tian J and Yu W. Comprehensive analysis to identify the neurotransmitter receptor-related genes as prognostic and therapeutic biomarkers in hepatocellular carcinoma. *Front Cell Dev Biol*. 2022;10:887076.
72. Miranda K, Girard-Dias W, Attias M, de Souza W and Ramos I. Three dimensional reconstruction by electron microscopy in the life sciences: An introduction for cell and tissue biologists. *Mol Reprod Dev*. 2015;82:530–47.
73. Venkataramani V, Tanev DI, Strahle C, Studier-Fischer A, Fankhauser L, Kessler T, Korber C, Kardorff M, Ratliff M, Xie R, Horstmann H, Messer M, Paik SP, Knabbe J, Sahm F, Kurz FT, Acikgoz AA, Herrmannsdorfer F, Agarwal A, Bergles DE, Chalmers A, Miletic H, Turcan S, Mawrin C, Hanggi D, Liu HK, Wick W, Winkler F and Kuner T. Glutamatergic synaptic input to glioma cells drives brain tumour progression. *Nature*. 2019;573:532–538.
74. Venkatesh HS, Morishita W, Geraghty AC, Silverbush D, Gillespie SM, Arzt M, Tam LT, Espenel C, Ponnuswami A, Ni L, Woo PJ, Taylor KR, Agarwal A, Regev A, Brang D, Vogel H, Hervey-Jumper S, Bergles DE, Suva ML, Malenka RC and Monje M. Electrical and synaptic integration of glioma into neural circuits. *Nature*. 2019;573:539–545.
75. Zeng Q, Michael IP, Zhang P, Saghafinia S, Knott G, Jiao W, McCabe BD, Galvan JA, Robinson HPC, Zlobec I, Ciriello G and Hanahan D. Synaptic proximity enables NMDAR signalling to promote brain metastasis. *Nature*. 2019;573:526–531.
76. Salavatian S, Beaumont E, Longpre JP, Armour JA, Vinet A, Jacquemet V, Shivkumar K and Ardell JL. Vagal stimulation targets select populations of intrinsic cardiac neurons to control neurally induced atrial fibrillation. *Am J Physiol Heart Circ Physiol*. 2016;311:H1311-H1320.
77. Shulgach JA, Beam DW, Nanivadekar AC, Miller DM, Fulton S, Sciullo M, Ogren J, Wong L, McLaughlin BL, Yates BJ, Horn CC and Fisher LE. Selective stimulation of the ferret abdominal vagus nerve with multi-contact nerve cuff electrodes. *Sci Rep*. 2021;11:12925.
78. Oh TK, Lee WJ, Woo SM, Kim NW, Yim J and Kim DH. Impact of Celiac Plexus Neurolysis on Survival in Patients with Unresectable Pancreatic Cancer: A Retrospective, Propensity Score Matching Analysis. *Pain Physician*. 2017;20:E357-E365.
79. Zou XP, Chen SY, Lv Y, Li W and Zhang XQ. Endoscopic ultrasound-guided celiac plexus neurolysis for pain management in patients with pancreatic carcinoma reasons to fight a losing battle. *Pancreas*. 2012;41:655–7.
80. Lavu H, Lengel HB, Sell NM, Baiocco JA, Kennedy EP, Yeo TP, Burrell SA, Winter JM, Hegarty S, Leiby BE and Yeo CJ. A prospective, randomized, double-blind, placebo controlled trial on the efficacy of ethanol celiac plexus neurolysis in patients with operable pancreatic and periampullary adenocarcinoma. *J Am Coll Surg*. 2015;220:497–508.

81. Johnson CD, Berry DP, Harris S, Pickering RM, Davis C, George S, Imrie CW, Neoptolemos JP and Sutton R. An open randomized comparison of clinical effectiveness of protocol-driven opioid analgesia, celiac plexus block or thoracoscopic splanchnicectomy for pain management in patients with pancreatic and other abdominal malignancies. *Pancreatology*. 2009;9:755–63.
82. Lillemoe KD, Cameron JL, Kaufman HS, Yeo CJ, Pitt HA and Sauter PK. Chemical splanchnicectomy in patients with unresectable pancreatic cancer. A prospective randomized trial. *Ann Surg*. 1993;217:447–55; discussion 456–7.
83. Curveilhier J. *Anatomie pathologique du corps humain, descriptions avec figures lithographiées et colorisées des diverses altérations morbides dont le corps humain est susceptible.* Paris: J.B. Baillière; 1835.
84. Shi DD, Guo JA, Hoffman HI, Su J, Mino-Kenudson M, Barth JL, Schenkel JM, Loeffler JS, Shih HA, Hong TS, Wo JY, Aguirre AJ, Jacks T, Zheng L, Wen PY, Wang TC and Hwang WL. Therapeutic avenues for cancer neuroscience: translational frontiers and clinical opportunities. *Lancet Oncol*. 2022;23:e62-e74.
85. Zahalka AH and Frenette PS. Nerves in cancer. *Nat Rev Cancer*. 2020;20:143–157.

Correction to: Tools and Model Systems to Study Nerve-Cancer Interactions

Peter L. Wang, Nicole A. Lester, Jimmy A. Guo, Jennifer Su, Carina Shiau, and William L. Hwang

Correction to:
Chapter 11 in: M. Amit, N. N. Scheff (eds.), *Cancer Neuroscience*, https://doi.org/10.1007/978-3-031-32429-1_11

The chapter was inadvertently published with an error in Figure 11.1 and it has been corrected now.

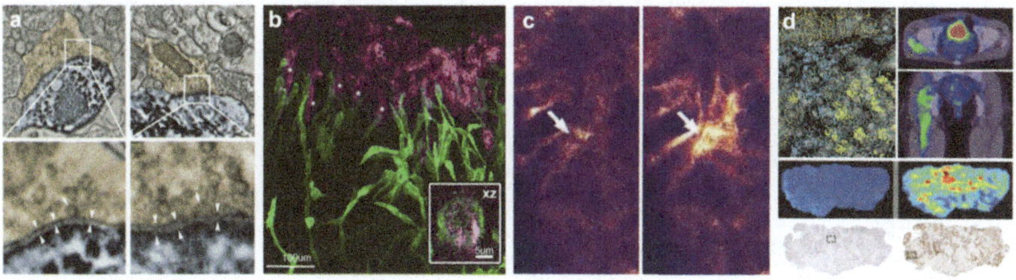

Fig. 11.1 Methods for visualizing tissue and cellular architecture in cancer neuroscience. (**a**) Synapses between presynaptic axons (sepia) and postsynaptic glioma cells (blue) visualized by immuno-electron microscopy [25]. (**b**) Confocal imaging of pancreatic cancer cells (magenta) associated with Schwann cells (green) in Matrigel [29]. (**c**) Intravital imaging of autonomous calcium activity in glioma cells [34]. (**d**) Multiplexed in situ and clinical imaging methods. (Top left) Spatial molecular imaging (SMI) provides mRNA and protein localization at single-cell/subcellular resolution [39]. (Top right) PET/CT detects perineural spread of the sciatic nerve [40]. (Bottom) Raman imaging reveals nanoparticle-stained cancer markers validated by IHC [38]

The updated original version of this chapter can be found at https://doi.org/10.1007/978-3-031-32429-1_11

M. Amit, N. N. Scheff (eds.), *Cancer Neuroscience*, https://doi.org/10.1007/978-3-031-32429-1_13

Index

A

Activity, vii, 6–9, 11, 13, 28, 30, 38–40, 42, 45, 58, 59, 64, 77, 91, 102, 104, 105, 107, 123, 133, 151, 156–158, 161, 171, 173–175, 188–190, 193
Adrenergic, 57, 58, 70, 72, 86–88, 90, 91, 103, 105–108, 120–122, 150–152, 154, 155, 159, 160, 176, 194
Axonogenesis, 73, 75, 76, 135, 153, 154, 194

B

Behavior, 6, 13, 14, 35, 37, 41, 56, 58, 64, 71, 73, 74, 76, 77, 90, 136, 139, 141, 153, 161, 190, 191
β-adrenergic, 55–59, 70, 86, 87, 102, 103, 109, 122, 150, 158, 169, 173
β-adrenergic signaling, 150
Brain, 3, 27, 35, 59, 73, 86, 101, 120, 137, 149, 174

C

Cancer, 5, 27, 35, 51, 69, 101, 117, 131, 149, 169, 185
Cancer metastasis, 57, 70, 74, 139, 152
Carcinogenesis, 118, 123, 124, 172, 173
Central nervous system (CNS), 3, 6, 7, 12–15, 35–45, 54, 71, 72, 89, 151, 158, 170, 174, 185
Checkpoint protein, 89, 187, 190, 193
Cholinergic, 56, 58, 70, 71, 87–89, 91, 103, 121, 150–152, 173, 189, 194
Cognitive, 43, 44, 156–158
Communication, 13–15, 27, 28, 31, 41, 51, 54, 65, 90, 149, 172, 186, 191, 194
Crosstalk, vii

E

Electrochemical, 27–29, 39, 41

G

Glioma, 3, 27, 35, 54, 72, 149, 171, 188

I

Immune, 42, 56, 69, 84, 101, 117, 133, 152, 176, 186
Immunosuppression, 58, 188–190
Innervation, ix, 52, 70–77, 84–86, 105, 106, 121, 155, 159, 169, 170, 172, 173, 187, 193
Invasion, 13, 29, 40, 51, 74, 84, 119, 134, 149, 169, 187

M

Methodology
Models, 4, 29, 35, 56, 70, 85, 104, 120, 135, 150, 169, 185
Multiple myeloma, 101–109

N

Nerve density, 51, 52, 59, 64, 70, 72, 120, 154
Nerves, 10, 37, 51, 70, 86, 102, 117, 131, 149, 169, 185

M. Amit, N. N. Scheff (eds.), *Cancer Neuroscience*,
https://doi.org/10.1007/978-3-031-32429-1

Nervous system, vii–ix, 27, 30–32, 35, 43, 70–72, 75, 84, 89–91, 103, 105, 151–156, 159, 161, 169, 170, 186, 190, 194
Network, 12, 13, 28, 29, 31, 38, 41, 43–45, 87, 88, 91, 152, 159, 172, 175
Neuroimmune, 43, 77, 84, 87, 90, 91, 160, 188, 193, 194
Neuroimmunomodulation
Neuronal activity, 6–7, 9–15, 27–29, 31, 35–37, 39–42, 44, 45, 76, 156, 157, 161, 175, 192, 193
Neuronal cells, 37, 40, 54–59, 62, 64, 174
Neurotrophin, 36–37, 54, 73, 123, 149, 151, 154, 155, 159

P
Parasympathetic nervous system, 87, 186
Perineural invasion, vii, 51, 52, 56–59, 61–64, 74, 77, 78, 119, 123, 135, 138, 151, 153, 161, 169, 170, 194
Peripheral nerves, ix, 52, 53, 63, 70–74, 76, 91, 152, 171, 174, 175, 178, 186, 187, 192, 193
Peripheral nervous system, vii, 54, 56, 69–78, 102, 105, 118, 131, 170, 174, 185–187
Progenitor, 3–6, 38, 52, 122, 124, 132, 154, 155, 176, 190

R
Reprogramming, 75, 76, 131, 133–136, 140, 141, 154

S
Schwann cells, 52–54, 59–64, 71, 76, 77, 90, 118, 119, 122–124, 131–141, 151, 152, 154, 170–172, 176, 186, 192
Small extracellular vesicless (sEVs), 73–76
Stroma, 59, 64, 118, 121, 140
Sympathetic nervous system (SNS), 86, 105, 107, 108, 120, 157, 158, 169, 173

T
T lymphocytes, 58, 84, 85, 87, 88, 91
Tool, 161, 169–172, 174, 177, 179, 180
Tumor, 3, 27, 35, 51, 69, 87, 101, 117, 132, 149, 169, 185
Tumorigenesis, vii, ix, 61, 71, 84, 87–89, 91, 104, 150, 152, 156, 169, 173, 188, 190, 192–194
Tumor microenvironment, ix, 8, 12, 27–29, 31, 36, 37, 39, 42–45, 51, 52, 54, 56–59, 63–65, 69, 70, 75–77, 85, 102, 104, 117, 124, 138, 151, 160, 170, 185

V
Visualization, 77, 140, 172, 186

SPRINGER NATURE

GPSR Compliance

The European Union's (EU) General Product Safety Regulation (GPSR) is a set of rules that requires consumer products to be safe and our obligations to ensure this.

If you have any concerns about our products, you can contact us on ProductSafety@springernature.com

In case Publisher is established outside the EU, the EU authorized representative is:

Springer Nature Customer Service Center GmbH
Europaplatz 3
69115 Heidelberg, Germany